Ecocriticism and the Anthropocene in Nineteenth-Century Art and Visual Culture

In this volume, emerging and established scholars bring ethical and political concerns for the environment, nonhuman animals and social justice to the study of nineteenth-century visual culture. They draw their theoretical inspiration from the vitality of emerging critical discourses, such as new materialism, ecofeminism, critical animal studies, food studies, object-oriented ontology and affect theory. This timely volume looks back at the early decades of the Anthropocene to query the agency of visual culture to critique, create and maintain more resilient and biologically diverse local and global ecologies.

Maura Coughlin is Professor of Visual Studies at Bryant University, USA.

Emily Gephart is Lecturer in the Department of Visual and Material Studies at Tufts University, USA.

Cover image: Designed by Theodore Russell Davis, made by Haviland & Co., Limoges, France, "Fish Service Platter – The Shad," c. 1880–1887. Porcelain with chromolithograph, enamel, and gilt decoration, 9 x 24 1/2 inches, Philadelphia Museum of Art, Gift of the McNeil Americana Collection, 2006.

Routledge Advances in Art and Visual Studies

This series is our home for innovative research in the fields of art and visual studies. It includes monographs and targeted edited collections that provide new insights into visual culture and art practice, theory, and research.

Play and the Artist's Creative Process
The Work of Philip Guston and Eduardo Paolozzi
Elly Thomas

Film and Modern American Art
The Dialogue between Cinema and Painting
Katherine Manthorne

Bridging Communities through Socially Engaged Art
Edited by Alice Wexler and Vida Sabbaghi

Abstract Painting and the Minimalist Critiques
Robert Mangold, David Novros, and Jo Baer in the 1960s
Matthew L. Levy

Arte Ambientale, Urban Space, and Participatory Art
Martina Tanga

Theory of the Art Object
Paul Crowther

The Digital Interface and New Media Art Installations
Phaedra Shanbaum

Ecocriticism and the Anthropocene in Nineteenth-Century Art and Visual Culture
Edited by Maura Coughlin and Emily Gephart

Popularization and Populism in the Visual Arts
Attraction Images
Edited by Anna Schober

For a full list of titles in this series, please visit https://www.routledge.com/Routledge-Advances-in-Art-and-Visual-Studies/book-series/RAVS

Ecocriticism and the Anthropocene in Nineteenth-Century Art and Visual Culture

Edited by Maura Coughlin and Emily Gephart

Routledge
Taylor & Francis Group
NEW YORK AND LONDON

First published 2020
by Routledge
52 Vanderbilt Avenue, New York, NY 10017

and by Routledge
2 Park Square, Milton Park, Abingdon, Oxon, OX14 4RN

Routledge is an imprint of the Taylor & Francis Group, an informa business

Library of Congress Cataloging-in-Publication Data
Names: Coughlin, Maura, editor. | Gephart, Emily, editor.
Title: Ecocriticism and the anthropocene in nineteenth-century art and visual culture / edited by Maura Coughlin and Emily Gephart.
Description: New York: Routledge, 2020. | Series: [Routledge advances in art and visual studies] | Includes bibliographical references and index. |
Identifiers: LCCN 2019025477 (print) | LCCN 2019025478 (ebook) |
ISBN 9780367180287 (hardback) | ISBN 9780429059193 (ebook)
Subjects: LCSH: Art and society—History—19th century. | Ecocriticism.
Classification: LCC N72.S6 E29 2020

ISBN: 978-0-367-18028-7 (hbk)
ISBN: 978-0-429-05919-3 (ebk)

Typeset in Sabon
by codeMantra

Contents

Figures

Color Plates

Contributors

Alan C. Braddock is Ralph H. Wark Associate Professor of Art History & American Studies, and Director of Graduate Studies in the American Studies Program at The College of William & Mary. He is the author of *Thomas Eakins and the Cultures of Modernity* (2009) and coeditor of *A Keener Perception: Ecocritical Studies in American Art History* (2009, with Christoph Irmscher) and *A Greene Country Towne: Philadelphia's Ecology in the Cultural Imagination* (2016, with Laura Turner Igoe). He is also the co-curator and co-author, with Karl Kusserow, of *Nature's Nation: American Art and Environment* (2018–2019), a book and traveling exhibition organized by the Princeton University Art Museum. His essay "From Nature to Ecology: The Emergence of Ecocritical Art History" appears in *The Blackwell Companion to American Art* (2015), edited by John Davis, Jennifer Greenhill and Jason LaFountain. In 2016–2017, Braddock was the Currie C. and Thomas A. Barron Visiting Professor of Environment and Humanities at Princeton. He has taught interdisciplinary courses on American art and environmental history, ecocriticism and animal studies since 2002.

Kelly P. Bushnell is Visiting Instructor in the Department of English at the University of West Florida. Her research focuses on oceans and animals in Victorian literature, particularly whales, whaling and material culture. Recent publications include the *Oxford Bibliography of Victorian Maritime Literature*, a state-of-the-field article on oceanic studies for *Victorian Literature and Culture* and chapters on early cetacean captivity and Tennyson's sense of aquaria, respectively. She has been the environmental humanities scholar aboard an all-women expedition to the Canadian High Arctic and Greenland, and her current project, co-authored with Inuk Traditional Knowledge Keeper Johnny Issaluk, considers ecological agency and colonialism in the Arctic. She lives just a short ferry ride from Seattle and teaches remotely.

Maura Coughlin is Professor of Visual Studies in the Department of English and Cultural Studies at Bryant University in Smithfield, RI, and is Vice President of the Nineteenth-Century Studies Association. Her research is focused on French Atlantic visual culture, coastal ecology, the rise of marine sciences in France and dialogues that have been struck between nineteenth- and twenty-first-century aesthetics and ecological ethics. Across her projects is a shared fascination with the material flows of fish and animals, seaweed, salt, people, sand, stones, boats and other actors that move across and through the tide line, and the ways in which the visual culture of the shore visualizes intensely local perceptions of tide, geology, beach morphology and marine botany.

Emily Gephart is a Lecturer in the Department of Visual and Material Studies at the School of the Museum of Fine Arts at Tufts University. Her work explores the ecological networks of mind in the long nineteenth century and addresses how material objects served as tools through which beholders comprehended discoveries about modern psychology, perception and aesthetics. Her work has appeared in the Routledge anthology *Architectures of Display* (with Michael Rossi) and in journals such as *New Directions in William James Studies, Cabinet* and *Imago Musicae.* She holds a PhD from the Massachusetts Institute of Technology.

Caroline L. Gillaspie is a PhD candidate in Environmental Design at The Graduate Center, CUNY, and a Visiting Instructor at the Pratt Institute. Her work examines the visual culture of the nineteenth-century Brazil-United States coffee trade, tracing the representations of American coffee culture from bean to cup through landscapes, cityscapes and genre scenes. She is interested in depictions of race, labor and environmental degradation, and the transport of commodities in these images. Caroline is the recipient of Early Research Initiative grants from the Graduate Center and the Patricia Phelps de Cisneros Fellowship for the study of Latin American art.

Polly Gould is a writer and an artist working between art and architecture. She holds a PhD in Art and Architecture from the Bartlett School of Architecture, University College London, and is currently Post-doctoral Research Fellow in Design-led Architectural Research with ARC Architectural Research Collaborative, Newcastle University, UK. Gould has shown in gallery spaces, botanic gardens and natural history museums internationally, and is represented by Danielle Arnaud. Gould's monograph, titled *Antarctica through Art and the Archive: Refractions of the Life of Edward Wilson*, takes the archival encounter with the watercolor practice of this Antarctic explorer as an opportunity to offer an ecocritical interpretation of our ecological entanglement with our environment. Gould has also devised and curated TOPOPHOBIA: Fear of Place in Contemporary Art, 2012, and taught at Central Saint Martins, London; the Bartlett School of Architecture, London; and KADK, Denmark.

Joan E. Greer (PhD Vrije Universiteit, Amsterdam) is Professor of History of Art, Design, and Visual Culture at the University of Alberta, Canada. She is a founding member of the U of A Environmental Studies and ongoing member of the Religious Studies and Science and Technology in Society Interdisciplinary Programmes. Her research engages with issues of artistic identity, the history of environmentalism and theories of nature and ecological envisioning, both historically (most particularly in the long nineteenth century) and in contemporary art and design. Her current major project is entitled *Visualizations of Nature in Late Nineteenth-Century Dutch Art and Print Culture: Religion, Science and Art,* and her most recent publication is "'To everything there is a season': the rhythms of the year in Vincent van Gogh's socio-religious world view," *Van Gogh and the Seasons*, ed. Sjraar van Heugten (exhibition catalogue, National Gallery of Victoria, Melbourne, 2017), republished by Princeton University Press, March 2018.

Jessica L. Horton is an Assistant Professor of Modern, Contemporary, and Native American Art History at the University of Delaware. She is the author of the book *Art for an Undivided Earth: The American Indian Movement Generation* (2017), which was awarded a Wyeth Foundation Publication Grant and the Mellon Art History Publication Initiative. Her essays about Indigenous art, space, ecology and globalization have appeared in publications such *as Art Journal, Art History, American Art, Third Text* and *The Journal of Transnational American Studies.* Her research has been supported by the Getty Research Institute, the Smithsonian Institution, the Center for Advanced Study in the Visual Arts, the Social Science Research Council and the Terra Foundation for American Art, among others. She is helping to design and build an earth sheltered, solar powered house in rural northern California.

Rosie Ibbotson is Senior Lecturer in Art History and Theory at Te Whare Wānanga o Waitaha University of Canterbury, Aotearoa New Zealand. Her research centers on the Anglophone long nineteenth century and concerns the entanglements of visual culture and environmental violence. Rosie is currently writing a book titled *Picturing the Imperial Anthropocene: Visual Representation and Environmental Change in Long Nineteenth-Century Aotearoa New Zealand*, and she has also published on the intersections of museology, visual and material culture, and de-extinction.

Laura Turner Igoe is the Curator of American Art at the James A. Michener Art Museum in Doylestown, Pennsylvania. She specializes in American art and material culture of the long nineteenth century, and its engagement with environmental conditions and change. She is the coeditor of *A Greene Country Towne: Philadelphia's Ecology in the Cultural Imagination* and has contributed essays to the journals *American Art*, *Panorama*, and *Common-place*, and the exhibition catalogue *Nature's Nation: American Art and Environment.*

Shana Klein is Assistant Professor of Art History at Kent State University. She is trained in the history of American art, with sub-specialties in African- and Native-American art. Klein's research on the visual culture of food will be published in her forthcoming book, "The Fruits of Empire: Art, Food, and the Politics of Race in the Age of National Expansion," which has been supported by several fellowships and published in journals such as *American Art, Southern California Quarterly* and *Southern Cultures.*

Jessica Landau is a PhD candidate in Art History at the University of Illinois at Urbana-Champaign. Her dissertation, "'Critical Habitat': Picturing Extinction, Conservation, and the American Animal in the Long Twentieth Century," seeks to challenge the idea of the nonhuman in the American imaginary, and to establish a better way to relate to the natural world while disrupting reliance on settler colonial notions of time, place and the animal. She is the managing editor of *Media-N, Journal of the New Media Caucus*, and has served as an Associate Curator of the Brinton Museum in Big Horn, WY, and as the managing editor of the *Journal of the Native American and Indigenous Studies Association.*

George Philip LeBourdais is a scholar and curator of American art, specializing in the history of photography, who is earning his PhD in Stanford University's Department of Art and Art History. His writing and curatorial work have earned awards from the Henry Luce Foundation, the Andrew W. Mellon Foundation and the U.S. Fulbright Commission. At Stanford, he has been a member of the Environmental

Humanities Project, an interdisciplinary research group focused on environmental history, ecocriticism and the place of nature in our lives.

Sarah J. Moore is Professor of American Art History at the University of Arizona. Questions regarding the shifting terrain of identities and geographies animate her work as a scholar and teacher of American art. Her research areas intersect with the global interdisciplinary arena of world's fair studies, considering in particular pre-World War I fairs in the United States and ecocriticism in nineteenth-century visual culture. Recent publications include "The Great American Desert is No More" in *Trans-Mississippi and International Exposition of 1898–99* (2018); "Mosquitoes, Malaria, and Cold Butter: Discourses of Health and Progress in the Panama Canal Zone, 1904–1915," *Panorama* (3.2 Fall 2017); *Empire on Display: San Francisco's Panama-Pacific International Exposition of 1915* (2013); and "Manliness and the New American Empire" in *Gendering the Fair* (2010).

Kimberly Rhodes is Professor of Art History and Director of the New York Semester on Contemporary Art at Drew University. Her publications include "Archetypes and Icons: Materialising Victorian Womanhood in 1970s Feminist Art" in *Neo-Victorian Studies* (2013); *Ophelia and Victorian Visual Culture: Representing Body Politics in the Nineteenth Century* (2008); "Double Take: Tom Hunter's The Way Home (2000)" in *The Afterlife of Ophelia* (2012); and "Degenerate Detail: John Everett Millais and Ophelia's Muddy Death" in *John Everett Millais: Beyond the Pre-Raphaelite Brotherhood* (2001).

Annie Ronan is an Assistant Professor of Art History in the School of Visual Arts at Virginia Tech. She earned her PhD from Stanford University's Department of Art and Art History, where she specialized in the art and visual culture of the United States. Her research, which largely focuses on the visual representation of animals during the long nineteenth century, has been published in *American Art* and supported by The Smithsonian American Art Museum's Douglass Foundation Fellowship, The Tyson Scholars Program at The Crystal Bridges Museum of American Art and the Iris & B. Gerald Cantor Center for Visual Arts at Stanford University. Her current book project, *Creature Comfort: Animal Art and the Reconstruction of American Whiteness*, investigates the racial politics of picturing animals and redefining the "humane" in the wake of the American Civil War.

Michael Rossi is a historian of science and medicine at the University of Chicago. His first book, *The Republic of Color: Science, Perception, and the Making of a Modern America* (2019), details the history of color science in the late nineteenth and early twentieth centuries. He also writes about natural history, the history of language and linguistics, and the history of anthropology, among other topics.

Naomi Slipp is an Assistant Professor of Art History at Auburn University at Montgomery and holds a PhD from Boston University and an MA from the University of Chicago. Her research in American art focuses on nineteenth-century intersections between art, science and identity. She has published in *Sculpture Journal* and the edited volumes *Beyond the Battlefield: New England and the Civil War* and *Bodies Beyond Borders: Moving Anatomies, 1750–1950*, and has forthcoming contributions in *Hyman Bloom: Matters of Life and Death* and *Victorian Science and Imagery: The Evolution of Form in Nineteenth-Century Visual Culture.*

Her current book project, *The Art of the Body: Medicine and Anatomy in American Culture, 1800–1880*, surveys connections between art and medicine in the study of human anatomy.

Corina Weidinger is an Art History Professor at Truckee Meadows Community College in Reno, NV. Her research on pollution, fatigue and industrial spectacle in fin de siècle French and Belgian representations of mines, ironworks and industrial landscapes has been supported by grants from the Fulbright Program, the Getty Research Institute, the Belgian American Educational Foundation and the Material Culture Institute at the University of Delaware. She has given talks on her research in the United States, Belgium, Spain and the United Kingdom, and has published articles in *Nineteenth-Century Art Worldwide* and *Art & Fact* (Liège). Weidinger holds a PhD from the University of Delaware.

1 Introduction

Maura Coughlin and Emily Gephart

> [I]t is but recently that ... public attention has been half awakened to the necessity of restoring the disturbed harmonies of nature, whose well-balanced influences are so propitious to all her organic offspring, of repaying to our great mother the debt which the prodigality and the thriftlessness of former generations have imposed upon their successors—thus fulfilling the command of religion and of practical wisdom, to use this world as not abusing it.
>
> —George Perkins Marsh (1864)

The long nineteenth century casts an even longer shadow across the twentieth century and into our present, with respect to the irrevocable transformations it set in motion. It brought about a geologic era of climactic change that many observers now broadly call the Anthropocene: a term whose anthropocentric hubris is not without its critics.[1] Yet it seems no coincidence that ecology, as both a science and a philosophy, also emerged during this century, amid intensifying imperialist colonization; capitalistic commodification; rising pollution; and global exploitation of human, animal and inorganic resources.

Given the period's prevailing Eurocentric, Christian and utilitarian faith in humankind's ultimate dominion over the earth, the ethical imperatives of a more secular, ecological mindset that acknowledged the mutual interdependence among planetary forces faced ambivalent acceptance, if not outright denial. Nonetheless, what many scholars today would call an emergent "ecocritical" consciousness also shaped the century, of the kind proposed by George Perkins Marsh in the oft-cited book *Man and Nature* (1864).[2] Even as Capitalocentric logics rose to ideological dominance, the changes that they set in motion fostered a slow-to-dawn acknowledgment of their results (Haraway 2015; Moore 2016). Amid rising demand for goods and resources that mandated wholesale reformulation of global networks, a few keen observers like Marsh began to recognize an even larger dynamic: the idea that all human beings were and had always been part of a "mesh" – a planetwide symbiotic community of interdependent entities (Morton 2011).

This book takes such a community as its core premise, and throughout its assembled essays, its authors speak to the non-hierarchical interconnectedness central to ecological thinking and environmental history. Bringing together a broad range of responses to the task of what Nicholas Mirzoeff has termed "Visualizing the Anthropocene," the chapters herein articulate a visual culture of ecology during its early development (Mirzoeff 2015). Embracing insights from the interdisciplinary environmental humanities, they echo Karl Kusserow's recent observation that ecocriticism is not merely a method but a habit of mind (Kusserow 2018; see also Chakrabarty 2009). Beyond merely applying ecocriticism to the art of the nineteenth century, they employ materials, images, technologies and epistemologies as productive "theory machines" (Galison 2003),[3] and in so doing, they seek to enrich the field of nineteenth-century visual culture. They read the century's seismic and world-wide shifts through pictures and with things, and by means of the ecological metaphors set in place during this formative era.

For instance, as both a concept and an organic exemplar, the tree proved a particularly effective image to think with. Among his many accomplishments, German biologist, philosopher, physician and artist Ernst Haeckel is celebrated today for coining the term "Oecologia," or "ecology," in his 1866 German-language publication "General Morphology of Organisms." Despite recognition of the "mutual relations of organisms" dawning across a host of inter-related discourses, Haeckel's neologism gained momentum slowly as a body of knowledge and model of thinking, earning widespread currency only in the twentieth century (Stauffer 1957; Kelsey 2014).[4] To his Anglophone readers, Haeckel was better known for his passionate defenses of Darwinian evolutionary theory, many of which he presented as ecological relationships. In the 1879 English edition of his text *The Evolution of Man*, Haeckel embraced the challenge of explaining Darwin's occasionally abstruse theories, providing helpful models through which his own equally complicated explanations might take root.

The guiding analogy undergirding Haeckel's defense of evolutionary processes compared them to the natural growth of trees, as demonstrated by plates in both his 1866 and 1879 texts (Figure 1.1) (Gontier 2011; Lima 2014; Miller 2016). Darwin

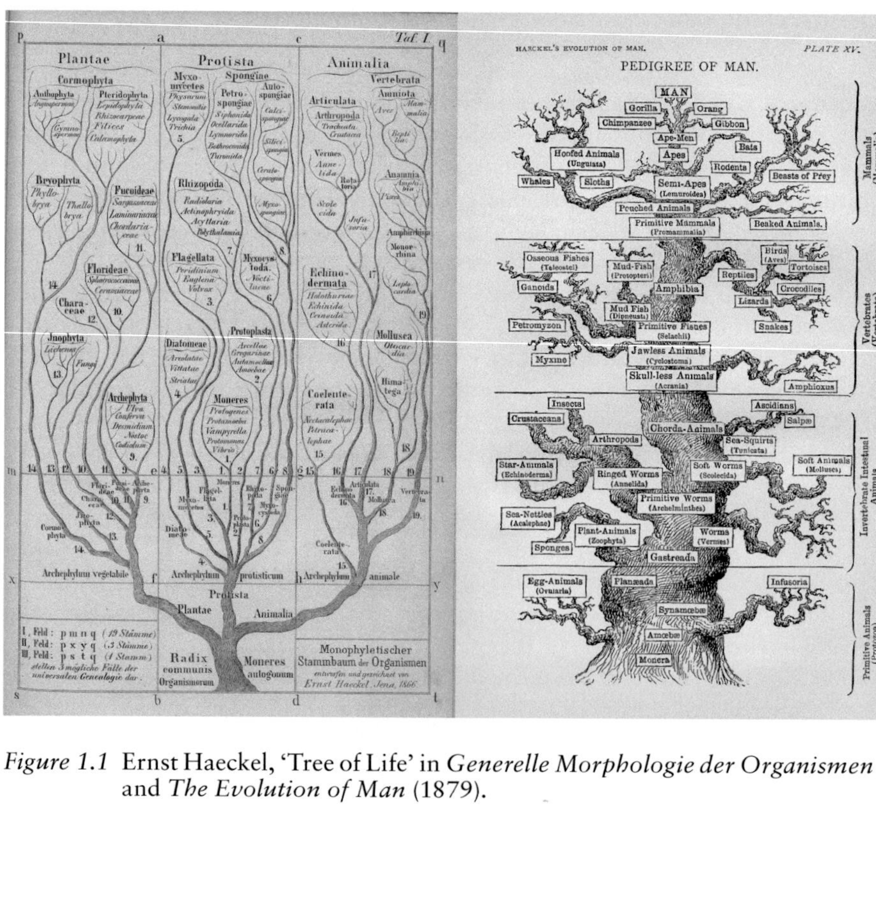

Figure 1.1 Ernst Haeckel, 'Tree of Life' in *Generelle Morphologie der Organismen* (1866) and *The Evolution of Man* (1879).

himself had suggested such a visualization in the only illustration gracing the 1859 edition of the *Origin of Species*, in which he charted life's proliferating diversity and evolution's endlessly bifurcating processes via an effective if rudimentary schematic.[5] It was Haeckel, however, who truly expanded upon the idea in his own diagrams, which took on increasingly elaborate resemblance to a tree. As a visualization of complex precepts, a tree's innate and immutable growth patterns proved equally significant to structuring information. Granting all biological development a form that was organically, not divinely, predestined – from taproot to trunk to ever more diverging branches and fine twigs – Haeckel pictured an originary, vitalized, sap-like stream of biotic entities from which all higher organisms grew upward, as if toward the sun's providential light.

While the tree-form that Haeckel helped popularize may have echoed nature's organic laws, it also simplified a complex and abstract system by means of pictorially convincing didactic principles.[6] To Haeckel, this metaphor had its own unassailable logic:

> The Natural System of Organisms, which classifies all the[ir] various forms in larger and smaller groups, ... can but be represented figuratively under the form of a tree with many branches. This tree is the genealogical tree of the groups related in form, and their relation in form really is their relation in blood. As no other explanation can be given of the fact that the system naturally assumes a tree-like form, we may regard this as an immediate and powerful proof of the truth.... (Haeckel 1879, 112)

His visualization not only literalized his comparison, but also proved morphologically persuasive to readers who might otherwise have remained skeptical of evolution: as symbols and as forms of 'natural' evidence, trees upheld an important if un-articulated association with God's underlying governance. Reconstituting references to Judeo-Christian genealogical trees in terms of nature's secular logic, Haeckel's diagram helped assuage lingering doubt that acceptance of Darwinian theory meant abandoning cherished beliefs (Gontier 2011).

Yet Haeckel's "pedigree of man" did not dislodge, but merely reconfigured the pre-existing scientific imperatives of classification, and his illustration upheld an anthropocentric and racially biased system of hierarchical categorization. For Haeckel, as for many of his fellow European and American biologists, white man's position atop an arboreal crown was unshakeable; the prejudice in this premise was conveniently naturalized by substituting models of ever-ascending dynamic growth for the ontologically fixed 'great chain of being.' Moreover, by giving the evolutionary struggle for existence the familiar shape of a pleasant sylvan refuge, rather than representing it as an arena for dark conflict, Haeckel envisioned the limbs of his mighty oak as a site of ecological mutualism. Under the shade of this three, the aesthetic appreciation of 'nature' could even be cast as a basic, universal or even 'instinctual' human activity (Figure 1.2).

A second visual example that enabled ecological thinking and afforded a profound understanding of the deep time of prehistory emerged in the public consciousness at about the same time as Haeckel's tree. The mammoth engraved on mammoth ivory, found by archeologist Edouard Lartet, was a nine-inch-long, fragmentary Ice Age image that was displayed at the Paris Universal Exposition of 1867 and was reproduced in Lartet's widely

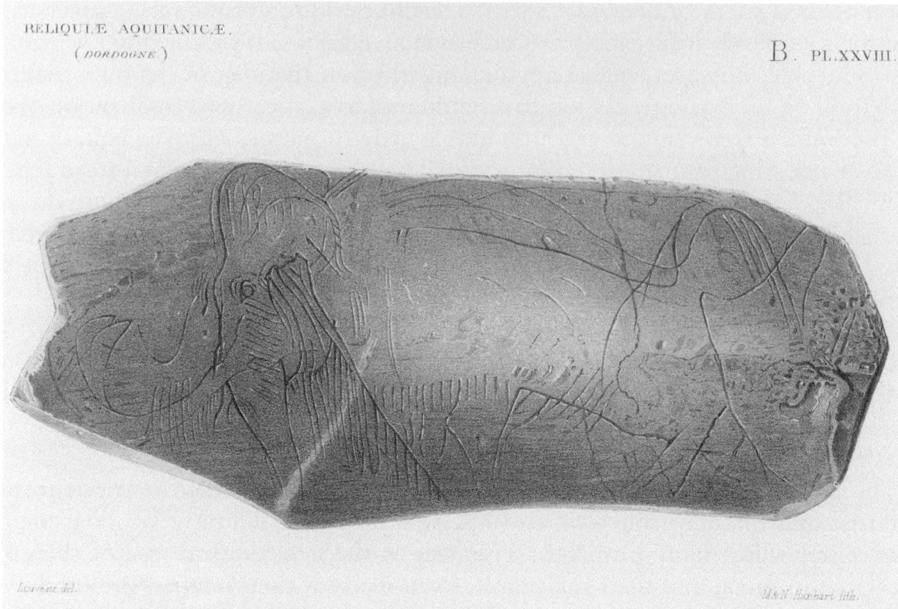

Figure 1.2 Image of a mammoth carved onto an elephant's tusk. *Reliquiæ aquitanicæ: being contributions to the archæology and palæontology of Périgord and the adjoining provinces of southern France*, by Edouard Lartet and Henry Christy; edited by Thomas Rupert Jones.

published text *Reliquiae Aquitanicae* (1865–1875). By 1864, Lartet had already uncovered much evidence that "aboriginal" humans and then-extinct mammals had inhabited the same, very ancient French landscape, but the discovery of the mammoth at the rock shelter of La Madeleine in the Dordogne (a site that had been steadily inhabited for 17,000 years) sealed the case, incontrovertibly demonstrating that this extinct creature was a contemporary of the "primitive" who engraved it.

Bearing both representational and indexical witness to prehistory, Lartet's mammoth opened the door to a hitherto unimaginably vast epoch, prior to Biblical time; he mused on "the coexistence of Man with the fossil Elephant ... in the earlier phases of the Quaternary Period" that the "truth of retrospective evidence is deduced now-a-days from so great a number of concordant observations, and of material facts of so clear a significance, that minds the least prepared to admit it are not slow to accept it in all its reality, when they will but take the trouble to look and then judge conscientiously" (Lartet 1875, 207–208).

Lartet's thoughts on the force of "retrospective evidence" are strangely symmetrical with the words penned 100 years later by prehistory enthusiast and Land Artist Robert Smithson, who wrote in 1968 that it is the passage of time that allows something to be understood as an artwork: "[w]hen a *thing* is seen through the consciousness of temporality, it is changed into something that is nothing....The object gets to be less and less but exists as something clearer. Every object, if it is art, is charged with the rush of time" (Smithson 2000, 112). Visual culture, such as the mammoth engraved by a prehistoric artist, "charged with the rush of time" has the capacity to initiate such connective insights.

This image thus facilitated the development of an ecological understanding of a shared human and extra-human history. Such conceptualization anticipates Donna Haraway's proposition that nonhuman animals and humans have co-constituted each other over the course of our millennia together, in an always-interconnected "natureculture" and that "none of it can be approached if the fleshly historical reality of face-to-face, body-to-body subject making across species is denied or forgotten in the humanist doctrine that holds only humans to be true subjects with real histories" (Haraway 2008, 66). Like Haeckel's tree, this striking image is an apt point of departure for thinking about how nineteenth-century visual culture might do (or might already have done) more than merely picture our origins or imaginatively place us in the past: it has powerfully produced twentieth- and twenty-first-century relationships to the other-than-human world.

Just as the human influence on the earth's climate, or what Haraway terms the "anthropogenic processes" (Haraway 2015, 160) of the Anthropocene accelerated, widespread curiosity about the natural world fueled a concomitant desire to picture nature and humanity's place within it. This collection seeks to demonstrate the range and reach of such curiosity, and to reveal the contradictions that underwrote the work of artists, scientists, historians and philosophers whose combined efforts pursued knowledge about the earth's fundamental function and structure. Although many of these nineteenth-century scholars found, in the words of English critic John Ruskin, that "the truths of nature are one eternal change—one infinite variety" (Ruskin 1844, 65), they also initiated and perpetuated spurious pseudo-scientific beliefs that are no longer supportable in the twenty-first century. Other ironies proliferated as modern industrialization and transnational expansion called for a new grasp of global interconnectivity: even as resources were *extricated* irreversibly from their surroundings, scientists studied evidence of *inextricable* ecological relationships and postulated comprehensive narratives of the world's deep time, vast scale and dynamic flux.

Scientific and documentary motivations coordinated with art and with emerging and progressive image technologies such as chromo-lithography, photography and microscopy, and with habits of collecting, mounting and displaying plants, minerals and animals as aesthetic evidence; all of these added to the visual terrain occupied by drawing, painting and other traditional art mediums in the promotion of inquiry. Visual culture rose to the challenges of stilling time, dissecting organisms and preserving vanishing species and communities. But it also asserted the imaginative human agency to reconstruct fossil creatures, to connect organisms in an ecotone, to re-green damaged terrain and to nourish the spirit. Thus, both well-known and less familiar items of art historical inquiry serve as entry points into an expanded ecocritical analysis of this period.

Contemporary art has frequently been the topic of ecocritical studies that foreground, for instance, the potential of art work to resist the extractive logic of globalization, to reveal the scope of the crisis, to inspire positive change or to imagine the relational potentials for creating new dialogues about living in the Anthropocene (see, for example, Kurt 2004; McKee 2007; Kraynak 2010; Weintraub 2012; Davis and Turpin 2015; Scott and Swenson 2015; or Demos 2016). In contrast, the field of nineteenth-century global art history has not seen as much of a focus on either the ecological agency of the image or the material ecology of art-making, notwithstanding the contributions of a host of scholars in the past decade whose work has advanced important and less humanistic counternarratives.

The State of the Field

Because the art historical literature of ecocriticism has been outlined in several recent publications, most notably Alan Braddock's 2015 survey essay "From Nature to Ecology: The Emergence of Ecocritical Art History," or Suzaan Boettger's 2016 "Within And Beyond the Art World: Environmentalist Criticism of Visual Art," what follows briefly summarizes relevant precedent texts, indicating the groundwork upon which this present scholarship rests and from which it departs. As new environmentally oriented histories have taken hold, the most conventional mode of reckoning with environmental attitudes has been via European and American landscape painting – one of the century's most popular artistic genres and an enduringly vital locus of study.

Ecocritical examinations of European art emerged in the late 1980s in the wake of Social Art History's focus on the rural landscape. With a hefty dose of Foucauldian analysis, Marxist anthropocentric approaches to the natural world perhaps reached a pinnacle in Nicholas Green's highly influential book *Spectacle of Nature* (1995). Green persuasively argued that landscape (whether a painting, a park or a piece of real estate), as a monetized and commodified discourse, is a construct of "nature" for the urban consumer. This model rendered landscape experience a marker of human social class, a stage for the enacting of bourgeois identity performance or the primitive other against which to measure one's "modernity." The materiality of landscape was central to Ann Bermingham's work on John Constable's landscapes, first published in the 1980s; she argued for the "material proximity" and "physicality of the landscape painter's encounter with his object," countering pervasive assumptions that an "experience of the materiality of things...created associations which in turn could be communicated through paintings in a way that was universally legible" (Bermingham 1987, 41 and 51). Greg Thomas, in his groundbreaking work on Théodore Rousseau (2000) was the first to connect an emerging nineteenth-century popular scientific ecological consciousness to an artist's landscape imagery. Thomas posited "earth narrative" readings of the Barbizon landscapes of the famously "rejected" Romantic painter and proposed new possibilities for reading the embodied and grounded, more-than-optical ecological understanding of the landscape and its biological constituents. Responding to the "animal turn" in the humanities, Stephen Eisenman's wide-ranging *The Cry of Nature – Art and the Making of Animal Rights* considered the ethical agency of images by both nineteenth-century illustrators and canonical painters. Eisenman positioned exploitative violence against non-human animals as a form of "class struggle" endemic to the alienating forces of capitalist modernity (Eisenman 2013, 65).

Art historical and ecocritical scholarship has converged in particularly revelatory ways in examining the visual culture of the United States. Historians, art historians and literary scholars have long attended to the close affiliation between the representation of North American space and the changing environment in the nineteenth century, given the significance so many artists and critics ascribed to the 'wilderness ideal,' to expansionist discourses, and to progressive models of national development (Miller 1967; Cronon 1995). Yet the artists most cherished by nineteenth-century beholders, like Thomas Cole or Frederick Edwin Church, were not insensitive to the costs of ecological upheaval: many scholars have found merit in exploring the deep ambivalence that informed Cole's celebrated pictures of sublime mountain vistas and pastoral farmland (as just one example, see Wallach 2002). Nonetheless, most artists and their audiences failed to recognize millennia of anthropogenic change that had occurred across the pre-industrial and pre-colonial continent (Mann 2005).

Ecocritical scholarship *avant le lettre* took shape in the 1980s and into the 1990s, as scholars queried how art affirmed or unsettled the dominant premises of American exceptionalism and Manifest Destiny, and reevaluated long-standing beliefs about the relationship between landscape art and nation-building. Barbara Novak, Nancy Anderson, Angela Miller and Alan Wallach directed attention to the ideological work performed by American landscape painters who invited beholders into dialog with their environmental surroundings (Novak 1980; Anderson 1991; Miller 1993; Wallach 2002). In 1991, the Smithsonian American Art Museum sparked controversy with its impressively revisionist exhibition *The West as America, Reinterpreting Images of the Frontier, 1820–1920*, in which scholars regarded many of the environmental concerns that underwrote the nation's claim to Western land long occupied by indigenous peoples (Truettner 1991). A subsequent generation responded to ecological challenges in concert with scholars in other disciplines such as Leo Marx and William Cronon, taking note of the formative role visual culture played in sustaining contradictory wilderness ideals and mythologizing America's abundant garden-scapes (Marx 1964; Cronon 1995).

This present collection would not have taken shape without the essential precedent of Alan Braddock and Christoph Irmscher's groundbreaking 2009 anthology *A Keener Perception: Ecocritical Studies in American Art History*, which addressed North American topics spanning four centuries. The example set by that book's laudable range nourished complementary scholarship (Dunaway 2005; Braddock and Igoe 2017). A special 2014 *Commentaries* feature in *American Art* embraced varied aesthetic and artistic responses to Art in the Anthropocene, including Robin Kelsey's thought-provoking essay "Ecology, Sustainability, and Historical Interpretation," in which he observes how the natural sciences were set apart from the moral and aesthetic imperatives of a divinely sanctioned, anthropocentric plan for resource use, instrumentalized by forms of culture (Braddock and Ater 2014; Kelsey 2014). Longer essays published in that journal have also embraced new approaches to familiar American topics, such as Maura Lyons's perceptive examination of the symbolism of wounded trees in Civil War imagery (Lyons 2012).

At the time of writing this introduction, a vigorous crop of writing about new materialist and ecocritical philosophies has taken root in and beyond academia, and in recent exhibitions such as *Nature's Nation: American Art and Environment*, organized by Karl Kusserow and Alan Braddock for the Princeton University Art Museum in 2018. Many emerging models, especially in literary studies, have affirmed overlap or interchange between the cognitive/semiotic realms of meaning making, aesthetics, materiality and sensuality, as exemplified by the redounding impact of Timothy Morton's proposal of "Ecology Without Nature" and "The Ecological Thought" (Morton 2009, 2012). So has interdisciplinary scholarship that takes the dispersed interactions of what Karen Barad terms "posthuman performativity" as a fundamental premise (Barad 2003; Coole and Frost 2010; Cohen 2013; Mentz 2015). Yet, although these topics lie near to the interests of art history, they have been slow to propagate widely. They are, in a sense, resistant to (if not incompatible with) traditional concerns with artistic biography and influence, style or provenance. Dismantling art historical business as usual requires a decolonized examination of globally dispersed networks and actants that shaped contradictory responses to environmental incursions (Latour 2004, 2005). To mobilize these goals, the authors in this book put into practice methodologies that have already invigorated the environmental and energy humanities, and those that have brought overdue attention to parallel histories of social injustice. Together, we seek new modes of scholarly ethics and interdependency that will prove the "retrospective evidence" of our own work (Lartet 1875).

Methodologies

Ecocriticism is not a uniform practice, but an array of multiple perspectives that share a common attitude toward mutuality, relatedness and non-hierarchical thinking. Applying ecocritical and environmentally engaged methodologies to visual culture can reshape traditional histories, coordinating historical intermediaries with the most pressing, activist demands of the present crisis, and attending to the global economic "supply chain" that produced wealth for some but misery for many. These methodologies are also emphatically process-oriented, examining production across categories, across borders, and with the exploitation of resources alongside extractive models of knowledge in mind. Yet this collection is not undertaken to lament the blindness of our forebears or to castigate their failure to recognize the consequences of their intemperate desires. Instead, we hope to relearn or retrace the logic they followed, in order to comprehend the trajectories that objects and images have taken across time. Made for a variety of reasons and embedded in a Euro-American dominated capitalist culture, these artifacts remain signifying agents which we 'read' both with and against the grain of their initial purpose.

The analytical methods in this collection have been selected for their appropriate utility to the particular sites of excavation, in the spirit of Linda Nochlin, who described herself in 1999 as an "ad-hoc historian" of nineteenth-century art (Nochlin 1999, 7). As a result, these essays offer multiple and transnational responses to the imperatives of a changing climate; as Haraway notes, "we need stories (and theories) that are just big enough to gather up the complexities and keep the edges open and greedy for surprising new and old connections" (Haraway 2015, 160). The essays gathered here avoid supercilious "criticality" or nihilistic pessimism, in order to explore a period whose inconsistencies far exceed the hubristic certainties of its key players, foreclosing any tidy descriptions or linear narratives.

Comprehensive coverage of nineteenth-century ecological sensibilities is beyond the scope of this – or any – collection. Since ecocritical scholarship is inherently interdisciplinary, a comprehensive summary would be cumbersome; the array of sources cited by the contributors to this volume attests to ever-new and amplified awareness of the intersecting arts, sciences and humanities. In these essays, the systems of extraction and fabrication required to bring an object into being exceed standard chronologies of making and beholding, interweaving with social dynamics in which unequally distributed privileges delimited possibilities. In working beyond long-standing dichotomies separating nature from culture, human from nonhuman animal and venerated 'high' art products from more ephemeral visual or material ones, we avoid imposing anachronistic sensibilities on our subjects of study, hoping instead to move the discipline of art history toward a more capacious, generous and socially engaged practice.

A de-centered ecocritical approach to the long nineteenth century not only reformulates well-trodden historical narratives, but also invites an enriched exploration of expanding colonialism, capitalism and science. As these discourses reshaped relations between humans, other-than-human species and a world of vibrant matter, they brought about permanent alterations and negative effects on global ecosystems that these assembled scholars collectively explore. This volume's essays have been organized into thematically inter-related clusters, even if their authors vary in approaches and interests. (They could well be re-organized very differently, since many ideas are held in common and several themes overlap). These themes concern the political ecologies and the movement of things; material ecologies of climate and extractive logics; the depletion and conservation of natural resources; animal agencies and natural histories; and agriculture and resource husbandry.

Throughout this collection, the words of anthropologist Timothy Ingold resonate, emphasizing the entanglements among which we find ourselves: "Like all other creatures, human beings do not exist on the 'other side' of materiality, but swim in an ocean of materials. Once we acknowledge our immersion, what this ocean reveals to us is not the bland homogeneity of different shades of matter but a flux in which materials of the most diverse kinds … undergo continual generation and transformation" (Ingold 2011, 24). We accept his invitation to a performative pictorial analysis "not as spectators, but as participants," and look to the continuities that link twenty-first-century artists, viewers and philosophers to their antecedents, finding meaning in objects whose materials bear witness to anthropogenic change (Ingold 1993, 159).

Acknowledgments

The editors have benefited from the help of numerous people whose assistance has been inestimable. Their contributions to this volume have included guidance, advice, encouragement and archival research, and in the ecocritical spirit of collegiality we have encountered along the many stages of this project, we extend our heartfelt appreciation. Particular thanks are offered to the anonymous reviewers of the volume in its preliminary form, and to Ted Geier; Julia Medina; Jean-Marie Carey; Sayandeb Chowdhury; Malcolm McNeill; Timothy Morton; Nina Amstutz; Michaela Rife; to the organizers and participants in Princeton University's Symposium *Picture Ecology: Art and Ecocriticism in Planetary Perspective* – Karl Kusserow; Alan Braddock; Mónica Domínguez Torres; Andrew Patrizio; Rachael Z. DeLue; De-Nin Lee; Gregory Levine; Sugata Ray; Stephen Eisenman; Greg Thomas; James Nisbet; T. J. Demos; Finis Dunaway; Anne McClintock; Faizal Sheikh; to Jeffrey Jerome Cohen and Stacy Alaimo; to the members of the Babel Working Group and the Material Collective; to Karen Overbey; Ellery Foutch; Joanne Lukitsh; Silvia Bottinelli; and Bryant University librarians Sam Simas and Maura Keating.

Notes

1 Climate scientists and historians continue to debate the precise moment at which anthropogenic change was initiated, dating it either to the invention of the steam engine, as Dutch chemist Paul Crutzen and his colleagues have done, or to a prior era of European colonial incursions in the New World which caused the catastrophic collapse of human societies, as others have proposed (Crutzen 2002; Kolbert 2011; Steffen, Grinevald, Crutzen and McNeill 2011; Lewis and Maslin 2015). Since Crutzen's early twenty-first-century observations were published, the very term 'anthropocene' has been subject to revisionist inquiry, positing capitalist extractive regimes and exploitative agricultural practices as alternate causes of climactic change (Haraway 2015; Moore 2016; Demos 2017).

2 For helpful summary assessments of the emergence of ecocriticism across the humanities, see Garrard (2011) and Braddock (2015).
3 This idea, embodied by a thing in the world that inspires new hypotheses and invites new models of thought, is Peter Galison's. Stefan Helmreich has more recently expanded upon this concept to include protean, planetary and ecologically connective substances such as water (Galison 2003; Helmreich 2011).
4 The idea that a web of ecological relationships had been forged across the world's deep time pre-dated Haeckel's conceptualization, taking early form in the work of scientists from Carolus Linnaeus to Charles Lyell, Charles Darwin, George Perkins Marsh and Alfred Russell Wallace, but the term 'ecology' didn't gain more widespread use until the twentieth century.
5 At the present time, art historian Rachel Ziady DeLue is preparing a book on the subject of Darwin's sole diagram in the *Origin of Species.*
6 While tree diagrams pre-dated Darwin and Haeckel's revisionary models of the great chain of being, they have proven a durable way of illustrating the development of eukaryotic life forms, and have proliferated across disciplines, rendering visible complex relational patterns and associational histories (Lima 2014).

Works Cited

Anderson, Nancy. 1991. "The Kiss of Enterprise: The Western Landscape as Symbol and Resource." In W.H. Truettner, ed., *The West as America: Reinterpreting Images of the Frontier, 1820–1920.* Washington, DC: Smithsonian Institution Press: 238–283.

Barad, Karen. 2003. "Posthumanist Performativity: Toward an Understanding of How Matter Comes to Matter." *Signs.* 28, 3: 801–831.

Bermingham, Ann. 1987. "Reading Constable." *Art History.* 10, 1: 38–58.

Braddock, Alan. 2015. "From Nature to Ecology the Emergence of Ecocritical Art History." In John Davis, ed., *A Companion to American Art.* Chichester: John Wiley & Sons: 447–467.

Braddock, Alan, and Christoph Irmscher, eds. 2009. *A Keener Perception: Ecocritical Studies in American Art History.* Tuscaloosa: University of Alabama Press.

Braddock, Alan, and Laura Turner Igoe. 2017. *A Greene Country Town: Philadelphia's Ecology in the Cultural Imagination.* University Park, PA: Penn State University Press.

Braddock, Alan, and Renée Ater. 2014. "Art in the Anthropocene." *American Art.* 28, 3: 2–8.

Boettger, Suzaan. 2016. "Within and Beyond the Art World: Environmentalist Criticism of Visual Art." In Hubert Zapf, ed. *Handbook of Ecocriticism and Cultural Ecology.* Berlin: Walter de Gruyter: 664–681.

Chakrabarty, Dipesh. 2009. "The Climate of History: Four Theses." *Critical Inquiry.* 35, 2 (Winter): 197–222.

Cohen, Jeffrey Jerome. 2013. *Prismatic Ecology: Ecotheory beyond Green.* Minneapolis: University of Minnesota Press.

Coole, Diana, and Samantha Frost, eds. 2010. *New Materialisms: Ontology, Agency and Politics.* Durham, NC: Duke University Press.

Cronon, William. 1995. "The Trouble with Wilderness, or, Getting Back to the Wrong Nature." In William Cronon, ed., *Uncommon Ground: Rethinking the Human Place in Nature,* New York: W. W. Norton & Co.: 69–90.

Crutzen, Paul. 2002. "Geology of Mankind." *Nature.* 415, 6867: 23.

Davis, Heather, and Etienne Turpin, eds. 2015. *Art in the Anthropocene: Encounters among Aesthetics, Politics, Environments and Epistemologies.* London: Open Humanities Press.

Demos, T. J. 2016. *Decolonizing Nature: Contemporary Art and the Politics of Ecology.* Berlin: Sternberg Press.

———. 2017. *Against the Anthropocene.* Berlin: Sternberg Press.

Dunaway, Finis. 2005. *Natural Visions: The Power of Images in American Environmental Reform.* Chicago, IL: University of Chicago Press.

Eisenman, Stephen. 2013. *The Cry of Nature – Art and the Making of Animal Rights*. London: Reaktion Books.
Galison, Peter. 2003. *Einstein's Clocks, Poincare´'s Maps: Empires of Time*. New York: W. W. Norton.
Garrard, Greg. 2011. *Ecocriticism*. 2nd edition. New York: Routledge.
Gontier, Nathalie. 2011. "Depicting the Tree of Life: the Philosophical and Historical Roots of Evolutionary Tree Diagrams." *Evolution Education and Outreach*. 4, 3 (September): 515–538.
Green, Nicholas. 1995. *The Spectacle of Nature: Landscape and Bourgeois Culture in Nineteenth Century France*. Manchester: Manchester University Press.
Haeckel, Ernst. 1866. *Generelle Morphologie der Organismus*. Berlin: G. Reimer.
———. 1879. *The Evolution of Man*. New York: Appleton & Co.
Haraway, Donna. 2008. *When Species Meet*. Minneapolis: University of Minnesota Press.
———. 2015. "Anthropocene, Capitalocene, Plantationocene, Chthulucene: Making Kin." *Environmental Humanities* 6: 159–165.
Helmreich, Stefan. 2011. "Nature/Culture/Seawater." *American Anthropologist*. 113, 1: 132–144.
Ingold, Tim. 1993. "The Temporality of the Landscape." *World Archaeology*. 25, 2: 159.
———. 2011. *Being Alive: Essays on Movement, Knowledge, and Description*. New York: Routledge.
Kelsey, Robin. 2014. "Ecology, Sustainability, and Historical Interpretation." *American Art*. 28, 3 (Fall): 8–13.
Kolbert, Elizabeth. 2011. "Enter the Age of Man." *National Geographic*. 219, 3: 60–85.
Kraynak, Janet. 2010. "The Land and the Economics of Sustainability." *Art Journal* 69(4). (Winter): 16–25.
Kurt, Hildegard. 2004. "Aesthetics of Sustainability." In Heike Strelow, ed., *Ecological Aesthetics, Art in Environmental Design, Theory and Practice*. Berlin: Birkhauser: 238–240.
Kusserow, Karl. 2018. "Welcome Remarks." *Picture Ecology: Art and Ecocriticism in Planetary Perspective Symposium*. December 7. Princeton University.
Kusserow, Karl, and Alan Braddock. 2018. *Nature's Nation: American Art and Environment*. Princeton, NJ: Princeton University Art Museum.
Lartet, Edouard. 1875. "On a Piece of Elephant's Tusk Engraved with the Outline of a Mammoth, from La Madeleine, dep. Dordogne." *Reliquiae Aquitanicae: Being Contributions to the Archaeology and Palaeontology of Périgord and the Adjoining Provinces of Southern France: 1865–75*. London: Williams & Norgate: 309–311.
Latour, Bruno. 2004. *The Politics of Nature*. Cambridge, MA: Harvard University Press.
———. 2005. *Reassembling the Social: An Introduction to Actor-Network-Theory*. Oxford: Oxford University Press.
Lewis, Simone L., and Mark A. Maslin. 2015. "Defining the Anthropocene." *Nature*. 519, 7542: 171–180.
Lima, Manuel. 2014. *The Book of Trees, Visualizing Branches of Knowledge*. Princeton: Princeton University Press.
Lyons, Maura. 2012. "An Embodied Landscape: Wounded Trees at Gettysburg." *American Art*. 26, 3: 44–65.
Mann, Charles. 2005. *1491: New Revelations of the Americas before Columbus*. New York: Alfred A. Knopf.
Marsh, George Perkins. 1864. *Man and Nature, or, Physical Geography as Modified by Human Action*, David Lowenthal, ed. (reprint, Seattle: University of Washington Press, 1965).
Marx, Leo. 1964. *The Machine in the Garden: Technology and the Pastoral Ideal in America*. New York: Oxford University Press.

McKee, Yates. 2007. "Art and the Ends of Environmentalism: from Biosphere to the Right to Survival." In Michel Feher, Yates McKee, and Gaelle Krikorian, eds., *NonGovernmental Politics*. New York: Zone Books: 539–561.

Mentz, Steve. 2015. *Shipwreck Modernity: Ecologies of Globalization, 1550–1719*. Minneapolis: University of Minnesota Press.

Miller, Angela. 1993. *The Empire of the Eye: Landscape Representation and American Cultural Politics, 1825 to 1875*. Ithaca, NY: Cornell University Press.

Miller, Carolyn R. 2016. "Genre Innovation: Evolution, Emergence, or Something Else?" *The Journal of Media Innovations*. 3, 2: 4–19.

Miller, Perry. 1967. *Nature's Nation*. Cambridge: Harvard University Press.

Mirzoeff, Nicholas. 2015. "Visualizing the Anthropocene." *Public Culture*. 26, 2: 213–232.

Moore, Jason W., ed. 2016. *Anthropocene or Capitalocene? Nature, History and the Crisis of Capitalism*. Oakland, CA: PM Press.

Morton, Timothy. 2009. *Ecology without Nature: Rethinking Environmental Aesthetics*. Cambridge, MA: Harvard University Press.

———. 2011. "The Mesh." In Stephanie LeMenager, Teresa Shewry, and Ken Hiltner, eds. *Environmental Criticism for the Twenty-First Century*. New York: Routledge.

———. 2012. *The Ecological Thought*. Cambridge, MA: Harvard University Press.

Nochlin, Linda. 1999. *Representing Women*. London: Thames and Hudson: 7–33.

Novak, Barbara. 1980. *Nature and Culture: American Landscape and Painting, 1825–1875*. New York: Oxford University Press.

Ruskin, John. 1844. *Modern Painters*, 2nd edition. London: Smith, Elder and Co.

Scott, Emily Eliza, and Kristin Swenson, eds. 2015. *Critical Landscapes: Art, Space, Politics*. Berkeley: University of California Press.

Smithson, Robert. 2000 [1968]. "A Sedimentation of the Mind: Earthworks." Reprinted in Smithson, Robert, and Jack D. Flam. *Robert Smithson, the Collected Writings*. Berkeley: University of California Press.

Stauffer, Robert C. 1957. "Haeckel, Darwin, and Ecology." *The Quarterly Review of Biology*. 32, 2 (June): 140.

Steffen, Will, Jacques Grinevald, Paul Crutzen, and John McNeill. 2011. "The Anthropocene: Conceptual and Historical Perspectives." *Philosophical Transactions of the Royal Society A*. 369, 1938: 842–867.

Thomas, Greg. 2000. *Art and Ecology in Nineteenth-Century France*. Princeton: Princeton University Press.

Truettner, William, ed. 1991. *The West as America: Reinterpreting Images of the Frontier, 1820–1920*, Washington, DC: Smithsonian Institution.

Wallach, Alan. 2002. "Thomas Cole's River in the Catskills as Antipastoral." *The Art Bulletin*. 84, 2: 334–350.

Weintraub, Linda. 2012. *To Life! Eco Art in Pursuit of a Sustainable Planet*. Berkeley: University of California Press.

Part 1

Political Ecologies and the Movement of Things

The Land Acknowledgements that frequently open public events in Canada, Australia and New Zealand (and increasingly elsewhere) draw the attention of participants to indigenous presences and histories that have been mistranslated, threatened or willfully erased by settler-colonial violence. The four essays in this section inquire how an ecocritical art history can likewise provide a lens through which to identify and acknowledge the scale and long-lasting planetary effects of anthropogenic change in the nineteenth century, arriving at a range of examples and conclusions.

Collectively, these scholars examine how visual culture is implicated in shaping and resisting contentious, geo-political, transnational exchanges. An apt designation for the concerns they share is located in Haraway's speculative periodization of the "Plantationocene" (rather than Anthropocene), in which planetary ecologies are seen through the lens of the transatlantic plantation system that commodified the natural world (Haraway 2016). These essays share the global ecological perspective encapsulated in Rob Nixon's important theorization of the "slow violence" of environmental degradation: a process that compounds the suffering of human populations who are relegated to "nature," and cast to the side of modern "progressive" narratives of globalization (Nixon 2011).

In his analysis of the "political ecology" of the American Civil War, Alan Braddock re-casts North America's internecine conflict as a matter of trans-Atlantic ecological concern, well beyond its immediate social and environmental effects. Like the other three writers of this section, he considers how the vast movements of people, biota and materials in the nineteenth century amplified the social inequalities and environmental deterioration that accompanied settler colonialism and enslavement. Re-reading familiar paintings by Winslow Homer, Thomas Moran and Édouard Manet, the narratives Braddock traces exceed traditional lineages of production and reception; they reveal the war's unparalleled loss of human and nonhuman animal life and devastated agro-ecologies.

Like Braddock, who invokes philosopher Timothy Morton's concept of the "hyperobject" to shift perspective from local preoccupations to the war's globally encompassing consequences (Morton 2010, 2013), Rosie Ibbotson envisions the Anthropocene as (in part) a scalar crisis of representation. She turns to another of Morton's pithy ecological paradigms: "ecology without nature" – to prize apart the loaded cultural constructions of "nature" from an ecological analysis (Morton 2007). Unpacking the colonial ecological violence enacted upon Aotearoa, New Zealand, she shows that British-instituted design principles of the Arts and Crafts movement were problematized by transnational adaptation. While this movement's initial emphasis on locally appropriate "natural" motifs in the UK is well documented, Ibbotson asks her readers to see New Zealand Arts and Crafts imagery as a project of the "imperial Anthropocene" which weaponized introduced English biota, and legitimized a view of "local nature" that acclimatized introduced species, even those detrimental to pre-colonial ecologies.

Urging the adoption of "animism" to an expanded vocabulary of ecocriticism, and noting its harmony with Bruno Latour's famous dictum "we have never been modern" (Latour 1991), Jessica Horton reassesses what she calls the "transcultural history of ecological thought" in the late nineteenth-century Indian craze. Reframing this aesthetic valorization of Indigenous culture affirms the ethical responsibilities endemic to ecological storytelling, Native American concepts and social justice movements across the Americas. Like Ibbotson, Horton considers the displacement of Indigenous peoples and posits "ecolonial relations" as a new tool with which to examine visual culture, within a history of colonization that is still unfolding, as, for example, in contemporary Indigenous activism in the face of environmental devastation. Horton examines two interlinked products of such entangled relations in the reservation era: a portrait of American businessman Edward Everett Ayer in his Chicago "Indian Room" painted in 1887, and a muslin drawing of the *Battle of the Little Bighorn* by the Lakota artist Standing Bear, attending to how Indigenous environmental knowledges intersected with colonial practices and formative ecological awareness.

Considering the ecological violence of the interoceanic canal's incursion through Panama, Sarah Moore de-centers the postcolonial human narrative of this technological monument to globalism to consider the "agentive presence" of climate and geography (Patrizio 2018, 17) that produced the canal as a more-than-human hybrid zone. In her close reading of three images, Moore challenges the long-standing assumption that the Panama Canal was a discrete location and definable problem that could be compartmentalized, measured and transformed. Instead, she proposes the region as a resistant zone, whose inhabitants demonstrate various models of transgressive hybridization in their interaction with non-local geo-political agents. Like the swamp-scapes in which Braddock locates slave resistance, the Panamanian landscape's nonhuman inhabitants show agentive opposition to infrastructural attempts to subdue them: plants overgrew construction equipment, thwarted work and complicated travel, while proliferating insects migrated freely across spaces and bodies. Defying invasive attempts at enforcing control, Panama's flora and fauna propelled counterevidence that complicated the dominant discourses of imperial subjugation. Resonating with Ibbotson's focus on the "imperial Anthropocene" and Horton's "ecolonial relations," Moore's analysis of how the aestheticizing metaphors of sublime perspective and idealization ran up against proliferating resistance invites us to think of Panamanian images as relational media entering into mutually transformative relationships with events, places and forces.

Works Cited

Haraway, Donna Jeanne. 2016. *Staying with the Trouble: Making Kin in the Chthulucene*. Durham, NC: Duke University Press.

Latour, Bruno. 1991. *Nous n'avons jamais été modernes: Essai d'anthropologie symétrique (We have never been modern)*. Paris, Editions La Découverte.

Morton, Timothy. 2007. *Ecology without Nature: Rethinking Environmental Aesthetics.* Cambridge, MA: Harvard University Press.

———. 2010. *The Ecological Thought.* Cambridge, MA: Harvard University Press.

———. 2013. *Hyperobjects: Philosophy and Ecology after the End of the World.* Minneapolis: University of Minnesota Press.

Nixon, Rob. 2011. *Slow Violence and the Environmentalism of the Poor.* London: Harvard University Press.

Patrizio, Andrew. 2018. *The Ecological Eye: Assembling an Ecocritical Art History.* Manchester: Manchester University Press.

2 "A Demonstration to the World"

Art, Political Ecology and the Global American Civil War

Alan C. Braddock

Perhaps no event of the nineteenth century did more to modify the social and environmental fabric – the political ecology – of the United States than the American Civil War (1861–1865). The term "political ecology," coined by a Smithsonian Institution biologist named Frank Thone during the Dust Bowl and later disseminated across disciplines from cultural geography to development studies, refers to the inextricably collective and often contested nature of environmental relationships, countering utopian notions of scientific or social neutrality (Thone 1935; Robbins 2012). The American Civil War certainly demolished such notions. This conflict between pro-Union states and the secessionist Confederacy ended legal slavery but it also destroyed much life, land and property, especially in the American South, where most of the military battles were fought. According to recent accounts, as many as 750,000 soldiers died along with a million horses and mules as well as millions of trees (Ransom 2001; Widmer 2014). Economists calculate the conflagration cost North and South more than six billion dollars in combined government expenditures, physical destruction and human capital as well as more than $7 billion in indirect costs (according to 1860 value estimates) from lost consumption (Hacker 2011). The conflict militarized long-standing ideological differences over labor and land use, pitting the industrialized North against the largely agrarian, slave-holding South. In this essay, I offer an art historical perspective on the American Civil War by highlighting several creative works that reveal the global dimensions of its political ecology. Though disparate in their origins and media, the artifacts in question variously disclose an international – even planetary – sense of scope, either in the hostilities themselves or in their social and environmental implications.

On Thinking Big: The Limits of Granular Environmental History

The ecocritical philosopher Timothy Morton has argued for the need to "think big" in order to comprehend environmental problems of unprecedented magnitude in the Anthropocene such as global warming, which he describes as a "hyperobject" (Morton 2010, 2013). Contrary to a familiar habit of environmentalist thought that nostalgically celebrates what he calls "Hobbit-like worlds" of localized smallness and self-containment, Morton praises different cultural expressions that broach vast scales of space and time – Milton's *Paradise Lost*, Darwin's *Origin of Species* and Tibetan prayer wheels, to name a few – as better models for understanding the contested cosmic mesh of Earth's present predicament.

Morton's observations help us see the limits of many recent environmental histories addressing the American Civil War, which have often focused on local material realities of battlefield conditions and other domestic details about the conflict in the United States. For example, Mark Fiege has extensively examined the pivotal 1863 Battle of Gettysburg, which he says "turned on the inescapable ties between people and the material world in which they lived," the "ability—or inability—of each side to procure resources from nature," and "the use and control of terrain" (Fiege 2012, 201). Similarly, Kathryn Shively Meier has studied the environmental challenges faced by common soldiers during the 1862 Shenandoah Valley and Peninsular Campaigns in Virginia, showing how "strange terrain, tainted water, swarms of flies and mosquitoes, interminable rain and snow storms, and oppressive heat" led the combatants to create "informal networks of health care based on prewar civilian experience" and "a universal set of self-care habits" to combat "their deadliest enemy—nature" (Meier 2013, 99). Interpretations of the war by these environmental historians have revealed fascinating details, but they emphasize domestic contexts and circumstances, not the broader global political ecology of the conflict.

Using a somewhat wider regional perspective, other environmental historians have investigated the Civil War in terms of agriculture, supply chains and the impact of Northern scorched-earth military strategies on Southern landscapes. For example, historian Lisa M. Brady describes the conflict as a "war upon the land," driven by Northern cultural desires to "control nature" and perceptions of the South as a disorderly, unkempt society with an immoral plantation regime of enslaved labor (Brady 2012, 20). Such perceptions existed before the war and were famously articulated by the prominent Yankee landscape architect Frederick Law Olmsted, who toured the South during the 1850s. In his 1861 book *The Cotton Kingdom*, Olmsted observed disparagingly that

> for every mile of road-side upon which I saw any evidence of cotton production, I am sure that I saw a hundred of forest or wasted land. ... Coming directly from my farm in New York to Eastern Virginia, I was satisfied ... that the proportion of men improving their condition was much less than in any northern community, and that the natural resources of the land were strangely unused, or were used with poor economy. (Olmsted 1861, 12)

"Economy" and "ecology" are etymologically related terms describing complex systemic relationships, usually with political inflections. Although "ecology" was first articulated in 1866 by the Prussian naturalist Ernst Haeckel, we can count Olmsted among earlier writers who anticipated the concept, in this case by casting aspersions on Southern agriculture (Worster 1994, 192–193).

Responding to the perceived disarray of Southern landscapes, Union military leaders in the American Civil War decided they might as well be destroyed in order to annihilate slavery and ensure victory. Accordingly, Northern commanders adopted a strategy of total war, devastating what Brady calls Confederate "agroecology." Officially codified in General William T. Sherman's Special Orders No. 120 of November 9, 1864, the Northern strategy authorized massive foraging raids, called *chevauchées*, in enemy territory to sustain Union forces as they marched through the South, asserting overwhelming federal power and disrupting Confederate supply lines. Sherman expressed the import of this policy in letters from the field, saying "We have devoured the land" (Brady 2009, 49).

Art historical accounts of the Civil War have also emphasized American battlefields, camps, soldiers or home front scenes in various media, concentrating on the representation of regional differences, racial identities, slavery, emancipation, gender and other sociopolitical or military matters relating directly to the war, its domestic reverberations or its subsequent memorialization in the U.S. Numerous monographs explore relevant work by prominent American artists while recent museum exhibitions have interpreted the struggle as a watershed in national aesthetic mood, especially in landscape painting and photography (Harvey 2012; Brownlee et al., 2013). Rare scholarship addressing environmental issues in Civil War art has similarly focused on local, regional or national landscapes. For example, Maura Lyons has examined what she calls "new and disturbing representations of the relationships between humans and nonhuman nature" exclusively in "Northern landscapes" of the U.S. (Lyons 2015).

And yet the American Civil War was a conflict with global reach, the implications of which have so far eluded ecomaterialist and art historical inquiry. It is important to remember that the United States as a nation always already existed within an international political ecology. Moreover, despite the institution of slavery and many other obvious shortcomings, it had established a representative democracy that challenged the older European monarchical order of things and provided the world with an important model of modern, liberal enlightenment. Indeed, the American Civil War made aspiring republicans everywhere concerned about the fragility and survival of representative democracy, especially as aristocratic empires reasserted power in Europe following the 1848 revolutions. This helps explain why President Abraham Lincoln, in his famous Gettysburg Address of 1863, defiantly declared – in the face of growing international doubt – "that government of the people, by the people, for the people, shall not perish from the Earth." Likewise, General Sherman touted Northern military strategy as "a demonstration to the World, foreign and domestic, that we have a power which [Jefferson] Davis cannot resist," clearly indicating awareness of the international stakes and perceptions of the war in the highest military circles. The American foreign diplomat and environmental historian George Perkins Marsh also evinced such global awareness in an 1861 letter to Secretary of State William Seward, describing the conflict as "a contest between the propagandists of domestic slavery and the advocates of emancipation and universal freedom," warning that "our hold upon the sympathy and good will of the governments, *and still more of the people of Europe*, will depend upon the distinctness with which this issue is kept before them" (emphasis in original) (Doyle 2015, 132). In the preface to his *Man and Nature, Or, Physical Geography as Modified by Human Action* (1864), a landmark in early ecological thought written in Europe during the American Civil War, Marsh articulated his purpose and perspective in explicitly global terms:

> The object of the present volume is: to indicate the character and, approximately, the extent of the changes produced by human action in the physical conditions of the globe we inhabit; to point out the dangers of imprudence and the necessity of caution in all operations which, on a large scale, interfere with the spontaneous arrangements of the organic or the inorganic world; to suggest the possibility and the importance of the restoration of disturbed harmonies and the material improvement of waste and exhausted regions. . . . [M]odern ambition aspires to yet grander achievements in the conquest of physical nature, and projects are

> meditated which quite eclipse the boldest enterprises hitherto undertaken for the modification of geographical surface. (Marsh 1864, 3, 4)

Examining complex and far-reaching environmental relationships over time, Marsh's *Man and Nature* expressed concern about the implications of humanity's "modern ambition" to conquer "physical nature" on a planetary scale. He also forcefully critiqued myths about the abundance and inexhaustibility of the Earth as a natural resource. Although he did not mention the scorched-earth campaign waged against the South by Sherman and other Northern military commanders, Marsh saw a disturbing global trend toward environmental depletion and destruction wrought by humankind. In his view, imprudent "changes produced by human action ... on a large scale" were unprecedented in the history of the world, demanding creative engagement and "restoration." Accordingly, an important 2011 study by several scientists, including Nobel Prize-winner Paul Crutzen, acknowledged Marsh's *Man and Nature* as a conceptual antecedent for understanding the Anthropocene (Steffen et al. 2011, 844).

Sherman, Marsh, Lincoln and many others knew that the Union's failure to win the American Civil War could result in the permanent institutionalization of slavery, irrevocably shaping the U.S. not just internally but also in its international relations. With these considerations in mind, a few historians now view the conflict as a global phenomenon. For example, Don Doyle refers to the American Civil War as "the cause of all nations" (Doyle 2015), and Brian Schoen explores what he calls "the global origins of the Civil War" (Schoen 2009). Taking a cue from these historians, the present essay uses ecocritical art history to highlight examples of creative work pertaining to the American Civil War that made "a demonstration to the world," either through engagement with international audiences or through the deployment of far-reaching environmental and aesthetic knowledge. In some cases, the works in question traveled and even originated outside U.S. national borders. By thus approaching art of the American Civil War in an expanded field, the essay attempts to model a more global art history responsive to the transnational impulses of ecocriticism in the Anthropocene.

C'est plus franc

Probably the most celebrated example of environmental representation in American Civil War art is the battle-scarred background landscape of Winslow Homer's *Prisoners from the Front* (Figure 2.1). Produced in 1866, this famous painting provided a pictorial summa of the conflict and established Homer's reputation as a leading American artist when it garnered strong praise in the Northern press during a debut exhibition at New York's National Academy of Design in 1866. The picture commemorated events surrounding the capture of Confederate soldiers at the Battle of Spotsylvania, Virginia, in 1864 by Union forces under the command of Brigadier General Francis Channing Barlow, a distant relative of the painter. We see Barlow in the picture standing at right, facing three stereotyped Confederate prisoners guarded by Union troops. Confederate rifles and a broken branch of Virginia pine rest symbolically at the feet of these figures. More troops and truncated trees appear in the background (Cikovsky and Kelly 1995, 17–59; Griffin 2006, 19–41; Harvey 2012, 169–171).

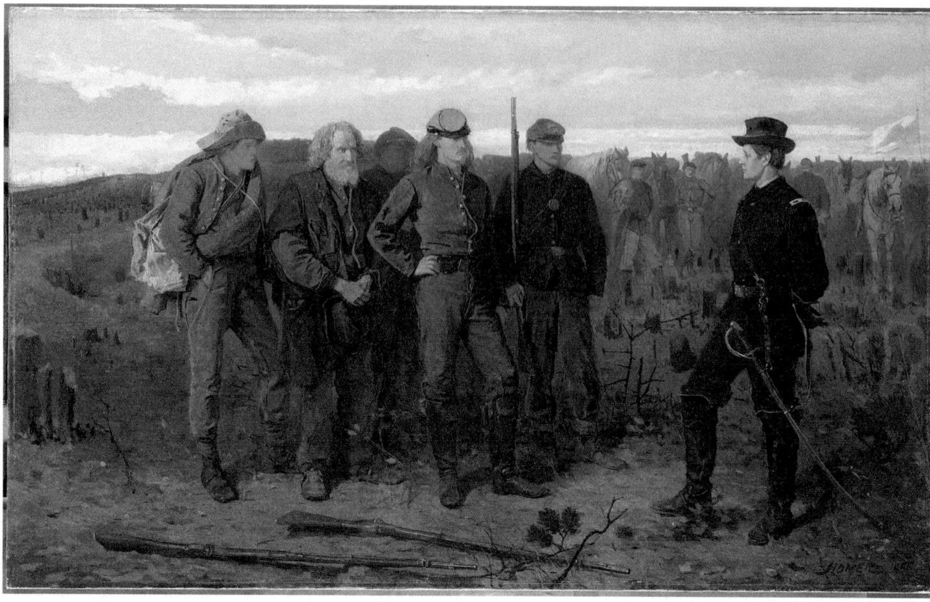

Figure 2.1 Winslow Homer, *Prisoners from the Front*, 1866. Oil on canvas, Metropolitan Museum of Art, New York.

New York art critics did not mention Homer's war-torn landscape, undoubtedly because its meaning seemed obvious and less important than the anthropocentric foreground drama. Generally sympathetic with the Union, they only addressed the human figures, praising the artist for epitomizing perceived regional differences between disciplined, upright Northern men and reckless, disorderly Southerners. As one critic observed in *The Round Table*, "It expresses, in a graphic and vital manner, the conditions of character North and South during the war." Another critic, writing in the *Evening Post*, said

> Mr. Homer shows us the North and South confronting each other; and looking at his facts, it is very easy to know why the South gave way. ... In our judgment Mr. Homer's picture shows the instinct of genius, for he seems to have selected his material without reflection; but reflection could not have secured a more adequate combination of facts, and they that think more must admit this natural superiority which enabled the painter to make his work at once comprehensive and effective. (Simpson 1988, 256, 257)

Analogous perceptions of "natural superiority" had informed Northern cultural and military views of the Confederacy as a morally bankrupt society incapable of imposing proper control over the land, making its destruction inevitable and appropriate. Visually acknowledging this political ecology of the war, Homer revealed the context and cost of battle in *Prisoners from the Front* by representing the Southern landscape of Virginia as a barren brown field "devoured" by Northern troops, with broken tree stumps extending into the distance as far as the eye can see. Such imagery appears in other Civil War pictures by Homer, whose job as an embedded Northern artist-journalist for the pro-Union New York magazine *Harper's Weekly* had regularly allowed him to witness such devastation.

Prisoners from the Front expressed Homer's Northern sense of political ecology both thematically and formally in terms of composition, color and technique. All the human figures and horses are enclosed within a pictorial field of drab earthen tones, except for the foreground protagonists, whose heads protrude into the gray sky, catching our eyes and focusing our attention. Barlow's dark blue uniform stands out above the ground (a technical term for the underlying layer of paint on a canvas), but the dirty grayish-brown tones of the Confederates' outfits fictively merge them with the broken land, suggesting that they "naturally" belong to it. Thus, while the composition renders both North and South as affected by the war, strategic formal and environmental cues clearly distinguish the victors from their vanquished adversaries, whose unruly character seems interchangeable with the Southern soil.

Both conceptually and logistically, Homer created *Prisoners from the Front* with an international audience in mind using cosmopolitan artistic knowledge. For example, its composition with opposing military forces recalls a long-standing tradition in the history of art, most famously exemplified in Diego Velázquez's *The Surrender of Breda* (1634–1635, Prado Museum, Madrid), a Baroque painting depicting the capitulation of Dutch soldiers to the Spanish army after a battle during the Eighty Years' War. The Velázquez picture was much admired by nineteenth-century realists and had circulated widely in photographic reproductions since the 1840s. Homer's first biographer, William Howe Downes, even compared the American artist with the Spanish Old Master, saying "Winslow Homer created his method of painting as truly as Velasquez created his method, that is to say, from the very ground up. ... Different as these men were, both dealt exclusively with realities—visible, tangible, material realities" (Downes 1989, 5).

Following his successful debut of *Prisoners from the Front* at the National Academy of Design in 1866, Homer sent the painting on a European tour to France and Belgium. At the Paris Exposition Universelle of 1867 – an enormous world's fair visited by Homer himself – the picture attracted the attention and approval of international art critics, all of whom praised the artist's realism. For instance, Théophile Thoré-Bürger declared Homer's picture to be better than the work of two prominent contemporary French artists in the same genre, Paul Alexandre Protais and Adolphe Yvon. According to Thoré-Bürger, "On trouve aussi ... une scène de la dernière guerre aux États-Unis, *Prisonniers confédérés* par M. Winslow Homer, de New-York. C'est plus franc que les idylles militaires de M. Protais ou que les épopées de M. Yvon" [One also finds ... a scene of the late war in the United States, *Prisoners from the Front*, by Mr. Winslow Homer, of New York. It is more honest than the military idylls of Mr. Protais or the epics of Mr. Yvon]. Such praise by this eminent French critic amounted to something more than mere technical appreciation. In addition to championing historical and contemporary realist painters like Johannes Vermeer and Édouard Manet for producing "art for the people" ("l'art pour l'homme"), Thoré-Bürger was well known for his republican political views, having founded the newspapers *La Vraie République* and *Le Journal de la Vraie République* in 1848 and 1849 only to see them both banned by conservative authorities in France. By expressing a preference for Homer over official French artists of the Second Empire such as Protais and Yvon, whose "military idylls" and "epics" served as propaganda for the reigning Bonapartist (and pro-Confederate) regime, Thoré-Bürger also signaled his approval of the "honest" American artist's Northern republicanism. Moreover, the critic's choice of wording contained a telling environmental dimension, for "idyll" connoted an excessively idealized, picturesque rendering of war. In contrast to such idealism, Thoré-Bürger probably admired the frank materialism of Homer's devoured Southern landscape. As an artistic "demonstration to the world," *Prisoners from the Front* revealed the environmental destruction of the Confederacy as a price worth paying to preserve the American republic (Thoré-Bürger 1870, 413).

Friendship to All and Every Nation

Americans obviously did not all share the same environmental experience or viewpoint about the Civil War. For enslaved African Americans, whom white Southerners generally considered to be only subhuman property, the political ecology of everyday life posed daunting existential challenges. Work usually lasted from dawn to dusk or longer, either in fields or in the master's home – environments of continual surveillance, intimidation, harassment, physical stress and deprivation. The slightest transgression could lead to torture or death. During the short periods of private time allotted to enslaved people, they attempted to raise families, build communities, grow their own food and maintain modest households, all the while subject to the whims of their owners, who could capriciously sell or abuse them with virtual impunity. The environmental historian Mark Fiege has described some of the ways in which enslaved African Americans negotiated their oppressive environments. For example, plantation field workers who understood the pace and life-cycle of cotton plant growth subtly and secretly adjusted their own activity accordingly to save energy, circumvent conflict, create extra time to cultivate private gardens or otherwise optimize their difficult predicament (Fiege 2012, 100–138).

The subtle, clandestine nature of such activity eluded artistic representation, but other environmental dimensions of black life surrounding the Civil War found expression in many works, including creative objects produced by African Americans themselves. For example, David Drake (1801–1870s), an enslaved man living on a plantation in Edgefield, South Carolina, signed his name, "Dave," prominently at the top of a glazed stoneware jug he made in 1857 (Figure 2.2). This was one of more than a hundred ceramic vessels created by Drake, a master potter, during the middle of the nineteenth century before emancipation. Like many of his pots, it also carries an extended verbal inscription, the mere existence of which defied systematic prohibitions against literacy among enslaved people. Though difficult to read in photographs, the inscription takes the form of a poetic couplet: "I wonder where is all my relation / Friendship to all and every nation" (Todd 2008, 239). In addition to hailing a world of nations beyond the borders of the United States and pondering the location of lost relatives, Drake's inscription here evokes a phrase from President Thomas Jefferson's first inaugural address referring to "Peace, commerce, and honest friendship with all nations—entangling alliances with none" (Woodson 2018, 181). Despite a repressive political ecology of slavery that conflated him and the pottery he made as property, Dave thus used the earthly materiality of clay to transcend his local space and express an avowedly global sensibility informed by American diplomacy. Moreover, the sturdy forms and articulate surfaces of his works even conjure living, speaking beings that defied bondage and retain an uncanny vitality today – amplified by an explosion

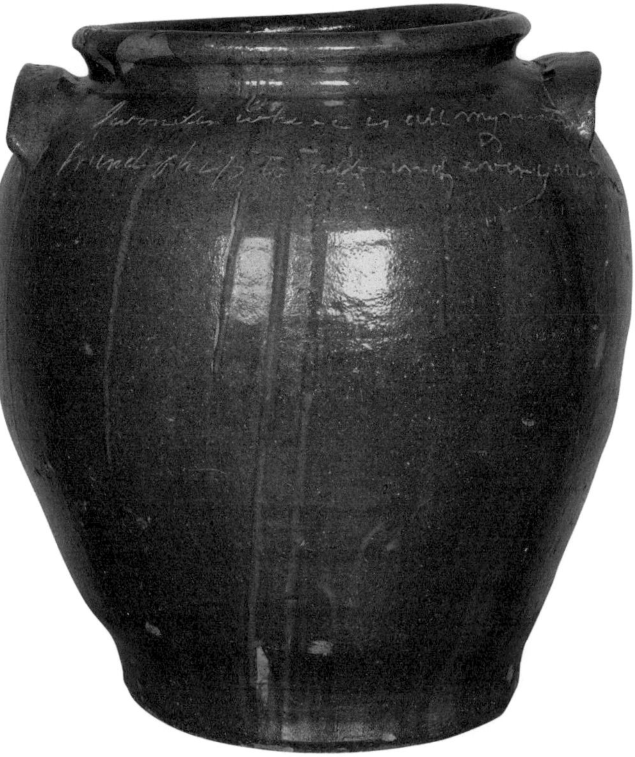

Figure 2.2 David Drake, *Food-storage Jar*, 1858, private collection, photo by David Greear, courtesy of the Folk Pottery Museum of Northeast Georgia.

of recent scholarship on the artist. An inscription on another pot made by Drake the following year alluded to the Big Dipper, a constellation pointing toward the North Star and offering an important beacon of freedom for enslaved people in the South (Burrison 2012). With his cosmic perspective, combining international awareness with environmental knowledge in the form of vernacular astronomy, Drake turned his pots into vehicles of global self-expression and instructional inspiration for runaway slaves.

The theme of the runaway, or fugitive slave, became a popular cultural sensation in all kinds of nineteenth-century media, thanks especially to the international fame of Harriet Beecher Stowe's antebellum abolitionist novel *Uncle Tom's Cabin* (1852). Stowe's celebrated book included a harrowing episode involving the winter escape of an enslaved woman named Eliza, who courageously carried her infant child across the frozen Ohio River from Covington, Kentucky, to freedom in the Northern city of Cincinnati (Stowe 1852, 98–117). In the visual arts, fugitive slave iconography varied widely, but one example giving particularly dramatic attention to environmental context is Thomas Moran's *Slave Hunt, Dismal Swamp, Virginia*, painted during the Civil War in 1862 (Plate 1). The picture represents a black man and woman with child in the left foreground of the marshy overgrown swamp, fleeing captors with hunting dogs in the distance at right. White Americans traditionally scorned swamps as

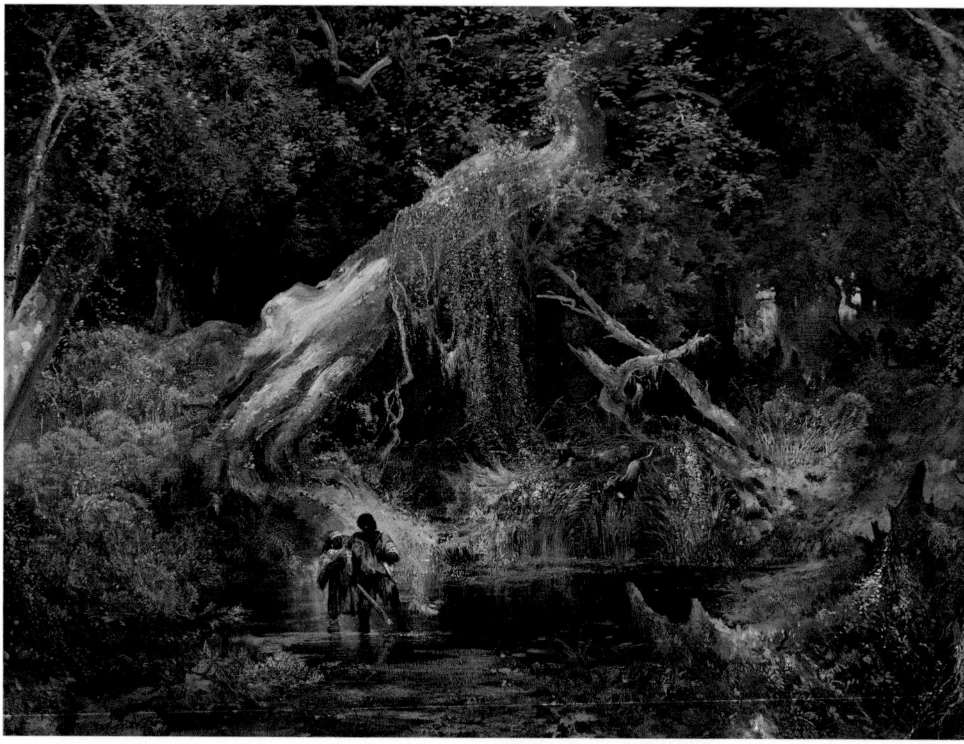

Plate 1 Thomas Moran, *Slave Hunt, Dismal Swamp, Virginia*, 1862, oil on canvas, Philbrook Museum, Tulsa.

wastelands that resisted economic development and nurtured malaria, among other unsavory things, giving rise to a host of negative associations. The very name of the Great Dismal Swamp, an enormous marshy area located on the border between Virginia and North Carolina, exemplified this cultural discourse (Miller 1989, 77–104; Harvey 2012, 195–198).

Yet Moran's picture projects a more complex environmental perspective inflected by the global political ecology of slavery. Whereas the white slave-trappers and their hunting dogs encounter the swamp as a series of obstacles to be overcome, the lush and colorful foreground vegetation of this semi-aquatic environment connotes something else for the fleeing African Americans, who often used such places as protective havens and passageways to freedom. As Stowe wrote in her 1856 novel *Dred: A Tale of the Great Dismal Swamp* – likely Moran's inspiration – "What the mountains of Switzerland were to the persecuted Vaudois, this swampy belt has been to the American slave," for "the near proximity of the swamp has always been a considerable check on the otherwise absolute power of the overseer" (Stowe 1856, vol. 1: 276). More recently, historian Daniel Sayers has called the Great Dismal Swamp "a desolate place for a defiant people." Accordingly, Moran's picture redeemed the benighted wetland as having value for African Americans, whose perspective the painting dramatically emphasizes by placing the fugitives directly before us. They even guide our gaze by turning to watch their pursuers through the dense forest. The fugitive man holds a bloody knife with which he has just killed one of the trapper's dogs.

A Northern artist of English origin, Moran sided with the Union and painted *Slave Hunt* for an English abolitionist patron while visiting London. This helps to explain why he depicted the swamp environment not as a negative wilderness or other sinister place, as had most previous artists, but rather as an exotic refuge of warmth, color and even visual allure. The swamp's strange beauty connotes resistance to Southern white control. In Moran's picture, cosmopolitan aesthetics thereby conspire with environmental vitality to connote "freedom." As the runaways advance into the marshy foreground, they confront the viewer with the political ecology and ethics of slavery. When we recognize that the fugitives comprise a family unit – recalling the biblical Holy Family on the Flight into Egypt – their empathetic appeal gains even greater moral and historical significance. This cosmopolitan reference appealed to anti-slavery sentiments that were particularly strong in Great Britain during the American Civil War.

Visualizing the Global Empire of Cotton

Not all art of the American Civil War was created by Americans. For instance, let us ponder a painting by the French artist, Édouard Manet, depicting *The Battle of the U.S.S. Kearsarge and the C.S.S. Alabama* (Figure 2.3). Manet produced this picture as part of a series of related works after a sea battle in which a Union warship sunk a Confederate raiding vessel off the coast of France in 1864. The artist did not witness the encounter but instead relied on news reports and his own aesthetic judgment in composing the series. In this example, the most dramatic of all, Manet placed the sinking *Alabama* in the upper center of the canvas, its damaged stern burning and partially submerged. The *Kearsarge* appears in the distance at left, obscured behind smoke billowing from the Confederate vessel. Two other unidentified boats sail nearby, one in the left foreground flying the French flag and another in the distance at right, ready to rescue survivors. The artist's animated brushwork and unusual vertical composition render the action with unusual drama for a maritime scene (Wilson-Bareau and Degener 2003).

Manet's picture highlights several important points about the American Civil War as a global phenomenon. By depicting a military encounter in European waters, it addressed hostilities that obviously extended beyond the boundaries of the U.S. and North America. Moreover, the painter's reliance on media reports in conceiving his series reveals how this "civil" conflict attracted international attention. As a republican artist living in imperial France, Manet (like Thoré-Bürger) opposed the aristocratic, Confederate-sympathizing government of Emperor Napoleon III. Other pictures by him similarly critiqued French colonialism, including several ensuing portraits of the *Kearsarge*, confirming his admiration for the Union vessel and the republican values it embodied (Herbert 1988, 60–65).

Furthermore, Manet's work underscores a fundamental truth about the global political ecology of the American Civil War. As a result of long-standing business relationships involving the production and international exchange of cotton from the American South, the *C.S.S. Alabama* was one of several Confederate military vessels built in the Birkenhead shipyards near Liverpool, England, with financing from British commercial textile interests. Confederate agent James Bulloch arranged for its construction there with funding from the Fraser Trenholm Company, a Liverpool cotton

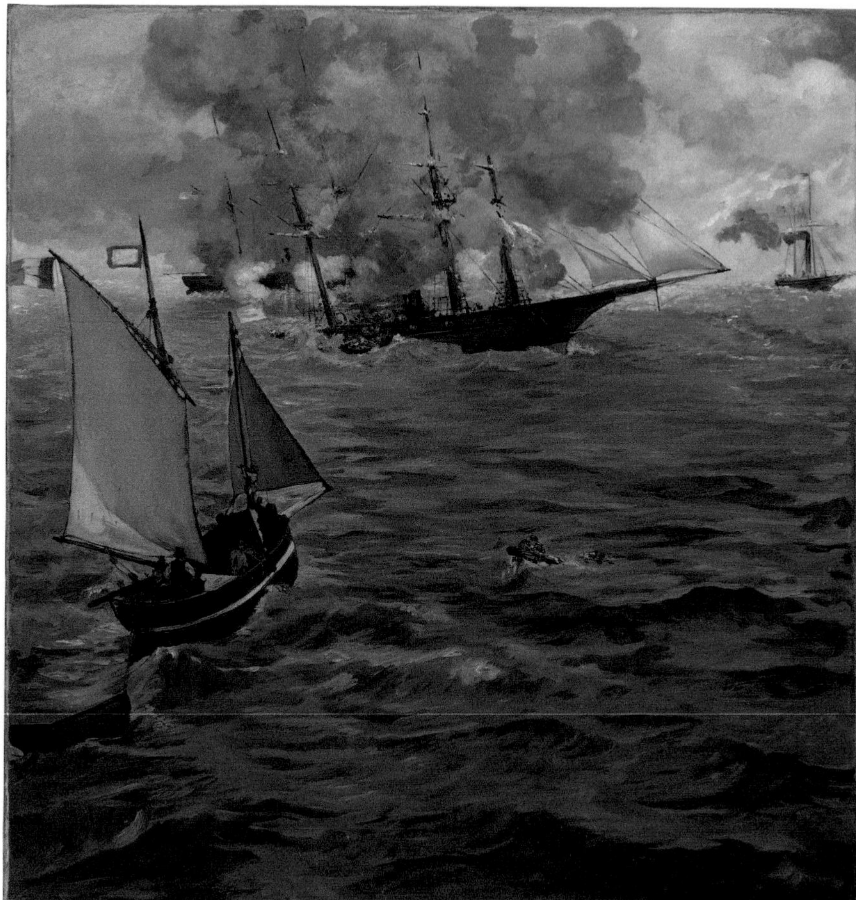

Figure 2.3 Édouard Manet, *The Battle of the U.S.S. Kearsarge and the C.S.S. Alabama*, 1864, oil on canvas, Philadelphia Museum of Art.

broker (Wilson-Bareau and Degener 2003, 22). As noted by historian Sven Beckert, "Liverpool, the world's largest cotton port, was the most pro-Confederate place in the world outside the Confederacy itself. Liverpool merchants helped bring out cotton from ports blockaded by the Union navy, built warships for the Confederacy and supplied the South with military equipment and credit" (Beckert 2014, 260). International cotton cultivation, trade and textile manufacture had fostered the Industrial Revolution, slavery and modern market capitalism – all of which were underlying economic causes of the American Civil War. The global "empire of cotton," as Beckert describes it, depended on the ecology and economics of a single plant species, which reshaped the world and precipitated a war with international consequences. By depicting a battle off the coast of France between a Union warship and a Confederate raider aligned with British textile interests, Manet broached this complex "empire of cotton."

Like other cosmopolitan republicans of the period, Manet probably knew of the *C.S.S. Alabama's* Liverpool cotton connections from reading newspaper accounts about British and French industrial ties to the Confederacy. These connections undoubtedly helped account for the ship's popular reputation among white colonial rulers of Cape Town, South Africa, a topic discussed recently by Tamar Garb (2016, 121–127). In celebrating the destruction of the *C.S.S. Alabama*, Manet affirmed American democracy and abolition. Ironically, by this time American cotton had insinuated itself into all sorts of modern consumer products, including artist materials. Cotton began to displace traditional linen (made from flax) in painters' canvases during the early nineteenth century. In a study of this material history, Anthea Callen observes "flax remained the most important vegetable textile fibre in Europe until the end of the eighteenth century, when cotton began to be imported on a large scale from the United States" (Callen 2000, 30). With the increasing abstraction and global flow of industrial commodities in modernity, painters of Manet's generation inadvertently became entangled in American cotton and slavery.

Cotton cultivation had already dramatically transformed landscapes and economies around the globe during the years leading up to the American Civil War. Historically, India was the world's leading cotton producer, but this role shifted to the U.S. in the early nineteenth century when Americans aggressively acquired aboriginal land for cultivation with low-cost slave labor. In 1830, U.S. President Andrew Jackson signed the Indian Removal Act, relocating Cherokee and other Native American communities from the Southeast to areas west of the Mississippi River, facilitating the expansion of cotton farming to feed rapidly growing British and American textile mills. America's ensuing sectional conflict therefore participated in a long-standing international contest over land and labor rooted in European colonialism. Beckert refers to the engine driving this contest as "war capitalism," involving the systematic European expropriation and clearing of indigenous territories for cotton production in various colonial-imperial contexts. Since cotton also rapidly leached nutrients from soil, its expanding growth increased demand for "virgin" land and slave labor across the South, exacerbating regional tensions and setting the stage for civil war (Beckert 2014, 20–97).

Well before the war erupted, environmental and social impacts of textile production based on American cotton had become impossible to ignore in other parts of the world. During an 1835 visit to Manchester – Britain's center of textile manufacturing – the eminent French writer-traveler Alexis de Tocqueville observed, "Black smoke covers the city. The sun seen through it is a disk without rays. A thousand noises disturb this damp, dark labyrinth. ... [From] this foul drain the great stream of human industry flows out to fertilise the whole world. From this filthy sewer pure gold flows" (Beckert 2014, 81). An 1852 painting titled *Manchester from Kersal Moor* (Royal Collection Trust, United Kingdom) by the British artist William Wyld testified to such conditions even as it attempted to naturalize them aesthetically. Representing the city as a distant forest of hazy industrial mills and smokestacks, it uneasily reconciled these with the picturesque rural landscape in the foreground. When Wyld created this work, the vast majority of cotton processed in Manchester came from plantations in the American South.

During the 1860s, Union naval blockades against Confederate shipping caused a "cotton famine," shuttering British factories and driving thousands of Manchester's wage laborers into soup kitchens. Speculators thrived from the economic instability by establishing a new futures market as textile manufacturers found alternative sources of cotton in India, Egypt, Brazil and Turkey. This production shift transformed local economies and environments in the latter countries by altering supply chains and clearing new land or replacing food crops with cotton. Beckert says that America's Civil War brought about "the world's first truly global raw materials crisis, and proved midwife to the emergence of new global networks of labor, capital, and state power" (Beckert 2014, 246). The pictures by Wyld and Manet testify to the effects of such networks.

The cotton crisis was temporary, however. Powerful postbellum forces of economic normalization led to the resumption and expansion of American cotton production. Slavery in the United States had been abolished, but a new global order emerged in which international market forces imposed an integrated system of wage labor for cultivating, processing, manufacturing and shipping cotton around the world. Lawyer and former Union military general Francis Channing Barlow – whom Winslow Homer depicted in *Prisoners from the Front* – put the matter bluntly in an 1865 letter to Henry Lee Higginson, discussing the viability of purchasing a cotton plantation in the South after the war: "Making money there is a simple question of being able to make the darkies work" (Beckert 2014, 281). In the American South and in other cotton-producing nations during the late nineteenth century, the predominant answer to this labor question involved a coercive system of sharecropping, compulsory contracts and debt peonage, effectively reinstating aspects of antebellum slavery under the guise of free market capitalism. African Americans in the former Confederacy had embraced emancipation, but the late nineteenth century brought daunting new challenges, including state-sponsored racial segregation laws – a repressive regime known as Jim Crow.

An artistic perspective on this postbellum political ecology appears in Winslow Homer's *The Cotton Pickers* of 1876 (Figure 2.4), representing cotton harvesting as a labor-intensive agricultural routine. Revisiting Virginia a decade after his wartime reporting there, Homer approached cotton harvesting as a form of economic captivity and ecological homogeneity, couched in cosmopolitan aesthetics. In *The Cotton Pickers*,

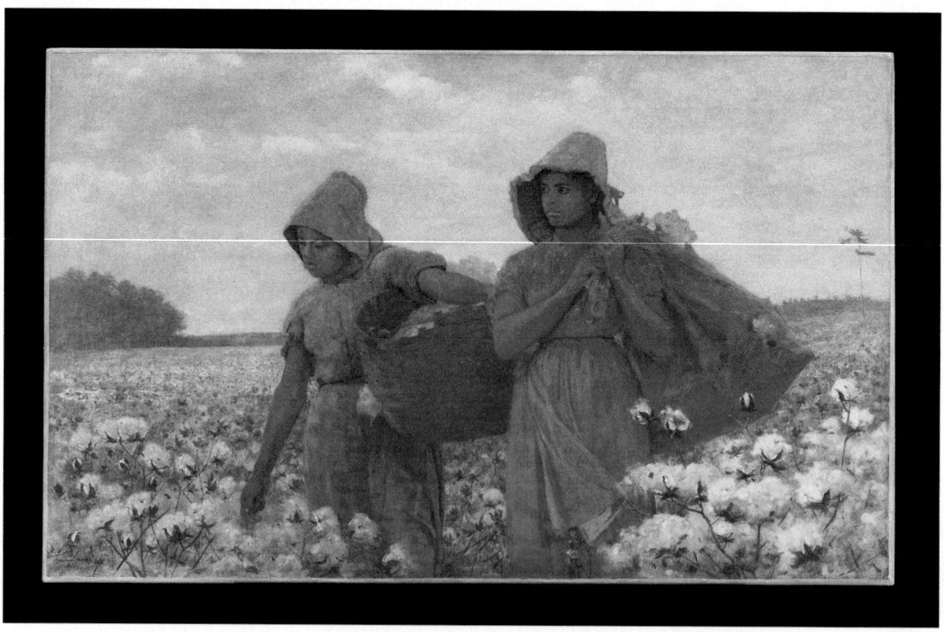

Figure 2.4 Winslow Homer, *The Cotton Pickers*, 1876, oil on canvas, Los Angeles County Museum of Art.

we see a pair of young African American women working an enormous field. The sea of ripe white cotton fiber around them provides a striking aesthetic motif that also acknowledges historical realities about large-scale monoculture during the late nineteenth century (Smith and Cothren 1999, 81). Here, the women's physical and emotional fatigue mirrors the exhaustion of Southern soils by this expansive, nutrient-sapping plant. A solitary tree in the distance at right alludes melancholically to the massive deforestation that made such immense cotton fields possible. Whereas Homer's *Prisoners from the Front* displayed a war-devoured Virginia battlefield with stumps stretching to the horizon, *The Cotton Pickers* presents a vast cash crop embodying postbellum economic normalization in the South. Slavery may have ended for these working women, but they remain trapped on the plantation by a reconstructed cotton empire that continued to deny them mobility or racial equality. Under this homogenizing regime, such laborers endured low wages, segregation, alienation and frustration. As one contemporary art reviewer observed about *The Cotton Pickers,* the women look "unhappy and disheartened," even "defiant and full of hatred." Another critic praised the foremost figure in *The Cotton Pickers* by saying, "The whole story of Southern slavery was written in every line of her patient, uncomplaining face" such that "It haunted me for days." But Homer's picture also reads somewhat ambiguously as a result of its cosmopolitan aesthetic beauty. Art historian Randall Griffin notes that the black women recall analogous field workers in nineteenth-century French paintings by Jean-François Millet and Jules Breton, while the cotton brings to mind fields of wheat and flowers in those same paintings. Shortly after Homer exhibited the painting in New York, the *American Art News* reported that a "wealthy English cotton spinner" acquired the work, which "is finely original and alluring, and when across the seas will do honor to the land that made it" (Cikovsky and Kelly 1995, 150; Griffin 2006, 86, 89).

Not unlike the empire of cotton, the art and political ecology of the American Civil War extended well beyond the borders of the United States, inviting scrutiny within an international context. In light of ongoing white nationalist racism and environmental injustice around the world, apparently the ideological landscape of this "nineteenth-century" "American" conflict surpasses its ostensible chronological parameters as well (Parrott 2017). Unfortunately, these expansive and enduring dimensions of the American Civil War seem to characterize not only the U.S. and Western modernity but also the Anthropocene, the new global geological epoch of human fashioning. Earth's inhabitants have left behind the Holocene era and entered an unprecedented historical condition, one in which the planet itself looks more and more like a built environment – an artwork of sorts – styled by the pervasive impacts of anthropogenic warming, pollution, warfare, inequality, migration, genetic modification, species extinction, mass slaughter and other human-dominated vectors (Ammons and Roy 2015). Under such conditions, the old distinctions between culture and nature, human history and natural history, have begun to collapse, merging ecology and art history in a global ethical struggle to curate survival. Ecocritical art history responds to this struggle by viewing art and "national" events like the American Civil War in a planetary perspective.

Works Cited

Ammons, Elizabeth, and Modhumita Roy, eds. 2015. *Sharing the Earth: An International Environmental Justice Reader.* Athens: University of Georgia Press.

Beckert, Sven. 2014. *Empire of Cotton: A Global History.* New York: Vintage.

Brady, Lisa M. 2009. "Devouring the Land: Sherman's 1864–1865 Campaigns." In Charles E. Closmann, ed., *War and the Environment: Military Destruction in the Modern Age*. College Station: Texas A&M University Press: 49–67.

———. 2012. *War upon the Land: Military Strategy and the Transformation of Southern Landscapes during the American Civil War*. Athens: University of Georgia Press.

Burrison, John A. 2012. "South Carolina's Edgefield District: An Early International Crossroads of Clay." *American Studies Journal*. 56. doi:10.18422/56-09.

Brownlee, Peter John, Sarah Burns, Diane Dillon, Daniel Greene, and Scott Manning Stevens. 2013. *Home Front: Daily Life in the Civil War North*. Chicago, IL: University of Chicago Press.

Callen, Anthea. 2000. *The Art of Impressionism: Painting Technique & the Making of Modernity*. New Haven: Yale University Press.

Cikovsky, Jr., Nicolai, and Franklin Kelly. 1995. *Winslow Homer*. Washington, DC: National Gallery of Art.

Downes, William Howe. 1911 [1989]. *The Life and Works of Winslow Homer*. Reprint New York: Dover.

Doyle, Don H. 2015. *The Cause of All Nations: An International History of the American Civil War*. New York: Basic Books.

Fiege, Mark. 2012. *The Republic of Nature: An Environmental History of the United States*. Seattle: University of Washington Press.

Garb, Tamar. 2016. "Revisiting the 1860s: Race and Place in Cape Town and Paris." In Hollis Clayson and André Dombrowski, eds., *Is Paris Still the Capital of the Nineteenth Century?: Essays on Art and Modernity, 1850–1900*. New York: Routledge: 115–130.

Griffin, Randall. 2006. *Winslow Homer: An American Vision*. New York: Phaidon.

Hacker, J. David. 2011. "Recounting the Dead." *New York Times*. September 20.

Harvey, Eleanor Jones. 2012. *The Civil War in American Art*. Washington, DC: Smithsonian American Art Museum.

Herbert, Robert L. 1988. *Impressionism: Art, Leisure, & Parisian Society*. New Haven: Yale University Press.

Lyons, Maura. 2015. "Nature Defamiliarized: Picturing New Relationships between Humans and Nonhuman Nature in Northern Landscapes from the American Civil War." *Panorama* 1, 1. https://editions.lib.umn.edu/panorama/article/nature-defamiliarized-picturing-new-relationships-between-humans-and-nonhuman-nature-in-northern-landscapes-from-the-american-civil-war/

Marsh, George Perkins. 1864 [1965]. *Man and Nature, or, Physical Geography as Modified by Human Action*, David Lowenthal, ed. Reprint, Seattle: University of Washington Press.

Meier, Kathryn Shively. 2013. *Nature's Civil War: Common Soldiers and the Environment in 1862 Virginia*. Chapel Hill: University of North Carolina Press.

Miller, David C. 1989. *Dark Eden: The Swamp in Nineteenth-Century American Culture*. New York: Cambridge University Press, 1989.

Morton, Timothy. 2010. *The Ecological Thought*. Cambridge, MA: Harvard University Press.

———. 2013. *Hyperobjects: Philosophy and Ecology after the End of the World*. Minneapolis: University of Minnesota Press.

Olmsted, Frederick Law. 1862. *The Cotton Kingdom*. New York: Mason Brothers.

Parrott, R. Joseph. 2017. "How White Supremacy Went Global." *Washington Post*, September 19.

Ransom, Roger L. 2001. "The Economics of the Civil War." *EH.net Encyclopedia*, edited by Robert Whaples. http://eh.net/encyclopedia/the-economics-of-the-civil-war/

Robbins, Paul. 2012. *Political Ecology: A Critical Introduction*. Malden, MA: Blackwell.

Schoen, Brian. 2009. *The Fragile Fabric of Union: Cotton, Federal Politics, and the Global Origins of the Civil War*. Berkeley, CA: University of California Press.

Sherman, William Tecumseh. 1864 [2004]. "To Ulysses S. Grant." November 6. In Bob Blaisdell, ed., *The Civil War: A Book of Quotations*. Mineola, NY: Dover: 156.

Simpson, Marc. 1988. *Winslow Homer: Paintings of the Civil War.* San Francisco, CA: Fine Arts Museums of San Francisco.

Smith, C. Wayne, and J. Tom Cothren, eds. 1999. *Cotton: Origin, History, Technology, Production.* New York: Wiley.

Steffen, Will, Jacques Grinevald, Paul Crutzen, and John McNeill. 2011. "The Anthropocene: Conceptual and Historical Perspectives." *Philosophical Transactions of the Royal Society A.* 369: 842–867.

Stowe, Harriet Beecher. 1852. *Uncle Tom's Cabin; or, Life among the Lowly.* Boston, MA: Jewett.

———. 1856. *Dred: A Tale of the Great Dismal Swamp.* Leipzig: Bernhard Tauchnitz.

Thone, Frank. 1935. "Political Ecology/Nature Ramblings: We Fight for Grass." *Science News Letter* (January 5): 14.

Thoré-Bürger, Théophile. 1870. *Salons de W. Bürger, 1861 à 1868.* Paris: Librairie de V. Jules Renouard.

Todd, Leonard. 2008. *Carolina Clay: The Life and Legend of the Slave Potter Dave.* New York: W.W. Norton & Co.

Widmer, Ted. 2014. "The Civil War's Environmental Impact." *New York Times.* November 15.

Wilson-Bareau, Juliet, and David C. Degener. 2003. *Manet and the American Civil War.* New York: Metropolitan Museum of Art.

Woodson, Jon. 2018. "Beneath Notice: A Social Philology of Poetry of Dave the Potter." In Michael A. Chaney, ed., *Where Is All My Relation?: The Poetics of Dave the Potter.* New York: Oxford University Press: 177–191.

Worster, Donald. 1994. *Nature's Economy: A History of Ecological Ideas.* 2nd edition. Cambridge: Cambridge University Press.

3 Crafting "Nature"

Ecocriticism, Environmental Violence and the Transnational Arts and Crafts Movement

Rosie Ibbotson

In discussions of the Arts and Crafts Movement, both primary and secondary sources emphasize its concern with "nature." A transnational and highly influential design reform phenomenon spanning several decades from the second half of the nineteenth century into the twentieth, the Arts and Crafts was as contradictory as it was interdisciplinary, but its reverence for "nature" appears to have been one of its few constants. In art-historical literature on the Movement, the centrality of "nature" is widely acknowledged, yet the concept itself – and its range of implications for the Arts and Crafts – has received insufficient critical examination. In light of renewed scrutiny for the idea of "nature" in recent posthumanities scholarship, this chapter is premised upon a reconsideration of the Movement informed by ecocritical thinking. Exploring what the Arts and Crafts' focus on "nature" might have meant in environmental terms, I look particularly at developments in Aotearoa New Zealand, where in the nineteenth century, ecologies were rapidly transformed under white settler colonialism. While the Arts and Crafts itself expressed prescient ecocritical philosophies, this essay illuminates how the Movement could also have environmental ramifications, and positions it as one of many forms of visual culture implicated in the imperial Anthropocene. In foregrounding the substantial agency of images within this environmental violence brought about by colonization – as well as the way in which changes to land and ecologies were often guided at least in part by aesthetic notions – this research responds to historiographical calls for environmental histories to analyze together the material and cultural aspects of ecological change (see, for example, Beattie, O'Gorman and Melillo 2014; Beattie, Melillo and O'Gorman 2015).

As one of its leading protagonists described, the Arts and Crafts was "a movement in the main of ideas and not of *objets d'art*" (Cobden-Sanderson 1905, 29). While it admittedly gave rise to certain visual styles, this was not the intention of the Movement, and it was instead informed by a set of interconnecting principles – all of which claimed deep reverence for "nature," and many of which resonated with contemporary developments in socialism. Premised at least initially on a holistic critique of industrialism, the Arts and Crafts first arose in Britain, where the multifarious effects of a capitalism vastly accelerated by fossil fuel extraction were evident early on. These included dangerous mechanized working environments; exploitative employment terms; and cramped, unsanitary living conditions in rapidly urbanizing areas. Appalled by these social and human rights issues, thinkers such as John Ruskin and William Morris decried the associated systems of production and looked to handicrafts as a means to restore dignity to labor and artistry to the decorative arts.

Perhaps rather prophetically, the theories of Ruskin, Morris and others connected with the Arts and Crafts Movement hinged upon a recognition of the agency of art, and understood that material culture and its production had the potential to effect substantial changes to social structures and economic power relations. Morris, an indefatigable Victorian polymath and arguably the Arts and Crafts' most influential practitioner, recognized the environmental component of this equation too, and today is seen by some as a significant proto-ecocritical voice. However, this strand within his thought is substantially contested, and histories concerning the Movement's visual culture have largely overlooked it, or diminished its relevance: as design historian Kjetil Fallan notes, "Design historical readings of Morris and the Arts and Crafts Movement ... have tended to view these ideological concerns more as social critique and aesthetic inspiration rather than as some sort of proto-environmentalism" (Fallan 2016, n.p.). Notwithstanding debates around the extent to which Morris can be positioned as an environmental visionary, it remains notable that not only was he keenly aware of an accelerating ecological devastation that characterized the nineteenth century, but that he unequivocally connected this with both the products and structures of industrial capitalism. In a frequently quoted passage from his essay "How I became a socialist," for example, Morris imagined "a counting-house on top of a cinder-heap" as part of a dystopian vignette that sardonically speculated about what might ultimately result from the dominant economic system (Morris 1963a, 36). Morris also saw environmental protection and attempts to reform the applied arts as inseparable, and in his canonical lecture of 1877 titled "The lesser arts," he parodied the pursuit of profit, positioning it as ecologically destructive and the common denominator of all contemporary ugliness:

> Is money to be gathered? cut down the pleasant trees among the houses, pull down the ancient and venerable buildings for the money that a few square yards of London dirt will fetch; blacken rivers, hide the sun and poison the air with smoke and worse... that is all that modern commerce, the counting house forgetful of the workshop, will do for us herein. (Morris 1963b, 102–103)

Arts and Crafts theory and practice drew heavily from Morris's example, and while environmental concerns were implicit across many of the Movement's ideas, not all of its protagonists dealt as directly with this issue in their writing. Nonetheless, practices that today might be read as orientated toward sustainability were enshrined within Arts and Crafts principles, such as the use of local materials in designs, and the preference for "slow" labor such as handwork over industrialized systems of production. Furthermore, ideas about environments pervaded the Arts and Crafts, not least in its advocation of design that would promote "healthy" and beautiful spaces for living in. The idea that such surroundings would be "improving" for society appears to have been a given within Arts and Crafts theory, suggesting the reliance of aspects of the Movement on problematic assumptions connected with sinister notions such as environmental determinism.

However, it was in the Arts and Crafts' fixation with "nature" that perhaps the most consistent evidence of the Movement's engagement with the environment can be found. Ubiquitously celebrated in both theory and designs, and positioned as the ultimate source of good art, the concept of "nature" also appears to have been central to how the Movement perceived its own legitimacy, and it frequently invoked "the moral authority of nature" – to borrow a phrase from Lorraine Daston and Fernando Vidal (2003) – to naturalize its artistic and social theories. The Arts and Crafts' range of uses for the idea of "nature" – as well as what this even comprised – can be productively complicated by ecocritical perspectives, not least because "nature is not natural," as posthumanist thinkers have emphasized (Harman 2005, 251). Crucially, as ecological philosopher Timothy Morton has shown, it needs to be understood as a construct that is distinct from ecology (Morton 2007, 2010): far from being plants or nonhuman animals or even their enmeshed totality, "nature" can be thought of as a potentially malevolent epistemological category that serves to conceal humans' inextricable and often deleterious interactions with other species.

Indeed, in its promotion of "nature," the Arts and Crafts participated in these destructive ecological entanglements, and this is particularly legible in the regions of the world to which the Movement spread as a result of colonialism. In addition to its universal engagement with natural motifs, the Arts and Crafts specifically advocated for design inspiration to be "locally appropriate." This principle encompassed determining which architectural vernaculars should be looked to for influence, and was, of course, highly ideological: in its most troubling implications, the co-option of Arts and Crafts ideas in the service of rampant nationalisms and xenophobia can be glimpsed. However, the prescription to focus on "local nature" – which extended to materials as well as design imagery – was frequently ambiguous too, including in Britain where the principle had been formulated. Here, of course, the hybrid biota bore the imprints of centuries of travel and imperial activity – the ecological tracers of myriad geopolitical, commercial and cultural interactions – as well as of extensive horticultural doctoring. While this has usually been overlooked in secondary sources on the Arts and Crafts Movement (exceptions including Calhoun 2015, and Parry 2005), nineteenth-century designers might have been more aware of it. In setting down what would constitute appropriate design inspiration, for example, Richard Redgrave – a key figure in the establishment of the "South Kensington" model of art education, which was crucial in the prehistory of the Arts and Crafts – prescribed referencing plant species "indigenous to our soil and climate ... in their native growth without the artificial qualifications of culture," but acknowledged that this would be "no trifling consideration when we observe how plants are modified by the art of the horticulturalist" (quoted in Calhoun 2015, 53). As well as suggesting how a number of key design principles of the Arts and Crafts did predate the Movement, this quotation might also sound a note of caution around the need to be critical about the political implications of attempting to discern hard lines between native, introduced and cultivated plants. Nonetheless, it seems to be the case that – as ideological as this pursuit might be – potential ecological "baseline states" are easier to determine for some regions than for others. Ecological historian Marcus Hall argues that the so-called "New World" countries seem to offer more straightforward cases here, making determinations of where distinctions fall between indigenous and introduced species often easier to plot than for vast continental regions such as Europe and Asia (Hall 2010, 2). Aotearoa New Zealand's status as an archipelago further adds to this, and while a number of plant and animal species were introduced by Māori, notions of the country's ecological baseline tend to hinge – sometimes controversially – around the extreme differences to the land pre- and post-European settlement, which began systematically in the first half of the nineteenth century.

As this suggests, colonial ecological violence in New Zealand was substantially connected to the prolific importation of non-native species of plants and animals. This "acclimatization" was aimed in part at emulating Britain, the "mother country," and had a transformative effect on the environment: Walter Buller, a Pākehā lawyer and amateur ornithologist who himself played a major role in the extinction of various bird species, wrote in 1905 with barely concealed approval that "The Colonial 'rage for acclimatization', as it has been termed, is altering the face of everything" (Buller 1905, vol. 1, xlii). In this British colonial context, therefore, notions as to what might have counted as "locally appropriate" to the Arts and Crafts Movement were rather more elastic, meaning that this design principle was problematized by its transnational adoption. While Arts and Crafts practitioners in countries to which the Movement spread adapted the principle so that their work reflected impressions of their particular environments, and therefore tended to draw on motifs that were different to those featured in British Arts and Crafts designs (which were known internationally but seldom copied directly), in New Zealand the picture appears to have been particularly complex. Plants and animals remembered from Britain constituted cultural symbols lodged deeply within the iconography of the settler imaginary, and this was supported by transnational flows of visual and material culture – as well as the fact that many of these species were themselves starting to appear in Aotearoa.

Arts and Crafts attitudes to nature, beauty and the environment were therefore to resonate significantly with ecologically destructive practices when the Movement spread to Aotearoa New Zealand. Ideas and objects from the British Arts and Crafts Movement were quick to transfer to the so-called "Antipodes" on a relentless tide of rapid colonial settlement and its attendant commercialism. By the closing decades of the nineteenth century, this was reflected in the decoration of many interiors, as well as in colonial New Zealand's nascent systems of European art education. While the Arts and Crafts Movement in Aotearoa is notable for its successful endurance into the second half of the twentieth century, it began as a distinctive hybrid, characterized by a mixture of imported goods and domestically produced designs. As well as furniture and other applied arts traveling to New Zealand among the personal effects of British immigrants, businesses in the major urban centers such as Wellington and Christchurch continued to supply consumers with recent trends, and the wide appeal of Arts and Crafts products (and mass-produced imitations) ensured their significant infiltration of the colonial domestic sphere. This growing market, and the need for technical skills, also prompted the establishment of a number of publicly funded art schools, including in Dunedin (1870), Christchurch (1882), Wellington (1886) and Auckland (1889). Most of the teachers that staffed these schools in the early decades had relocated from Britain, bringing with them the Arts and Crafts-inspired approaches to design training that had often characterized their own artistic educations. While many of these artist-teachers settled permanently in New Zealand, they often maintained close ties to artistic developments happening at "Home" (as settlers often called Britain at the time) through travel, correspondence and the periodical press.

In addition to this, strong British influences were perpetuated by these New Zealand schools' adherence to the "South Kensington" model of design training. As well as having been devised in Britain, it required that work be sent to London for examination – a practice that continued until New Zealand schools' withdrawal from this system in 1913 (Calhoun 2015, 165). Awkwardly, in relation to Arts and Crafts principles, this approach saw prominent art educators in London attempting to evaluate design work that originated in a country most of them had not been to. A design paper written in 1898 by Canterbury College School of Art (CCSA) student Annie L Ford, for example, was examined by Walter Crane and Lewis Day – both luminaries of the British Arts and Crafts Movement – and a heavily stylized design for a linoleum print she made in the same year was awarded a South Kensington prize (Calhoun 2015, 233). This variously symmetrical design featured an intricate border enclosing the central dark blue and yellow variegated pansy motifs – significantly a flower that was neither native to Aotearoa, nor particularly "natural," having been extensively cultivated. While it is impossible to be certain about whether details of this sort would have been known to the examiners, or the extent to which they would have mattered, the success of Ford's design is possibly symptomatic of a system of evaluation that intentionally or not might have privileged motifs imported to Aotearoa, arguably at the expense of local ones which might have been devalued on account of their unfamiliarity. If nothing else, however, the transplantation to New Zealand of both the pansy flower and the South Kensington educational system were of a piece: both were symptomatic of this colony's particularly close emulation in the nineteenth century of its "mother country."

However, in perhaps a more faithful interpretation of the Arts and Crafts principle of "fitness to locality," indigenous plant species also featured prominently within the Movement's design from New Zealand. While historiographical distortions (including significant gaps in the record) make it difficult to assess the balance between representations of native and introduced species, and therefore any shifts over time, histories of other phenomena might hold clues. In environmental historian Paul Star's account of bird protection in Aotearoa, for example, a general pattern away from the dismissal of indigenous species and toward their greater appreciation is indicated, framed within the growth of New Zealand nationalism (Star 2002). Indeed even from an early stage in New Zealand Arts and Crafts design, there is evidence of attempts within art education ecologies specifically to promote native plants as the most appropriate exemplars. At the New Zealand Industrial Exhibition held in Wellington in 1885, for example, young Dunedin artist Isabel Hodgkins won first prize (Phillips 2014, n.p.) and an "Honorable mention for design for wall-paper, New Zealand foliage" (quoted in Calhoun 2000, 49) for a recent watercolor and ink drawing of tātarāmoa, an indigenous plant of the genus *Rubus* also known as "bush lawyer" (Figure 3.1). Based around a strictly limited palette mostly comprising grayish greens, this design appears to feature the leaves and flowers of a single plant, energetically composed around curving stems. The motif is lightly stylized yet strongly graphic, flattened against a plain background of a not dissimilar color to the serpentine leaves. Closer inspection of Hodgkins's image reveals a deeper sense of artifice, however: the sparsely scattered subtly pink-tinged flowers are not attached to the greenery but instead appear to float in the small gaps between it, pinning the composition. In its restrained colors and simplicity, Hodgkins's watercolor resonates strongly with British Arts and Crafts wallpapers – including many later designs. In its familiar formal logic, and the resemblance of the flowers to wild roses, the native tātarāmoa in Hodgkins's design appears absorbed into the hybrid imaginary of the transnational Arts and Crafts Movement, reflecting the same aesthetic that was the vehicle to Aotearoa for many images of introduced plants. No doubt partly in response to this, by 1898, in a report to the government on "manual and technical instruction" in education, British immigrant and founder of the Wellington School of Design Arthur Dewhurst Riley specifically discussed "Painting still-life and plant-form," instructing that "native foliage should receive particular attention" (*AJHR* 1898, 38–39).

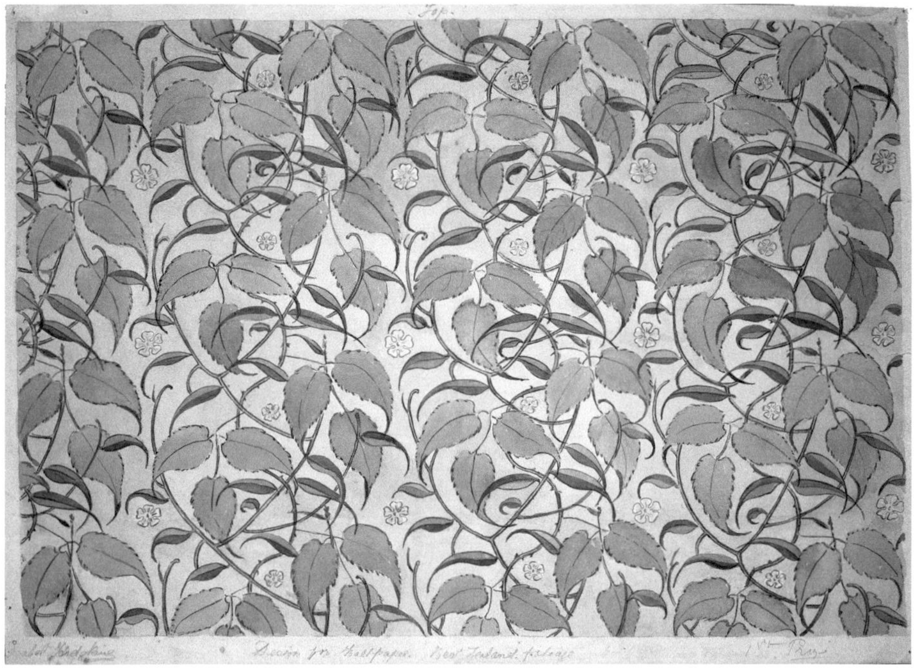

Figure 3.1 Isabel Jane Field, design for wallpaper, New Zealand foliage. 1st prize New Zealand Industrial Exhibition, Wellington, 1885. Alexander Turnbull Library, Wellington, New Zealand.

Despite this, however, the botanical volumes that Riley's report mentioned as potential "text books" were all British sources, and spanned from John Gerard's *Herbal* of the late sixteenth century to various nineteenth-century titles, including Anne Pratt's *Flowering plants of Great Britain*, first published in 1855 (*AJHR* 1898, 40). This heavy reliance on imported textual sources is all the more conspicuous given the existence by that date of various atlases documenting the indigenous flora of Aotearoa, including, for example, Thomas Kirk's *Forest Flora of New Zealand* of 1889. In addition, the numerous design manuals mentioned in Riley's report were imports too, and feature several British Arts and Crafts titles, including Crane's *The Claims of Decorative Art* of 1892, Day's *The Anatomy of Pattern* of 1887 and *Arts and Crafts Essays* of 1893, to which many leading figures of the Movement contributed, and which had a preface by Morris himself (Riley 1898). Books appear to have been pivotal within the design exercises practised in New Zealand art schools: in many cases, students seem to have commenced their training by working extensively from books, before progressing onto producing designs that referenced specimens drawn from life, including native plants. As described in the CCSA's 1894 prospectus by Samuel Hurst Seager, architect and lecturer in decorative design, "The designs at first will be simple modifications of the examples given, and later original designs based on New Zealand flora and fauna" (quoted in Calhoun 2015, 236).

The significance of these imported books should not be underestimated: as historian James Belich has observed in relation to culture in nineteenth-century New Zealand,

"Books, news, mail, and the like were the nervous system of Greater Britain, and ... could carry identity as well as information" (Belich 2009, 460–461). Visual cultures were, of course, a significant part of this too, and it should be noted that the interior designs the majority of European settlers would have been exposed to in this period did not necessarily reflect the ideas being explored in art schools. Despite Hodgkins's tātarāmoa imagery being designed for wallpaper, for example, it is unclear whether this ended up being manufactured. While stenciling was a technique also sometimes used in the decoration of interiors (Salmond 1986, 153), presumably increasing the likelihood of some walls having represented local indigenous flora, the vast majority of wallpapers were imported from Britain (Stewart 2002, 56; McCarthy 2009, 8–9) and did not reflect Aotearoa's native plants. Design historian Christine McCarthy, in her survey of the literature on domestic wallpapers in New Zealand, suggests the range of colors featured in imported British products, and indicates the ubiquity of flower motifs by the end of the nineteenth century. Flowers performed significant symbolic work within the colonial imagination, and elsewhere McCarthy highlights the nostalgia implied by the popularity of such designs, arguing that "Wallpaper was surely one of these 'little comforts' which enabled the British colonist to create an image of established domesticity in the antipodes" (McCarthy 2011, 101). As "the cheapest universally applied wall decoration throughout the nineteenth and early twentieth century" (Stewart 2002, 56), imported wallpaper was also very popular.

Therefore, despite the interest in motifs drawn from indigenous flora expressed in some New Zealand Arts and Crafts designs, other locally produced imagery – as well as the vast majority of patterns imported and consumed – reflected plant species from elsewhere. This would have had the effect of creating "interior landscapes," to borrow Ronald Rees's term (1993), that did not reflect Aotearoa's precolonial environments, and which instead featured potent agents of ecological change. As European settlers rapidly altered New Zealand landscapes – in part through the importation of plant and animal species, many of which were to become invasive – domestic interiors would have normalized these damaging environmental transformations. In addition to reflecting and promoting the "new nature" taking hold outside, such designs would have subtly destabilized perceptions of the precolonial ecological "baseline state," naturalizing introduced species and to some extent concealing the devastation many of these did to indigenous environments. Indeed, the idea of blurring the boundaries between interiors and exteriors was prominent within Arts and Crafts thinking around domestic design, suggesting that these colonial environmental transformations – while not expressly motivated by a desire for ecological destruction per se – were not an accidental part of the Movement's wider purview.

While species introductions were, of course, practised in many places – including back in Britain itself – the extent and consequences of "acclimatization" in Aotearoa New Zealand were compounded not only by the uniqueness and fragility of the ecosystem, but by the pervasive idea that these Pacific islands might become a "better Britain" (Belich 2001). For some, the desirable end goal was "a daguerreotype miniature of the mother country" as it was put in *The Lyttelton Times* in 1851 ("Canterbury settlement," April 5, 1851), a remark that illuminates the expressly pictorial terms in which the forging of the colony was imagined. Tāngata whenua (the indigenous Māori peoples) were confronted with vast tracts of their land being stolen, something that not only resulted in drastically diminishing resources, but also involved the cultural shocks that attended the widespread clearance of forests by European settlers – including the related extinction of several bird species. Animals and plants introduced by colonists keen to profit from Aotearoa's soils and nostalgic for their birth countries played key roles in this ecological violence: many of these species were to prove profoundly detrimental to endemic ones, and as environmental writer Philip Temple startlingly puts it, "The exploitation and destruction of New Zealand's nature and landscape gathered momentum as British settlers and their imported plants and animals spread like a virus through a body without an immune system" (Temple 1998, 77). While this metaphor has some potentially problematic implications, it speaks to the unusually fragile biota of Aotearoa, which up until the arrival of Polynesian settlers less than 1,000 years ago had evolved entirely in the absence of people and other mammals. While tāngata whenua had developed complex systems for ensuring the sustainability of most species, based on the principle of kaitiakitanga (a concept that might be translated as "guardianship"), British colonists' impact on the archipelago's ecology was highly deleterious and rapid.

The organized introduction of a profusion of plant and animal species by European settlers was framed by the idea of "acclimatization." As historian of science Michael Osborne has noted, this word in the nineteenth century "embraced an astounding range of uses and meanings," including pseudo-scientific worldviews (Osborne 2000, 137), and he also argues that "many perceived acclimatization to be the paradigmatic colonial science" (Osborne 2000, 135). In terms of New Zealand, the "systematic colonization" that had been promulgated by colonial theorist Edward Gibbon Wakefield and others was, of course, entirely dependent upon the establishment of introduced plants to supply enough food to support a rapidly growing population. While fruits and vegetables as well as other utilitarian considerations were therefore the initial priority, species selections became more ambitious as aesthetics and settler nostalgia became key driving motivations; as literary historian Chris Tiffin comments, "the strange phenomenon of the Acclimatization movement... provide[s] a unique window on the colonizing eye of imperial England" (Tiffin 2007, 175). However, in his essay "The empire of the rhododendron," environmental historian James Beattie has complicated previous accounts of acclimatization in Aotearoa such as Alfred Crosby's that subscribe to a simplistic "center-periphery" conceptualization of empire, instead suggesting "more of a spaghetti junction of congested roadways than an orderly one-way street leading from Europe directly to New Zealand" (Crosby 1986; Beattie 2013, 243). Beattie emphasizes in particular plants and gardening techniques introduced from China, as well as the export of cultivars from Aotearoa itself.

Many species' acclimatizations to New Zealand were conducted under the auspices of societies established with this particular aim, and they collaborated with other such groups overseas, including in Britain. The introduction of plants and animals in Aotearoa appears to have been as haphazard as it was extensive, and for most of the nineteenth century – up until 1895 – it was unregulated, largely driven by the acclimatization societies as well as by private individuals. Sometimes, the societies received governmental backing; at other times, attempts to introduce species were more informal, arising out of transnational "botanical friendships" (Beattie 2013, 251). Perhaps significantly, this varied and transnational associational framework was structurally rather similar to that of the Arts and Crafts Movement, and the timelines also closely coincided, both beginning in the second half of the nineteenth century and lasting – in New Zealand at least – significantly into the twentieth. Furthermore, both movements were motivated at least in part by aesthetics, ecological attitudes and the desire to forge "improved" environments.

Indeed, the rhetorical significance of images in envisioning the transformation of Aotearoa to a "better Britain of the South Pacific" suggests that the wider agency of visual sources within nineteenth-century political, cultural and environmental changes should not be underestimated. The aesthetic ideals that informed the ecological transformation of land under the control of European settlers were also grounded in nostalgia: London-born Julius Vogel, who was twice New Zealand Premier in the 1870s, claimed, for example, that "The settler found comfort... and was cheered by seeing the daisy, primrose, and other British flowers, and all the fruits of his native land flourish in luxuriance round his humble cottage" (quoted in Druett 1983, 40). The ostensible bucolic abundance of this vignette is the textual analogue of the supposedly "homely" visual imaginary of the Arts and Crafts Movement, and the quotation also suggests how flowers in particular loomed large in British colonizing aesthetics of landscape. Despite all the flowers native to Aotearoa, introduced blooms were frequently positioned in contrast to the supposedly "empty landscape," that pervasive myth which attempted to conceal the violence of colonization toward indigenous peoples by erasing their presence from representations and erroneously framing their lands as "untouched wildernesses." This weaponized invocation of pretty introduced plants is particularly explicit in a trite poem titled *A Colonist in His Garden* by politician William Pember Reeves, written shortly before he assumed office in 1896 as Agent-General to the United Kingdom. Here again, New Zealand is styled as a barren *terra nullius* in need of cultivation, something that is in turn positioned as a triumphant manly feat: "Who, matched against the desert's power, / Hath made the wilderness to flower, / Can turn, forsaking all? / Yet that my heart to England cleaves / This garden tells with blooms and leaves / In old familiar throng" (quoted in Alexander and Currie 1906, 24). Reeves's poem goes on to name-check various introduced trees, birds and flowers, and reiterates bizarre myths perpetuated by many colonists that claimed that, prior to such acclimatizations, there was no birdsong[1] or even color in the landscape: "On the silent waste, / In pigments not to be effaced, / We paint the hues of life" (quoted in Alexander and Currie 1906, 25). This idea of Aotearoa's indigenous land as a "silent waste" had been around at least since Charles Darwin's visit of 1835 aboard the *Beagle*; in a section of his journal discussing New Zealand – alongside repeated racist remarks he made about Māori – he commented disparagingly that "The surface appears from a distance as if clothed with coarse pasture, but this in truth is nothing but fern ... The general tint of the landscape is not a bright green ... an air of extreme quietness reigned over the whole district" (Darwin 1871, 417). For colonists, "bright green" appears to have been a ubiquitous signifier of landscape "improvement," and was often positioned in contrast to dark green, gloom and shadows. In a particularly revealing passage describing the environmental transformation of Horomaka/Banks Peninsula, British settler and journalist Howard Charles Jacobson wrote approvingly that "True gloomy Rembrandt-like shadows have disappeared" (Jacobson 1893, 229), suggesting the expressly art-historical referents through which landscapes were often perceived. Furthermore, Reeves's notion of painting the landscape "in pigments not to be effaced" implies an interesting inversion, in which the landscape is painted according to images rather than the other way around, suggesting circularities in art and ecologies' remediations of one another.

A similar identification of commonalities between painting and landscape transformation is also evident within Arts and Crafts thinking. In a discussion of Walter Crane's flower books, art historian Morna O'Neill notes that "Crane, for one, realized the parallels between the gardener and the artist" (O'Neill 2008, 297). In his 1892 essay "Art and Labour," for example, Crane mused about how "the labourer is engaged in moving, say, Earth or minerals from one place to another with his shovel. The painter is engaged in moving Earth or minerals (in the form of colors) from one place to another – from his palette to his canvas with his brush" (quoted in O'Neill 2008, 297). However, for ecologically manipulating the color of landscapes, settlers often required the nonhuman agency of species besides plants. In a passage of 1883 that further demonstrates the centrality of flowers and ideas about color in colonial discourses of "landscape improvement," Irish priest, naturalist and mountaineer William Spotswood Green assessed the ecology around Fairlie, in the foothills of the Southern Alps:

> here and there patches of vivid green on the hill-sides showed where the soil was gradually being reclaimed from its native state by the sowing of clover and grass. A strange fact concerning the clover is that red clover won't seed in New Zealand, owing to the absence of the humble-bee, which is the insect adapted by nature to fertilize its flowers. White clover, on the other hand, thrives amazingly, and in the last report of the [North Canterbury] Acclimatization Society I see that an invoice had been received from England of a consignment of humble-bees which had been shipped to the colony. I hope they may arrive in good condition after their long voyage, and be ready to "improve each shining hour" in the meadows of Fairlie Creek... If in time green becomes the pervading color, the landscape may be improved.... (Green 1883, 115–116)

As well as suggesting the colonial associations of different colors and the prominence of aesthetic judgments in the assessment of ecologies – even by biologists such as Green – this passage highlights the interspecies entanglements that constituted the mechanisms of acclimatization. However, it simultaneously implies how these were only initially – if really at all – within the domain of anthropocentric control. The bees that Green refers to did not arrive in good condition; of the 282 sent aboard the Tongariro under the care of the ship's surgeon, only 48 had survived the journey on arrival in January 1885. As recalled in the account subtitled "The first hundred years of the North Canterbury Acclimatisation Society," of the remaining bees most were "in a torpid state – except for two which were so lively that they flew away as soon as they were released" (Lamb 1964, 81). Forty-nine more of these insects (of an initial 260) arrived on a different ship, and three-and-a-half years later, one farm alone reported obtaining three tons of red clover seed in a single season (Lamb 1964, 81). While there were other shipments too, bringing to Aotearoa a range of different bee species for a range of different purposes, the sea voyage for bees was particularly treacherous. Indeed, the staggering proliferations of certain species must not obscure the devastating catalog of animal death that attended acclimatization in New Zealand. This encompassed cases where it was unsuccessful, including when animals died at sea, as well as the decimation of many native species caused by competition, or predation from introduced mammals such as mustelids.

The dramatic transformation under white settler colonialism of the face of the land in Aotearoa New Zealand, including the idea of "painting" it by harnessing the ecological agency of introduced species, might be productively interpreted by extending literature historian Elizabeth Carolyn Miller's notion of an "aesthetics of surface"

(Miller 2015). In discussing Morris's fraught relationship with mining and its polluting externalities, Miller identifies in his thought a celebration of the riches to be found on the surface of the Earth, which stand in contrast to those deep within valued by – and (still) driving – industrial capitalism. Furthermore, Miller demonstrates Morris's anxiety at his sense that what used to be contained deep within the Earth was now encroaching on its previously beautiful surface, citing his lament at "whole counties of England, and the heavens that hang over them, disappeared beneath a crust of unutterable grime" (Morris 1902, 172). In Aotearoa New Zealand, however, the environmental violence apparent at the surface was of a different sort: it was less a layer of industrial pollution accumulating on the land and leeching into the atmosphere, and more an expanding carpet of destructive colonial ecologies, informed at least in part by the same English country garden aesthetics that Morris had looked to for his image of environmental redemption. Indeed, a colonial aesthetics of surface had arguably blinded the colonizing eye to indigenous ecologies, resulting in a perception of the country's "gloominess" that motivated European settlers' violent campaign of landscape redecoration. As ecologist Geoff Park described settlers' responses to their new environment, "the wild of the new land [was] punished for its unfamiliarity" (Park 2006, 9).

The notion of an "aesthetics of surface" also resonates with the design and consumption of wallpapers. Like the face of the land, the ecological agency of decorated walls was not just a matter of what their imagery promoted, but also relates to the environments and atmospheres that they themselves created. The popularity of wallpapers in colonial homes in New Zealand was no doubt substantially due to a desire to beautify damp and poorly lit rooms, to provide something colorful to counter the polluted atmospheres of the many nineteenth-century domestic spaces blighted by poor ventilation, smoky fires and gas lighting. However, wallpapers in this period contributed to domestic atmospheres not just by their enveloping imagery of "nature" and other motifs, but also as a result of the materials from which they were made diffusing into the air. As scholars have recently highlighted (see, for example, Hawksley 2016), many nineteenth-century wallpapers used pigments containing arsenic, which when inhaled could have very harmful consequences for the health of rooms' inhabitants. While arsenic was used in the manufacture of a range of colors, green pigments tended to contain the highest concentrations – an irony, Fallan notes, "given the color's current connotations as the visual shorthand for all things sustainable and ecological" (Fallan 2016, n.p.). Fallan claims that in the nineteenth century, however, "green symbolised poison and sickness as much as anything else" (Fallan 2016, n.p.). In a detail rich with symbolism, imported toxic wallpapers – as they had elsewhere – caused problems in New Zealand; *The Lyttelton Times* reported on the issue in 1879, for instance, concluding: "Fortunately, the measures necessary in these Colonies for preventing the use of these papers, need only be of the simplest kind. A short Act of Parliament prohibiting their importation, under a heavy penalty, will effectually put a stop to their sale when the present stocks are exhausted" (*The Lyttelton Times* 16 October 1879).[2] Indeed, wallpapers made by Morris's firm had initially contained arsenic too (see, for example, Hawksley 2016, 133); one such example is his *Trellis* design of 1862. Said to have been inspired by species he saw in his garden, the design's image of harmonious interspecies co-operation – in which birds, insects (seemingly including bees) and a rambling rose interact around the human-made trellis structure – reads unsettlingly in light of the poisonous material from which the scene is delineated. Compounding this uneasy revelation further is the fact that by 1870, not long after Morris designed *Trellis*, the mine on which his family's wealth was based, Devon Great Consols, was the source of half the arsenic extracted on (or from) the planet (Boos and O'Sullivan 2012, 15) (Figure 3.2).

While it might at first seem ironic that, in Morris's "desire to turn rooms into gardens ... [that] lay behind all his papers and textiles" (Rees 1993, 128), he for a time at least participated in the designing of toxic spaces, such an interpretation would rest on the notion that gardens are ecologically beneficial, or at the very least innocuous.

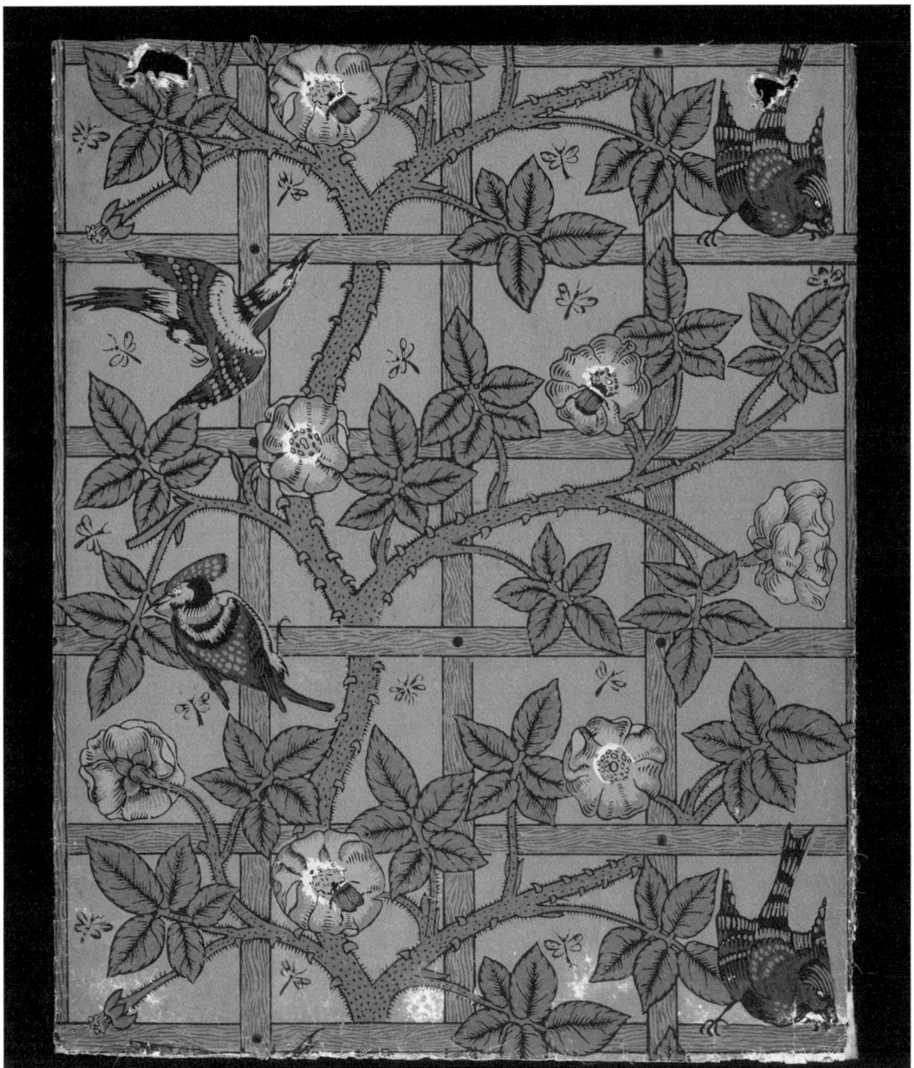

Figure 3.2 William Morris, (designer), Philip Speakman Webb, 1831–1915 (designer), Jeffrey & Co. (printer), Morris & Co. (publisher): *Trellis* wallpaper. 1862 (designed) and 1864 (design registered and manufactured). Printed in London. V&A, London. © Victoria and Albert Museum, London.

However, as the widespread crafting of "nature" in Aotearoa suggests, this was often not the case. Furthermore, in addition to the role that Arts and Crafts designs in New Zealand played in normalizing environments incorporating imported plants, gardens themselves were a significant area of endeavor in the Movement – and, of course, had direct ecological implications. In Aotearoa as in Britain, garden design in the decades around the *fin de siècle* was frequently informed by Arts and Crafts ideas, and a number of the Movement's leading architects – including John Dando Sedding, Edward S Prior and Reginald Blomfield – published books or essays on this topic. So too did Gertrude Jekyll, who remains the Arts and Crafts' best recognized horticulturalist. As Ann Calhoun has noted, the Movement's gardening principles in New Zealand followed British precedents both in terms of layouts and plantings, and – tellingly – William Robinson's text *The English Flower Garden* (first published in 1883) was particularly influential (Calhoun 2015, 154–155). Robinson, an Irish gardener and journalist, had advocated for a profuse and naturalistic appearance in the arrangement of gardens, arguing for "gardens full of life and change, and of such beauty as is nowhere to be found in deadly formalism" (Robinson 1893, 2). This unruly and seemingly organic aesthetic might have further aided a misleading impression of belonging for introduced plants, in part due to its ostensible "naturalness," which stood in marked contrast with the aggressively structured style in Victorian garden design that Robinson denounced.

As well as sometimes being Arts and Crafts objects in their own right, gardens full of imported flora in late nineteenth-century New Zealand also became sources for plant specimens for art students to copy. This was part of the curriculum that appears to have been for a long time central at such as the CCSA and the Wellington School of Design. Later renamed Wellington Technical College, its director William La Trobe (who had recently returned to Aotearoa after a decade in Britain) wrote about these "plant studies" in a 1908 report, describing how "Although it is always possible to obtain cut flowers for these classes, it is advisable to have as many specimens of growing plants in the school as possible, and a small glasshouse would greatly facilitate the keeping of such plants" (*AJHR* 1908, 52). Such a detail adds to a picture of the profound entanglements between visual culture and "nature" in this period, and the circularities in their remediations[3] of one another: while design played a role in influencing the assemblages of flowers that became prevalent in New Zealand, these ecologies went on to become the basis of many artworks. La Trobe's reflections also gesture toward the significance of glasshouses and related technologies as the life support system of acclimatization's extensive traffic in plants. In this, botanical gardens – a number of which were established in Aotearoa in the 1860s (including in Dunedin, Christchurch and Wellington) – were supplied in significant part by the transportation of flora in Wardian cases, sealed glazed containers that protected growing plants that otherwise would not have survived long sea voyages. Microcosms to the macrocosm of the glasshouse, these two symbiotic atmospheres were instrumental in the nineteenth-century proliferation of an international network of botanical gardens, which in turn precipitated new cultivars and hybrids, and had implications for surrounding ecologies. While the transportation of live plants aboard ships was costly and had a high failure rate (Beattie 2013, 244), it had a hopeful resonance to many at the time, evoking the biblical idea of the ark (Druett 1983). Indeed, acclimatization abounded with Christian symbolism (Druett 1983), and drew significantly on Edenic imagery – a prevalent trope through which botanical gardens, as well as the wider colonial project, were frequently framed in the New Zealand settler imaginary. This was also an important metaphor within the Arts and Crafts: as O'Neill has noted in relation to the Movement, "the ideal of the Garden of Eden held great symbolic currency within the rhetoric of socialism" (O'Neill 2008, 293). However, botanical gardens can also be seen as sites of violence – ecological and otherwise. With optics redolent of the cognitive dissonance that characterized the "utopian" settler imaginary, the construction in the early twentieth century in several New Zealand gardens of coal-powered hothouses suggests the environmental impacts that underpinned these spaces' extreme coercion of "nature." Furthermore, the forced labor of imprisoned men – including many Māori arrested during Ngā Pākanga Whenua O Mua/the Land Wars, the majority of which took place during the 1860s – had been used extensively in the landscaping of some botanic gardens, including those in Dunedin and Christchurch (Matchett 2013, 20).

Indeed, the fraught politics around the very idea of a "garden" might provide some additional context for the Arts and Crafts Movement's conflicted ecocriticality. While its reverence for "nature" and concern for the environment at first seem at odds with how it is implicated in environmental violence, these things can, in fact, be seen as intertwined phenomena which to some extent co-construct one another. For example, as Park argued in a seminal essay on the New Zealand landscape, certain forms of environmentalism are part of the same colonizing gaze that projected the idea of "wilderness" onto the land, and leveraged this in an attempt to justify transforming it. Highlighting the problematic nature of certain approaches to landscape preservation, Park described how

> Modern environmentalism... emerged as a direct response to the ecological mistakes and excesses of imperial rule... Its apparent universality, in the form of state-owned national parks, wilderness areas and wildlife reserves, has little more to do with the fact that by 1914, when New Zealand's conservation laws... were put in place, 85 percent of the Earth's land was under direct European control. (Park 2006, 82)

Indeed, in her essay "Wilderness as a walled garden," Jacinta Ruru speaks of the exclusion of indigenous peoples by "conservation estate boundaries," which, in Aotearoa, draw "lines on land through the gardens of iwi and hapū."[4] Critiquing how the European notion of wilderness has functioned within this, Ruru reminds readers that, for Māori prior to colonization, "Aotearoa New Zealand was their large garden estate" (Ruru 2011, 172).

* * * * *

In illuminating some of the ecological entanglements of the transnational Arts and Crafts Movement, particularly as it manifested itself in Aotearoa New Zealand, this chapter has sought to suggest aspects of the agency of interior design within environmental change. Arts and Crafts patterns not only drew on organic motifs based on both indigenous and imported plants, but played a role in shaping the aesthetics that informed these local ecologies. Images and ecologies therefore to some extent constructed one another, and this is something also seen in other regions that fell within the reach of Britain's imperial Anthropocene. It might indeed be said that in the hybrid environments of colonial New Zealand, "nature" itself acquired the role of visual representation, as landscapes were curated to emulate places elsewhere. This arguably extends and complicates ecocritical interpretations of the Anthropocene that position it as in part a crisis of representation (see, for example, discussions in Clark, 2015). This representational crisis is often framed in relation to the planetary scale – as Ed Ayres puts it, "[w]e are being confronted by something so completely outside our collective experience that we don't really see it, even when the evidence is overwhelming" (Ayres 1999, 6). However, this essay has sought to offer a related but more local perspective, suggesting that the links between environmental violence and representation extend to aesthetic modes of reading – and transforming – indigenous land.

Notes

1 This idea is revealing of both hermeneutic erasure and ecological violence, especially when considered in the context of Joseph Banks's observations of birdsong in Aotearoa early in 1770. While aboard Captain Cook's ship *Endeavour*, which was moored at the north end of Te Waipounamu (the South Island), Banks heard "the most melodious wild musick I have ever heard, almost imitating small bells but with the most tuneable silver sound imaginable." https://teara.govt.nz/en/speech/10106/joseph-bankss-journal. Accessed October 14, 2017.

2 Reporting on this issue some years later, other newspapers also took the opportunity to discuss "the influence of [wallpaper] colors on the temperament of the inmates" ("Wallpaper affecting health" 1905, n.p.), suggesting another way in which environments and wellbeing were believed to relate. See, for example, the article titled "Wallpaper affecting health" in the *Wairarapa Daily Times* of 10 April 1905, and the article titled "Wall-paper and hygiene: An important matter" in the *Clutha Leader* of 24 June 1910.

3 For discussions of the notion of "remediation," see Bolter and Grusin (1999).

4 Iwi is sometimes approximately translated as "tribe(s)," and hapū as "subtribe(s)."

Works Cited

Alexander, William Frederick, and Archibald Ernest Currie, eds. 1906. *New Zealand Verse.* London and Felling-on-Tyne: The Walter Scott Publishing Co.

Ayres, Ed. 1999. *God's Last Offer: Negotiating for a Sustainable Future.* New York: Four Walls Eight Windows.

Beattie, James. 2013. "The Empire of the Rhododendron: Reorienting New Zealand Garden History." In Eric Pawson and Tom Brooking, eds., *Making a New Land: Environmental Histories of New Zealand.* New edition. Dunedin: Otago University Press: 241–257.

Beattie, James, Emily O'Gorman, and Edward Melillo. 2014. "Rethinking the British Empire through Eco-cultural Networks: Materialist-Cultural Environmental History, Relational Connections and Agency." *Environment and History.* 20, 4: 561–575.

———. Editors. 2015. *Eco-cultural Networks and the British Empire: New Views on Environmental History.* New York and London: Bloomsbury.

Belich, James. 2001. *Paradise Reforged: A History of the New Zealanders from the 1880s to the Year 2000.* Harmondsworth: Penguin Books.

———. 2009. *Replenishing the Earth: The Settler Revolution and the Rise of the Anglo-World, 1783–1939.* Oxford and New York: Oxford University Press.

Bolter, Jay David and Richard Grusin. 1999. *Remediation: Understanding New Media.* Cambridge: MIT Press.

Boos, Florence S., and Patrick O'Sullivan. 2012. "Morris and Devon Great Consols." *The Journal of William Morris Studies.* 19, 4: 11–39.

Buller, Walter. 1905. *Supplement to the Birds of New Zealand.* London: The Author.

Calhoun, Ann. 2000. *The Arts and Crafts Movement in New Zealand 1870–1940: Women Make Their Mark*. Auckland: Auckland University Press.

———. 2015. *Arts and Crafts Design: Like Yet Not Like Nature – Sources for a New Zealand Story*. Wellington: The Author.

"Canterbury settlement." 1851. *The Lyttelton Times* 1, 13. 5 April.

———. 1879. *The Lyttelton Times* 52, 5816. 16 October.

Clark, Timothy. 2015. *Ecocriticism on the Edge: The Anthropocene as a Threshold Concept*. London: Bloomsbury Academic.

Cobden-Sanderson, Thomas James. 1905. *The Arts and Crafts Movement*. Hammersmith: Hammersmith Publishing Society.

Crosby, Alfred W. 1986. *Ecological Imperialism: The Biological Expansion of Europe, 900–1900*. Cambridge and New York: Cambridge University Press.

Darwin, Charles. 1871. *Journal of Researches into the Natural History and Geology of the Countries Visited during the Voyage of HMS Beagle Round the World*. New edition. New York: D. Appleton and Company.

Daston, Lorraine, and Fernando Vidal, eds. 2003. *The Moral Authority of Nature*. Chicago, IL: University of Chicago Press.

Druett, Joan. 1983. *Exotic Intruders: The Introduction of Plants and Animals into New Zealand*. Auckland: Heinemann.

Fallan, Kjetil. 2016. "Learning from Nowhere? Locating William Morris' Eco-fiction in Design History." In *Back to the Sustainable Future*. Accessed October 28, 2018. www.backtothesustainablefuture.net/learning-from-nowhere-locating-william-morris-eco-fiction-in-design-history/

Green, William Spotswood. 1883. *The High Alps of New Zealand: or, a Trip to the Glaciers of the Antipodes with an Ascent of Mount Cook*. London: Macmillan and Co.

Hall, Marcus, ed. 2010. *Restoration and History: The Search for a Usable Environmental Past*. New York and London: Routledge.

Harman, Graham. 2005. *Guerilla Metaphysics: Phenomenology and the Carpentry of Things*. Peru, IL: Open Court.

Hawksley, Lucinda. 2016. *Bitten by Witch Fever: Wallpaper & Arsenic in the Victorian Home*. London: Thames & Hudson.

Jacobson, H. C. 1893. *Tales of Banks Peninsula*. 2nd edition. Akaroa: The Author.

La Trobe, William Sanderson. 1908. "Extract from the Report of the Wellington Education Board". *Appendix to the Journals of the House of Representatives (AJHR), E-5*. Wellington: New Zealand Parliament: 49–56. https://paperspast.natlib.govt.nz/parliamentary/AJHR1908-I.2.3.3.10

Lamb, R. C. 1964. *Birds, Beasts and Fishes: The First Hundred Years of the North Canterbury Acclimatisation Society*. Christchurch: North Canterbury Acclimatisation Society.

Matchett, Alan. 2013. "The 2013 Banks Memorial Lecture: 150 Years of Botanic Gardens in New Zealand." *New Zealand Garden Journal*. 16, 2: 15–22.

McCarthy, Christine. 2009. "Domestic Wallpaper in New Zealand: A Literature Survey." Wellington: Centre for Building Performance Research, Victoria University of Wellington.

———. 2011. "Before Official Statistics: The Early Commerce of Wallpaper in New Zealand." *Fabrications: The Journal of the Society of Architectural Historians, Australia and New Zealand*. 20, 1: 96–119.

Miller, Elizabeth Carolyn. 2015. "William Morris, Extraction Capitalism, and the Aesthetics of Surface." *Victorian Studies*. 57, 3: 395–404.

Morris, William. 1902. "Art under Plutocracy" (Lecture first delivered in 1883). In William Morris, ed., *Architecture, Industry & Wealth*. New York: Longmans, Green and Co.

———. 1894 [1963a]. "How I Became a Socialist." In Asa Briggs, ed. *William Morris: Selected Writings and Designs*. Harmondsworth: Pelican Books: 33–37.

———. 1877 [1963b]. "The Lesser Arts" (Lecture First Delivered in 1877). In Asa Briggs, ed., *William Morris: Selected Writings and Designs*. Harmondsworth: Pelican Books: 84–105.

Morton, Timothy. 2007. *Ecology without Nature: Rethinking Environmental Aesthetics*. Cambridge MA: Harvard University Press.

———. 2010. *The Ecological Thought*. Cambridge: Harvard University Press.

O'Neill, Morna. 2008. "Walter Crane's Floral Fantasy: The Garden in Arts and Crafts Politics." *Garden History*. 36, 2: 289–300.

Osborne, Michael A. 2000. "Acclimatizing the World: A History of the Paradigmatic Colonial Science." *Osiris Nature and Empire: Science and the Colonial Enterprise*. 15: 135–151.

Park, Geoff. 2006. *Theatre Country: Essays on Landscape & Whenua*. Wellington: Victoria University Press.

Parry, Linda. 2005. *Textiles of the Arts and Crafts Movement*. London: Thames & Hudson.

Phillips, Jock. 2014. "Exhibitions and World's Fairs – New Zealand Exhibitions, 1865 to 1900." In *Te Ara – the Encyclopedia of New Zealand*. Accessed November 11, 2018. www.TeAra.govt.nz/en/artwork/45429/wallpaper-design-from-the-new-zealand-industrial-exhibition-wellington-1885.

Rees, Ronald. 1993. *Interior Landscapes: Gardens and the Domestic Environment*. Baltimore, MD: Johns Hopkins University Press.

Riley, Arthur Dewhurst. 1898. "Education: Manual and Technical Instruction". *Appendix to the Journals of the House of Representatives (AJHR), E-5B*. Wellington: New Zealand Parliament: 38–39. https://paperspast.natlib.govt.nz/parliamentary/AJHR1898-I.2.2.3.11

Robinson, William. 1893. *The English Flower Garden: Style, Position & Arrangement*. 3rd edition. London: John Murray.

Ruru, Jacinta. 2011. "Wilderness as a Walled Garden." In Mick Abbott and Richard Reeve, eds. *Wild Heart: The Possibility of Wilderness in Aotearoa New Zealand*. Dunedin: Otago University Press: 172–179.

Salmond, Jeremy. 1986. *Old New Zealand Houses 1800–1940*. Auckland: Reed Publishing.

Star, Paul. 2002. "Native Bird Protection, National Identity and the Rise of Preservation in New Zealand to 1914." *New Zealand Journal of History*. 36, 2: 123–136.

Stewart, Di. 2002. *The New Zealand Villa Past and Present*. New Edition. Auckland and London: Penguin Books.

Temple, Philip, ed. 1998. *Lake, Mountain, Tree: An Anthology of Writing on New Zealand Nature & Landscape*. Auckland: Godwit Publishing.

Tiffin, Chris. 2007. "Five emus to the King of Siam: Acclimatization and Colonialism." In Helen Tiffin, ed. *Five Emus to the King of Siam: Environment and Empire*. Amsterdam and New York: Editions Rodopi BV: 165–176.

"Wallpaper Affecting Health." 1905. *Wairarapa Daily Times*. 29, 8112. 10 April.

"Wall-Paper and Hygiene: An Important Matter." 1910. *Clutha Leader*. 36, 98. 24 June.

4 An Ecolonial Reassessment of the Indian Craze

Elbridge Ayer Burbank and Standing Bear

Jessica L. Horton

The ecocritical turn in humanities scholarship has grown rapidly alongside a resurgence of Indigenous justice movements across the Americas. What new perspectives on the past might their confluence opportune, even demand? Conversely, what overlooked past knowledges may be recovered and revalued as a result of their union? The words of Native participants point to a specific historical consciousness informing contemporary political struggles in a manner that resonates with the aims of this volume. Athabasca Chipewyan activist Eriel Deranger expressed the motivations for Idle No More, a movement begun by First Nations women in Canada in 2012: "Our people and our mother earth can no longer afford to be economic hostages in the race to industrialise our homelands. It's time for our people to rise up and take back our role as caretakers and stewards of the land."[1] Lakota historian Nick Estes similarly described No Dakota Access Pipeline (NoDAPL), a global effort to stop construction of the Dakota Access Pipeline across waterways and burial grounds deemed sacred by Oceti Sakowin citizens (the seven bands of Lakota, Dakota and Nakota which form the Great Sioux Nation) in 2016: "Like our ancestors' wars of the nineteenth century, our current war is also defensive—it is to protect water and land from inevitable spoliation in the name of profit" (Estes 2016, n.p.). No frontier closure. No settled ground. Deranger and Estes trace the present unfolding of human and ecological damage to the same colonial-industrial processes that drove the orchestrated dislocation of Native people in the nineteenth century. Likewise, they underscore a resurgence of Indigenous land-based knowledges that survived Indian removal and assimilation, offering compelling sources for reconnection amid ongoing conditions of extractive capitalism.

I propose that behind the global spread of movements such as Idle No More and NoDAPL lies an already thoroughly transcultural history of ecological thought that is expressed with special clarity in select works of art. Centering the creative production of Indigenous actors within nineteenth-century studies is a step toward grasping its diverse – non-European – dimensions. This essay will highlight several ways in which Indigenous and colonial environmental knowledges were imbricated in the reservation era. I build on scholarship that locates the origins of American environmentalism in the colonial displacement of Native people and the emergence of ecology in biological sciences that studied "nature" apart from humans (Jacoby 2001; Bibby 2006; McKee 2007). Yet critiques of these historical formations, taken on their own, do little to illuminate how Indigenous philosophies are implicated in the past and future of ecological thought. A different set of possibilities emerge from a reassessment of the well-documented Indian craze, an art market fueled by Euro-American nostalgia for handmade Native American objects that spanned the late nineteenth and early twentieth centuries. I will briefly discuss two interlinked products of this history, a painted portrait of American businessman Edward Everett Ayer (1841–1927) in his Chicago "Indian Room" completed by his nephew, Elbridge Ayer Burbank (1858–1949) in 1887, and a muslin drawing of the Battle of the Little Bighorn by the Lakota artist Standing Bear (1859–1933), which Burbank likely commissioned in 1899. Each in its own way distills what I call "ecolonial relations," rendering sensible the mutual transformation of diverse peoples and environments throughout a history of colonization that is still unfolding. Bridging a justice-oriented critique of the transformations wrought by colonial violence with attention to the creative survival of Indigenous land-based knowledge, "ecolonial" encompasses processes that "ecocritical" has only begun to address.

This brief study of ecolonial relations in nineteenth-century art helps to illuminate how and why much current critical work in the environmental humanities echoes enduring tenets of Indigenous thought. While ecocriticism is a diverse and growing field, its proponents share a denunciation of capitalist approaches to the environment, namely the division and hierarchical ordering of Euro-American culture over nature and the reduction of complex systems to discrete, consumable "resources" extracted by means of imperialist expansion. A branch concerned with environmental justice underscores how colonization relegates a majority of humans to the side of nature and renders them particularly vulnerable to the "slow violence" of ecological devastation (Nixon 2011). At the same time, ecocritics have envisioned alternative paradigms in which humans are embedded in relationships of dependency with each other and everything else, agency is distributed among nonhuman beings and matter, and mutual survival is contingent on granting political status to the earth. For some in Native Studies, the resemblance of these frameworks to long-standing Indigenous lifeways is disingenuous. Métis artist and scholar Zoe Todd has stated that powerful voices in the environmental humanities neglect to "credit Indigenous thinkers for their millennia of engagement with sentient environments, with cosmologies that enmesh people into complex relationships between themselves and all relations, and with climates and atmospheres as important points of organization and action" (2016, 6–7).[2] She underscores how the politics of citation bears on the utility of theory – who and what it serves and excludes.

Such critiques notwithstanding, this essay builds upon the foundational work of Joni Adamson to establish that "ecocriticism and Indigenous studies...emerge out of the same long, entangled, historical roots" (Adamson 2001; Adamson and Monani 2016, 4). Considering paintings by Burbank and Standing Bear side by side poses the difficulty of culturally isolating ideas, in particular notions of complex relationality that concern the intellectual genealogy of ecocriticism as much as the systems it seeks to understand. While this essay is tightly focused on only two objects, a similar reassessment of a wide variety of nineteenth-century artifacts would have broad methodological implications for the field of art history. Depending on their specialization, scholars may be adept at assessing how colonial power is exercised through Euro-American-made objects, or uncovering the transcultural processes shaping Indigenous-authored materials. An ecolonial reading necessarily works across these concentrations to parse the tangled threads of colonial, Indigenous and other-than-human forces shaping all sorts of objects and milieus. Attention to the material expressions of ecolonial relations will expose the co-presence of Native knowledge in far more locations than are currently addressed within the purview of Indigenous studies. It should be possible, for example, to simply state that American art history *is* Indigenous art history.

Burbank's portrait of his uncle specifically reveals how disparate cultural approaches to land collided amid westward expansion and its art historical corollary, the Indian craze (Figure 4.1). Following an early start in mining, Ayer built a fortune supplying timber to the railroads, implicating him in the extractive industries driving the federal project of Indian removal. He was also an avid collector of art and literature, particularly concerning North American Indians. He presented a vast collection of Native objects to the Field Columbian Museum in Chicago in 1894, followed by the announcement of a donation of roughly 50,000 books to the Newberry Library in Chicago in 1897 (Kastner 2003, 138; 144). Burbank accordingly depicted the businessman surrounded by wooden shelves of leatherbound volumes and Native material culture, signaling the multiple sources of his erudition and patronage. Ayer's collection was among the most prominent products of a fad that "almost amounts to a disease" to acquire and display diverse objects of Indigenous manufacture that swept across the United States following the completion of the transcontinental railroad (Mason 1904, 187). As art historian Elizabeth Hutchinson has emphasized, the creation of "Indian corners" in American homes was dependent on an influx of innovative artforms produced by Native people who were newly confined on reservations

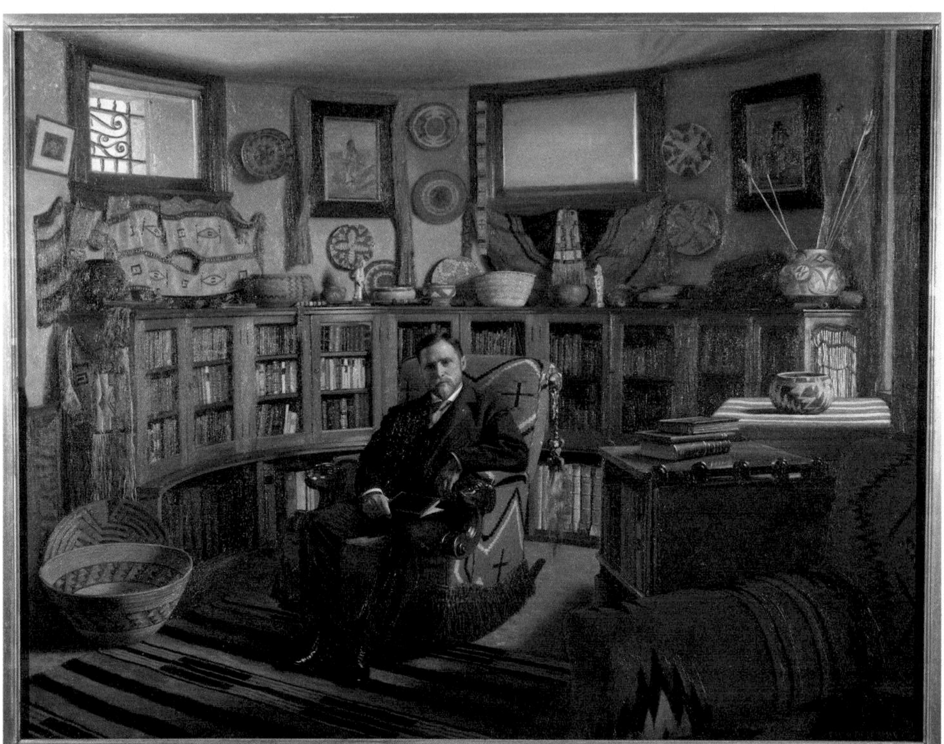

Figure 4.1 Elbridge Ayer Burbank, *Edward Everett Ayer*, 1897. Oil on canvas. Courtesy of the Edward E. Ayer Collection, The Newberry Library.

and forced to transition to a cash economy (2009, 1–50; see also Lears 1981; Cohodas 1997). Plains leather beadwork, Diné (Navajo) textiles, Pueblo pottery and baskets from the Indigenous Southwest spill across every available surface, visually overwhelming the inscrutible imprints of Ayer's books.

Burbank's artistic career was similarly tied to the financial rewards of westward expansion. After graduating from Chicago Academy of Design, he produced illustrations for *The Northwest Illustrated Monthly* to encourage homesteading along the path of the Northern Pacific Railroad (Wolfe 2000, 13). In 1897, Ayer sponsored him to embark on a new project that ultimately yielded over 2,000 portraits of Native Americans from 125 tribes in the western United States. Scholars have attributed the encyclopedic impulse to record and preserve Native cultures to "imperialist nostalgia," in which "the agents of colonialism long for the very forms of life they intentionally altered or destroyed" (Rosaldo 1989, 107–108). Carolyn Kastner has interpreted Burbank's painting of Ayer as "a visual metaphor of his power over the collected cultures. ...[The] Native meaning and identity of each object was ruptured and written over by Ayer as he removed it from its cultural context and made it part of his collection" (2003, 142). In this vein of critique, the extraction of lumber, beadwork and likenesses equally serve the endgame of conquest, the total possession of the Indigenous Americas.

While such insights are warranted, it is the unplumbed ecological dimensions of the Indian craze – specifically, Native frameworks for organizing human-earth relations that were present throughout – that interest me here. A hint can be found in Burbank's essays, collected and published in *Burbank Among the Indians*:

To these Children of the Sun each animal, each plant, each rock or wind or other object of nature was animated by a spirit which might be either beneficial or injurious to them personally according to the way it was propitiated or offended by the Indian's conduct. To the Indians, certain forces of nature were especially potent, particularly the sun, as were also fire and water, the rain clouds, the eagle, the rattle-snake and the buffalo. Changing conditions and the influence of the white man were rapidly undermining these simple beliefs, but they had not changed the Indian entirely (1972, n.p.).

Awash in paternalism, Burbank nonetheless described philosophical tenets that have lately resurfaced as innovative insights amid the contemporary ecocritical and materialist turns. In his account, Indigenous people ascribed agency to nonhuman plants, animals and weather, sought kinship and reciprocity with other beings, and envisioned their embeddedness in systems that were vulnerable to colonial damage. Burbank was reiterating popular ideas about Native people emerging from the discipline of anthropology, which produced the term, "animism," to describe irrational Indigenous commitments to an inspirited, responsive landscape (Tylor 1871). Contemporaneous with the emergence of "ecology" in the biological sciences, "animism" proposed a comparatively thorough confusion of "nature" and culture in the late nineteenth century. As Burbank's words reveal, proponents of the Indian craze simultaneously admired this union as an antidote to industrial alienation and dismissed it as "simple beliefs." I urge the inclusion of animism in an expanded genealogy of ecocriticism, beyond its capacity to illuminate how colonizers have "never been modern," in Bruno Latour's famous formulation (1993).[3] Mired in Euro-American notions of primitivism, it nonetheless names a site of transcultural transit where Native approaches to the land were translated – and just as often mistranslated – into new forms that affected a shared history. In particular, the Indian craze created a context for daily, intimate interaction with powerful objects that collectors valued in part for their ability to confer "the healthful influence of the natural world" (Hutchinson 2009, 31).

In an essay for *Brush and Pencil* in 1900, Burbank specifically credited Indigenous artworks such as those that filled his uncle's study as sites where environmental learning was made manifest: "The Indians...are to-day as loyal as ever to their first teacher. Their artistic work was copied direct from the hills and plains over which they roved...They used, not merely as models, but as materials, nuts, berries, bones, elk-teeth, furs, skins, feathers, porcupine-quills and the like" (1900, 75). More than a century later, literary theorist Timothy Morton argued in a much-quoted passage that art is meaningfully ecological not because it is "*about* something (trees, mountains, animals, pollution, and so forth," but because it "hardwires the environment into its *form*" (2010, 11).[4] In art historian James Nisbet's allied formulation, "the work of art as an ecological object" materially condenses and makes sensible relationships among political, technological and "natural" systems that are otherwise so diffuse as to elude comprehension (2014, 3). Approached in this manner, the Indigenous objects in Ayer's Indian Room could not be wholly decontextualized, for they carried their contexts with them. More than props for the identity of the industrialist, the Native materials drew the collector into a tangible relational fold through bodily contact and much as intellectual stimulation.

Perhaps most interesting in this regard is Burbank's choice to represent Ayer cradled in a plush Victorian rocking chair upholstered with a vivid red Diné weaving and decorated with what Kastner has identified as a Plains fly-chaser (2003, 142). Bold diagonal lines and crosses made of spun wool, plant pigments and aniline dyes realize a Diné ethical imperative to create *hózhó* (harmony, goodness, wellness, beauty). Weaving, a materialization of disciplined thought, helps to forge and maintain positive relationships between humans, plants, animals, mountains, stars and other elements of Dinétah, the Diné homeland in the southwestern United States (Berlo 2009, 237–253). Such symbolic and metonymic associations remain embedded in the textile as it travels beyond its community of origin. At the same time, its deformation into a comfortable rest for Ayer is a deceptively benign analogue for his involvement in the colonization of Dinétah, an arid region that outsiders valued primarily for its mineral wealth. As a young man in the army, he participated in a federal campaign to forcibly relocate Diné and Apache men, women and children from their lands to Fort Sumner in New Mexico, resulting in thousands of deaths between 1864 and 1868 (Kastner 2003, 156). Reconstituted in a painting that "collects" that same man's expansive pursuits, the upholstery temporarily arrests the transit of ideas and practices that comprised the Indian craze during the subsequent reservation era. Although its provenance is unknown, it resembles those developed for sale at Hubbell Trading Post in Ganado, Arizona where, at the encouragement of his uncle, Burbank set up a temporary studio in 1898. He compensated his host, trader John Lorenzo Hubbell, by producing studies of older Diné patterns that have been utilized by weavers for more than a century (Burbank 1972, n.p.; McLerran 2006, 14–15, 17). The rocker thus distills a complex chain of interactions across cultures, mediums and landforms that accelerated in the latter nineteenth century.

This reading insists that Native environmental values and practices were not written out of the scene of Euro-American erudition. Rather, the pictured furniture envelops Ayer's black-suited body in a patterned web of Euro-American, Indigenous and "natural" design.[5] Burbank presents us with an irreducible assemblage of human flesh, carved wood, spun wool, plant pigments and commercial dyes, procured and refashioned through disparate cultural practices that collided through colonization. It can be understood as a seat of ecolonial relations that unsettles the portrait's presumption of sovereign knowledge and authority.

While Burbank's painting is rich in relationships, attention to specific Native artworks in their contexts of production yields markedly different understandings of the ecological dimensions of the Indian craze. Whereas a generic anthropological conception of animism guided Burbank's approach to Native art, the work of his associate, Standing Bear, is better understood as a specific embodiment of the Lakota phrase, *mitákuye oyás'iŋ*, or "all our relations." The prayer, which acknowledges and thanks humans, animals, plants, rivers, mountains and other beings as common kin in a framework of reciprocity, has lately become a shorthand in Native Studies for Indigenous approaches to ecology. It has touched diverse publics around the world through contemporary art and environmental activism (Deloria, Jr. 1999b, 223–229; LaDuke 1999; de Zegher and McMaster 2012). The expanded intellectual genealogy of ecocriticism that I envision must account for a historical relationship between animism and *mitákuye oyás'iŋ*, without reducing one to the other. Put another way, an ecolonial analysis attends equally to the embeddedness of Native thought within the "material 'mesh' of meanings, properties, and processes" that characterize ecocritical inquiry (and Ayer's rocker), and the variegations and mistranslations of worldviews that underlie colonial dispossession and ongoing calls for Indigenous justice (Iovino and Oppermann 2014, 102).

Standing Bear's monumental rendering of the Battle of the Little Bighorn and the ceremonies that preceded was probably commissioned by Burbank, who wrote that he "was eager to get the Indians' account of the Custer Massacre" during his trip to paint portraits of Lakota people in 1899 (Burbank 1972, n.p.; Powell 1993, 82) (Figure 4.2). Yet the famous general is nowhere to be seen in Standing Bear's version. Instead, the muslin reveals how other-than-human sources of power shaped the most legendary conflict of Indian removal and aided the material and cultural survival of the people in its wake. In 1874, George Armstrong Custer supervised the largest of the nineteenth-century federal surveys mapping western topographies for the purposes of military and industrial expansion. He reported "lumber sufficient for all time to come" and "gold in paying quantities" in the Black Hills, a sacred center to the Oceti Sakowin people that was protected in the 1868 Treaty of Fort Laramie. The survey was instrumental to the Indian Appropriations Act of 1876 and the Black Hills Act of 1877, which, respectively, canceled treaty obligations and annexed the coveted land for mining (1974 [1874], 246).[6] At the age of 17, Standing Bear took part in the great battle between the U.S. cavalry and Lakota, Dakota and allied Northern Cheyenne and Arapaho forces beside the Little Bighorn River in Montana (Amiotte, Warren and Berlo 2014, 3–5). Twenty-three years later, he drew images from memory in pencil and painted them with watercolor on a six-foot square of muslin.

Standing Bear's battle scene continues a Plains graphic tradition in which men recorded events of individual bravery and collective importance on tanned animal skin. However, the shift from hide to industrial cloth marks a site of negotiation between two starkly different systems of valuing other-than-humans. Skins were more than a dead ground upon which human histories were inscribed; animals were spiritual intermediaries in the events whose memory they hosted. Given their importance to nearly every aspect of Plains culture, N. Scott Momaday has likened the U.S. government's

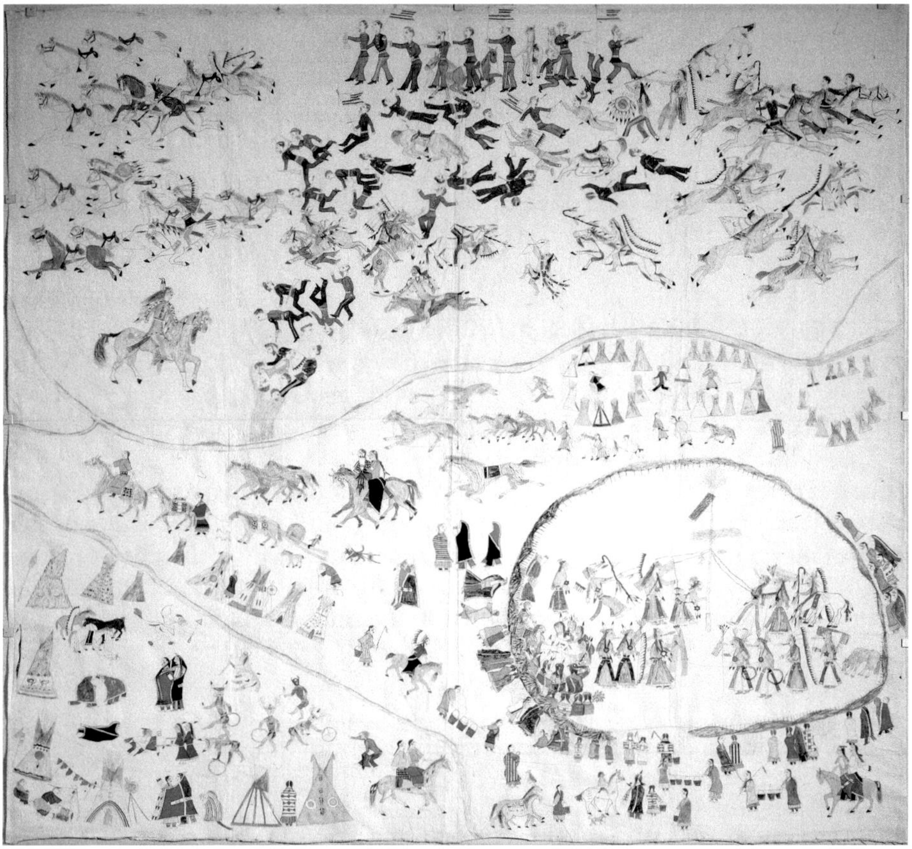

Figure 4.2 Standing Bear, *Events Leading to the Battle of the Little Bighorn*, ca. 1899. Muslin, pencil, red, blue, yellow, green and black pigment. Courtesy of the Foundation for the Preservation of American Indian Art and Culture, Inc., Chicago, Illinois.

orchestrated decimation of buffalo herds as a "deicide" (1969, 10). Standing Bear joined countless other Native artists who used manufactured pens, pencils, paints, paper and cloth procured at reservation trading posts, military prisons and boarding schools, throughout the nineteenth century. By choice or necessity, artists who replaced the uneven contours of animal bodies with rectangles of manufactured paper or cloth were drawn into a capitalist revaluation of their environments. Standing Bear completed four known large-scale works on muslin and countless drawings on paper before his death in 1933. Most left the reservation, sold for needed cash to traders, tourists or collectors spurred by the Indian craze (Berlo 1996, 12–18). He and his Austrian wife, Louise Renwick, further participated in the new cash economy by farming, ranching, making caskets and providing medical services for others in the community (Amiotte, Warren and Berlo 2014, 3–5).

In contrast to the battle scene, Standing Bear's inclusion of ceremonial imagery has no deep precedent in tribal annals. Art historian Janet Berlo has indicated that the first known Lakota examples appeared in the 1880s, when the "Code of Indian

Offences" banned the most important of the annual gatherings, the Sun Dance (2000, 40, 18). The ceremonies involved the redistribution of material goods throughout the community, which violated the logic of private capitalist accumulation driving Indian removal and assimilation. Some Lakota people nonetheless continued to dance and gift covertly (Berlo and Amiotte 2016, 32, 48–50). In the lower left, Standing Bear depicts black-tailed deer and elk dreamers, who danced during a multi-tribal gathering on nearby Rosebud creek prior to the battle. There, powers bestowed upon humans by animal guardians who visited them in dreams were publicly displayed and tested. The dreamers' hoops concentrate the power of everything that is wakan, or sacred, as expressed in the Lakota phrase *Wakan Tanka*, "the great mysterious" (Powers 1975, 45–47). Eagle feathers, horns and star and cloud symbols adorn the dreamers' headdresses. Such signs, gifted to individuals by their other-than-human caretakers, were reproduced on clothing and belongings as marks of personal distinction and protection during conflict (Blish 1967, 63). Their "active power," verified during the Battle of the Little Bighorn, would, in turn, enliven industrial cloth and the social contexts from which the muslin calls forth stories.

At the heart of the muslin is a circular Sun Dance lodge with a tall cottonwood pole and an entrance facing the sunrise, an emblem of the eternal interconnectedness of all things (Deloria, Jr. 1999a, 33). Men are shown attached by their flesh to the tree, a practice denigrated by colonizers as a form of torture. Rather, the ceremony merges body, sun, earth and sky, in a recapitulation of Lakota origins. Standing Bear's grandson, the artist Arthur Amiotte, who participated in later Sun Dance revivals, writes,

In the mythic beginning of the Lakota world, its sacred and temporal dimensions were one, and the Lakota still recognizes himself as a microcosmic reflection of that macrocosm. If he can live in concert with the holy rhythm of that which causes all life to move, he is then assisting in the ongoing process of creation. To maintain his participation in this process, he needs annually to make the journey to the Center of the World, which is the place of his beginning and the origin of all things. There he can renew his relation with the sacred rhythm in the ceremony known as the Sun Dance (1978, 46).

Amiotte describes an approach to the earth that differs in significant ways from Burbank's description of animist beliefs among the "Children of the Sun." All things are related not because of a naïve anthropomorphism – that is, the projection of human characteristics onto mute others – but because humans share with others a common system of energy, or animating power, that moves through the world, "a universe that people experience as alive and not as dead or inert" (Deloria, Jr. 1999a, 34). Because interactions with others can affect the distribution and intensity of this force, maintaining positive social relationships is paramount to harmonious coexistence. Humans are not the most powerful beings in this system. They are simply an incomplete part of it, and bear a responsibility to help complete the life-cycles of others – to assist in "the ongoing process of creation" – just as other animals look after human well-being. However, as the Dakota philosopher and legal scholar Vine Deloria, Jr. once cautioned, "some relationships are antagonistic to begin with" (1999b, 226–227). It is while dancing with bloodied arms during the 1876 Sun Dance that the Lakota chief leader of the resistance, Sitting Bull, received a prophetic vision of blue-coated soldiers raining from above (Powell 1993, 91–92). Standing Bear merges his premonition with the battle itself, as the soldiers cascade downhill, or perhaps, given the minimal topography, from the sky. In contrast to animism's associations with a premodern holism that is degraded by contact with Euro-Americans, his work reveals *mitákuye oyás'iŋ* to be a powerful framework for affecting, interpreting and withstanding colonial violence.

The muslin finally helps to clarify a social function for artworks within the broader ecolonial relations that set them in motion. To borrow the words of Timiskaming First Nation scholar Sherry Farrell-Racette, it possesses its "own life force" and is capable of "nudging memory, calling for a story" (2011, 41). Standing Bear's daughter, Christina Standing Bear (1894–1987), recalled that before his works left the community, they occasioned communal gatherings in the family's log cabin. The furniture was pushed aside. Guests sat on the floor, smoked pipes and enjoyed a lavish feast, while the artist unrolled his yet-unfinished work. A process of verification and storytelling among age mates continued late into the night. Such gatherings functioned as a means of maintaining Indigenous knowledge and redistributing material goods through the most aggressive period of assimilation (Berlo and Amiotte 2016, 14). This social dimension is anticipated in the formal arrangement of the Sun Dance circle, surrounded by rivers, "life-giving and healing forces," that course like dynamic lines of energy to the edges of the muslin (Howe and Tallbear, 2006, 79). It calls to mind winter counts, mnemonic calendars in which pictographic representations of a single important event of each year are arranged in a spiral. It is the job of the count keeper to forge a relationship between past and present during oratory, integrating other-than-human sources of power into the telling of Oceti Sakowin histories (Walker 1982, 113).

It is not clear whether the muslin ever joined the profusion of Native objects in Ayer's Indian Room, prompting accounts of Indian removal and assimilation that differ greatly from those in nearby books. Today, it resides in the Foundation for the Preservation of American Indian Art and Culture in Chicago, overseen by Father Peter J. Powell, Director of the Newberry Library where Burbank's painting is held. As "Native struggles for land and life" continue to unfold on the Plains and across the Americas, these works can occasion new past-present stories that expand the cultural and intellectual parameters of both ecocriticism and art history (LaDuke 1999). By contemplating their interwoven histories, we can invite them to participate anew in transforming the ecolonial relations – antagonistic and nurturing, human and other-than-human – that they picture.

Acknowledgments

I am grateful to Maura Coughlin, Emily Gephart, Janet Catherine Berlo, Kristine Ronan, Christopher Heuer and three anonymous reviewers for their incisive feedback on this essay. I thank Arthur Amiotte for sharing perspectives on his grandfather's life and Father Peter Powell for his scholarship and supportive letters. The Clark Art Institute has granted me permission to include some material previously published as "'All Our Relations' as an Eco–Art Historical Challenge: Lessons from Standing Bear's Muslin" in *Ecologies, Agents, Terrains*, eds. Christopher P. Heuer and Rebecca Zorach (Clark Art Institute with Yale University Press, 2018), 73–93.

Notes

1 Deranger is quoted on the official website for Idle No More, where elaborations of the movement's history and purpose can be found: http://www.idlenomore.ca.

2 Todd responded to Bruno Latour's position on climate change, outlined in a series of public lectures (2017). It is worth noting, as Todd does, that Indigenous Studies scholars *are* explicitly engaging with ecological subjects and have done so for a long time (see, for example, Deloria, Jr. 1972; Wildcat 2009).
3 Latour's text is at the center of curator Anselm Franke's reassessment of animism as a valuable tool for understanding Euro-American modernity in a two-part exhibition and catalog (Franke, 2010; Franke and Folie 2011). The exclusion of Native knowledge from this framework is addressed in Horton and Berlo (2013) and Horton (2017, 152–196).
4 On the adoption of this idea by art historians, see Braddock (2015, 450).
5 I am thinking here of anthropologist Alfred Gell's description of the "cognitive stickiness" of pattern, a material web designed to entrap and enthrall viewers by creating attachments to things and by extension, the people who made them (1998, 80–81, 86). Also relevant is Morton's theorization of the "mesh" of ecological relations in which strange beings collide (2010, 33–38).
6 For more on the importance of the Black Hills to Oceti Sakowin history and thought, as well as relevant historical treaties, see Howe, Soldier, and Lee (2011).

Works Cited

Adamson, Joni. 2001. *American Indian Literature, Environmental Justice, and Ecocriticism: The Middle Place*. Tucson: University of Arizona Press.

Adamson, Joni and Salma Monani. 2016. "Introduction: Cosmovisions, Ecocriticism and Indigenous Studies." In Salma Monani and Joni Adamson, eds., *Ecocriticism and Indigenous Studies: Conversations from Earth to Cosmos*. London: Routledge: 1–23.

Amiotte, Arthur, Louise S. Warren, and Janet Catherine Berlo. 2014. *Transformation and Continuity in Lakota Culture: The Collages of Arthur Amiotte, 1988–2014*. Pierre: South Dakota State Historical Society Press.

Berlo, Janet Catherine, ed. 1996. *Plains Indian Drawings, 1865–1935: Pages from a Visual History*. New York: Harry N. Abrams.

———. 2000. *Spirit Beings and Sun Dancers: Black Hawk's Vision of the Lakota World*. New York: George Braziller.

———. 2009. "Alberta Thomas, Navajo Pictorial Arts, and Ecocrisis in Dinétah." In Alan C. Braddock and Christoph Irmscher, eds., *A Keener Perception: Ecocritical Studies in American Art History*. Tuscaloosa: University of Alabama Press: 237–253.

Berlo, Janet Catherine, and Arthur Amiotte. 2016. "Generosity, Trade, and Reciprocity among the Lakota: Three Moments in Time." In Jill Ahlberg Yohe and Janet Catherine Berlo, eds. *Plains Indian Art of the Early Reservation Era: The Donald Danforth Jr. Collection at the Saint Louis Art Museum*. St. Louis: St. Louis Art Museum: 32–73.

Bibby, Brian. 2006. "Native American Art of the Yosemite Region." In *Yosemite: Art of an American Icon*, ed. Amy Scott. Los Angeles: Autry National Center.

Blish, Helen H. 1967. *A Pictographic History of the Oglala Sioux*. Lincoln: University of Nebraska Press.

Braddock, Alan. 2015. "From Nature to Ecology: The Emergence of Ecocritical Art History." In John Davis, Jennifer A. Greenhill, and Jason D. LaFountain, eds. *A Companion to American Art*. West Sussex, UK: Wiley Blackwell: 447–467.

Burbank, Elbridge Ayer. 1972 [1944]. *Burbank among the Indians*, ed. Frank J. Taylor. Harvard Diggins Library and Caxton Press. www.harvard-diggins.org/Burbank/Burbank_Among_The_Indians/Burbank_Among_the_Indians.htm.

———. 1900. "Studies of Art in American Life—III: In Indian Tepees," *Brush and Pencil*. 7, 2: 75–91.

Cohodas, Marvin. 1997. *Basket Weavers for the California Curio Trade: Elizabeth and Louise Hickox*. Phoenix: University of Arizona Press.

Deloria, Jr., Vine. 1972. *God Is Red: A Native View of Religion*. Golden, CO: Fulcrum Publishing.

———. 1999a. *Singing for a Spirit: A Portrait of the Dakota Sioux*. Santa Fe: Clear Light Publishing.

———. 1999b. *Spirit & Reason: The Vine Deloria, Jr., Reader*, eds. Barbara Deloria, Kristen Foehner, and Sam Scinta. Golden, CO: Fulcrum Publishing.

de Zegher, Catherine, and Gerald McMaster, eds. 2012. *The 18th Biennale of Sydney: all our relations*. Sydney: Biennale of Sydney.

Estes, Nick. 2016. "Fighting for Our Lives: #NoDAPL in Historical Context," *The Red Nation*, September 18. https://therednation.org/2016/09/18/fighting-for-our-lives-nodapl-in-context/.

Franke, Anselm, ed. 2010. *Animism Volume I*. Berlin: Sternberg Press.

Franke, Anselm, and Sabine Folie, eds. 2011. *Animism: Modernity Through the looking Glass*. Berlin: Sternberg Press.

Gell, Alfred. 1998. *Art and Agency: An Anthropological Theory*. Oxford: Clarendon Press.

Horton, Jessica L. 2017. *Art for an Undivided Earth: The American Indian Movement Generation*. Duke University Press.

Horton, Jessica L., and Janet Catherine Berlo. 2013. "Beyond the Mirror: Indigenous Ecologies and 'New Materialisms' in Contemporary Art." In T. J. Demos, ed. *Contemporary Art and the Politics of Ecology*, 17–28. Special issue of *Third Text* 27, 1.

Howe, Craig P., and Kimberly TallBear. 2006. *This Stretch of the River*. Sioux Falls, S.D.: Pine Hill Press.

Howe, Craig P., Lydia Whirlwind Soldier, and Lanniko L. Lee, eds. 2011. *He Sapa Woihanble: Black Hills Dream*. St. Paul, MN: Living Justice Press.

Hutchinson, Elizabeth. 2009. *The Indian Craze: Primitivism, Modernism, and Transculturation in American Art, 1890–1915*. Durham, NC: Duke University Press.

Iovino, Serenella and Serpil Oppermann, 2014. *Material Ecocriticism*. Bloomington, IN: University of Indiana Press.

Jacoby, Karl. 2001. *Crimes against Nature: Squatters, Poachers, Thieves, and the Hidden History of American Conservation*. Berkeley: University of California Press.

Kastner, Carolyn. 2003. "Collecting Mr. Ayer's Narrative." In Leah Dilworth, ed., *Acts of Possession: Collecting in America*. New Brunswick, NJ: Rutgers University Press: 138–162.

LaDuke, Winona. 1999. *All Our Relations: Native Struggles for Life and Land*. Cambridge, MA: South End Press.

Latour, Bruno. 1993. *We Have Never Been Modern*. Trans. Catherine Porter. Cambridge, MA: Harvard University Press.

———. 2017. *Facing Gaia: Eight Lectures on the New Climatic Regime*. Cambridge: Polity Press.

Lears, T. J. Jackson. 1981. *No Place of Grace: Antimodernism and the Transformation of American Culture, 1880–1920*. Chicago, IL: University of Chicago Press.

Mason, Otis. 1970 [1904]. "Aboriginal American Basketry: Studies in a Textile Art without Machinery." In *Annual Report of the U.S. National Museum, Smithsonian Institution, 1902*. Reprinted as *Aboriginal Indian Basketry*. Glorieta, NM: The Rio Grande Press.

McKee, Yates. 2007. "Art and the Ends of Environmentalism: From Biosphere to the Right to Survival." In Michel Feher, ed. *Nongovernmental Politics*. New York: Zone Books: 539–583.

McLerran, Jennifer. 2006. "Textile as Cultural Text: Navajo Weaving as Autoethnographic Practice." In. Jennifer McLerran, ed. *Weaving is Life: Navajo Weavings from the Edwin L. & Ruth E. Kennedy Southwest Native American Collection*, 8–33. Athens, OH: Kennedy Museum of Art.

Momaday, N. Scott. 1969. *Way to Rainy Mountain*. Albuquerque: University of New Mexico Press.

Morton, Timothy. 2010. *The Ecological Thought*. Cambridge, MA: Harvard University Press, 2010.

Nisbet, James. 2014. *Ecologies, Environments, and Energy Systems in Art of the 1960s and 1970s*. Cambridge: MIT Press.

Nixon, Rob. 2011. *Slow Violence and the Environmentalism of the Poor*. Cambridge, MA: Harvard University Press.

Powell, Father Peter J. 1993. "Sacrifice Transformed into Victory: Standing Bear Portrays Sitting Bull's Sun Dance and the Final Summer of Lakota Freedom." In Evan M. Maurer, ed. *Visions of the People: A Pictorial History of Plains Indian Life*. Minneapolis, MN: Minneapolis Institute of Arts: 81–106.

Powers, William K. 1975. *Oglala Religion*. Lincoln: University of Nebraska Press.

Racette, Sherry Farrell. 2011. "Encoded Knowledge: Memory and Objects in Contemporary Native American Art." In Nancy Marie Mithlo, ed., *Manifestations: New Native Art Criticism*, 40–55. Santa Fe: Museum of Contemporary Native Arts.

Rosaldo, Renato. 1989. "Imperialist Nostalgia." *Representations*. 26: 107–122.

Todd, Zoe. 2016. "An Indigenous Feminist's Take on the Ontological Turn: 'Ontology' Is Just another Word for Colonialism." *Journal of Historical Sociology*. 29: 4–22.

Tylor, Edward. 1871. *Primitive Culture*, 2 vols. London: John Murray.

Walker, James R. 1982. *Lakota Society*, Raymond J. DeMallie, ed. Lincoln: University of Nebraska Press.

Wildcat, Daniel R. 2009. *Red Alert! Saving the Planet with Indigenous Knowledge*. Golden, CO: Fulcrum Publishing.

Wolfe, Melissa M. 2000. *American Indian Portraits: Elbridge Ayer Burbank in the West (1897–1910)*. Youngstown, OH: Butler Institute of American Art.

5 The Panama Canal Zone as a Hybrid Landscape

A Case Study[1]

Sarah J. Moore

In 1913, artist Jonas Lie (1880–1940) traveled to Central America to create images of the late construction phase of the Panama Canal. Inspired by the soaring colonialist rhetoric of mastery and triumph of civilization that characterized most contemporary accounts of America's technological transformation of the Panama Canal zone – journalist Willis J. Abbot referred to the enterprise as "changing the Isthmus of Panama from a pest-hole into a spot as fit for human habitation as any spot on the globe" (Abbott 1913, 254). Lie created a series of paintings that visually canonized the epic nature of the engineering feat with conventional colonial tropes of mastery. His aptly named *The Conquerors*, 1913, for example, utilizes a view from above – a widely used compositional strategy in nineteenth-century landscape paintings to disclose the viewer's privilege and dominance over a broad swath of land in order to convey the sublime enormity and complexity of the project. Lie invites the presumptively white, privileged viewer to survey the herculean task of transforming the vexing problem of a hostile environment *in progress*; men, machines and smoke move across and into the deep space of Culebra Cut. This compositional distance underscores the massive scale of the engineering project while obscuring the thousands of laboring bodies who actually dug the canal and their excruciating and dangerous toil in the most trenchant obstacle of the enterprise (Green 2009, 15). The title expresses the hubris shared by most observers in the nation's ability to subdue even the most recalcitrant of natural obstacles in the name of progress, indeed civilization itself.

At five by four feet in size, Lie's image rendered in material form what was widely assumed at the time: that the radical transformation of the land in Panama during the building of the interoceanic canal was the most magnificent modern expression of progress and civilization in its conquest of a formidable adversary, nature. Geographer R. H. Whitbeck's colonial rhetoric expresses a sense of battle and triumph; "It is an inspiring example of man's conquest of adverse nature; not man's response to a hostile environment but defiance of it and his subjugation of it; an example of so-called geographic control upon which is superimposed a still more impressive example of man's control" (Whitbeck 1921, 8). Moreover, the Panama Canal rendered in tons of concrete and steel the Cartesian dualism of nature and human beings, seeing the former as something out there – inert, empty, a resource for human use, a mere background – and into which civilization could and would encroach and triumph (Alaimo 2010, 2).

The implicit anthropocentrism and colonial tropes of this image and countless other contemporaneous tracts on the Panama Canal – articles and essays, government reports, illustrations and paintings, maps, health statistics and guidebooks – were not an invention of the period, however, but were an acquired supposition that can be traced back at least until the middle of the nineteenth century and the United States' first technological intervention in the Panamanian isthmus. Indeed, Americans cut their imperial teeth and anthropocentric ambitions in the tropics with the building of the Panama Railway in the 1850s. I argue that it was images from that time period that did the ideological heavy lifting to construct twentieth-century mindsets and the political ecology that made the completion of the canal, and images like Lie's, not only possible but inevitable.

In this essay, I consider what most contemporary viewers failed to observe, what I would call the political and visual ecology of the massive engineering projects – the railroad and the canal – and the fact that the tropical ecosystem was not only the most formidable and often lethal obstacle to their realization but also unstable, changing and capable of producing change. To be sure, the geography and climate of the region were widely recognized as compromising the productivity of the venture. However, each was seen as a discrete location or definable problem that could be compartmentalized, measured, mapped and otherwise contained and controlled so as to transform the tropics from a torrid zone to a temperate one. This attempted transformation of the tropics, however, was not a one-way street; the notion that transforming agent (United States technology) plus labor (largely West Indian) surmounts natural forces and objects (continental divide, torrential rains, mudslides, disease-carrying mosquitoes) was oversimplified and smacked of the hubris of colonialist privilege. Indeed, many of the objects against which the enterprise struggled – perhaps none more than the lowly mosquito – were transformative agents themselves that threatened the builders' (real and imagined) efforts to separate themselves from the natural world.

This essay foregrounds the ecocritical agency of hybridity in proposing that the Panamanian isthmus, in its prolonged contact with Euro-American incursions and colonial practices, became a hybrid and shifting zone, as much a process as a discrete place. Within it, the entanglements between human and nonhuman as well as economic, technological, political, colonial, racial and ecological systems interacted, overlapped and often contradicted one another. To be sure, the Panama Canal zone – its terrain, technology, documentation and human and more-than-human subjects – was managed by a powerful ideological and bureaucratic structure and an archive of visual images and written texts that valorized American ingenuity, progress, modernity, health and superior fitness in contrast to the marginalized and debased tropics. This discourse of progress and modernity, however, hinged on the faulty premise that the tropics were a "mere empty space" – a blank slate – in which white men could, using the calculus of reason and imperial authority, map, measure and otherwise gain firm ground (Alaimo 2010, 2). The Panama Canal zone, however, "complicated colonial efforts to make clear distinctions between human beings and the natural world" (Allewaert 2013, 34).

Despite material attempts to subdue the Panamanian isthmus – the construction of the railroad, physically cutting through the isthmus to make a canal, and the containment and eradication of disease – Panama's nonhuman inhabitants, alongside indigenous and racially mixed human inhabitants, proved to be stunningly resilient. Plants overgrew construction equipment, thwarted work and complicated travel, while insects, particularly disease-carrying mosquitoes, proliferated in tropical humidity and migrated freely across spaces and bodies. An unintended consequence was that the resulting terrain facilitated the spread of malaria, although this reality was largely invisible in contemporary discussions and views of the region. Nonetheless, even in the rare cases when nature was recognized as having the potential of agentive resistance, entrenched anthropocentrism inspired confidence that no matter how formidable, nature would inevitably fall to human mastery (Frenkel 1996, 317–333; White 2004, 557–564).

Thus, in order to shift the focus on the historical designation of the Panamanian isthmus from place to hybrid zone, and to demonstrate the work that visual culture performed in setting up counter-narratives to dominant discourses, I will consider four images in some detail – three from the mid-nineteenth century during and following the building of the Panama Railroad and one just before the completion of the Panama Canal – as case studies of how such depictions revealed cracks in the triumphant discourses of conquest and colonial mastery and challenged the long shadow cast by Cartesian dualisms. These images, in their attempt to transform the Panamanian isthmus into a landscape – a backdrop for human habitation and dominion – invite a reading of hybridization and ecological resistance. Panama, it seems, defied compartmentalization.

Traversing the Waste Spaces

Although the story of nineteenth-century westward expansion across the North and Central American continent is often conceptualized on an east-to-west trajectory, in fact the most traveled route between the east and west coasts was southerly, via Panama. One needs to remember that at mid-century, the territory west of the State of Missouri was often depicted as a blank space across which were printed the words, "Great American Desert" (Moore 2018, 240–275). And the four-to-five-month passage around Cape Horn – some 14,000 miles – was far too long for most impatient travelers and certainly for those seeking their fortune in California gold. By the late 1840s, the so-called Panama route was facilitated by two steamship lines operating between New York and San Francisco, respectively; with the discovery of gold in California in 1849, demand on both lines increased dramatically.

The 60-mile overland journey from the Caribbean to the city of Panama, on the Pacific, was not for the faint of heart. The first stage of the journey was made on the Chagres River, 40-odd miles, in wooden canoes propelled by indigenous men using steering poles. In addition to the staggering heat and disease-carrying mosquitoes, the river swarmed with alligators, venomous snakes and stretches of rough water. Following the harrowing river passage, the final stage of the passage across the isthmus was on foot or on the back of a mule. Paths were narrow and swampy, often filled with mud due to incessant heavy rains. Contemporary descriptions underscored the wretchedness of the passage, including that by Frank Marryat, British sailor, artist and writer, who traversed the Panamanian isthmus in 1850 on his way to California. His memoir of the journey, which was illustrated by the author, was published in 1855 in London and New York; his New York publisher, Harper and Brothers, printed a review of his book, including some of the original illustrations, in their periodical, *Harper's New Monthly Magazine*, assuring a wide readership.

Marryat's narrative actively produced perceptions of the future canal zone as the "foremost pest-hole of the earth, infamous for its fevers, and interesting only because of the variety of its malarial disorders and pestilences" (Adams 1911, 5). He described the beginning of the river passage: "The town of Chagres deserves notice, inasmuch as it is the birthplace of a malignant fever, that became excessively popular among California emigrants, many of whom have acknowledged the superiority of this malady by giving up the ghost a very few hours after landing" (Marryat 1855, 2). His sardonic tone can be understood as an acknowledgment of the resistance of the region to ready control; *Crossing the Isthmus* (Figure 5.1) depicts many of the journey's hardships he described. Against an impenetrable jungle backdrop complete with jagged rocky peaks, a precipitous crevasse, palm trees entangled with vines and incongruous cacti are vignettes of hapless foreign travelers, utterly ill-equipped for the journey across the jungle – some on foot, others on mules up to their bellies in mud, still others wrestling with or having been bucked off their obstinate mounts. The wry humor in the illustration is noteworthy. In contrast with the fully dressed Anglo travelers, including one woman – referencing the anticipated domestication of the region – four indigenous Panamanians, dressed in loincloths, conform to widely held racial stereotypes; the male figure in the background gestures violently, aligning his behavior with the unruliness of the jungle in which he is embedded.

The eye moves across the image from right to left, as do the Anglo travelers and their pack mules that struggle with their heavy parcels. One such creature has stumbled on the rocky terrain, and its parcel, clearly marked with the instructions "This Side Up," lists precariously. Two travelers depicted in small scale are being transported in makeshift chairs strapped to the backs of indigenous men. This rather perplexing

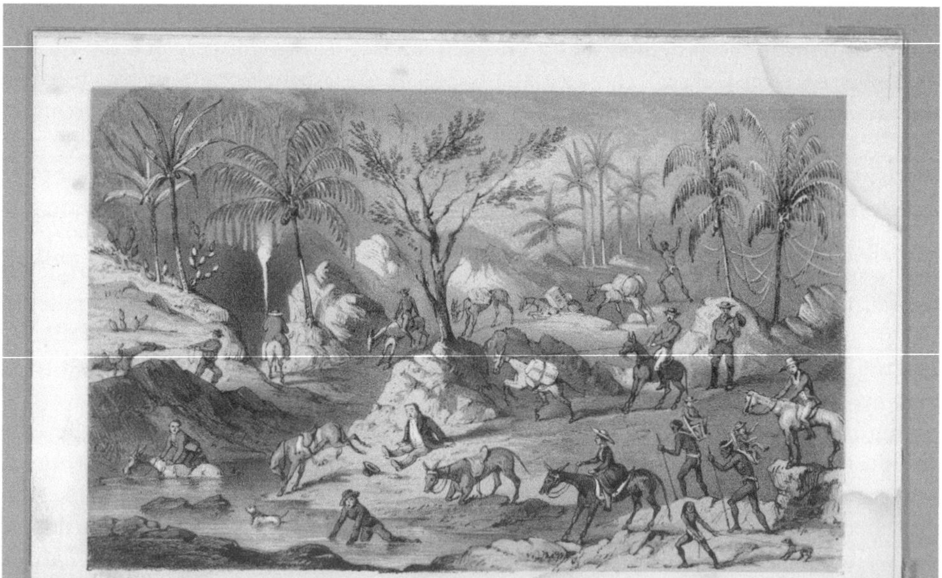

Figure 5.1 *Crossing the Isthmus*, as illustrated in Frank Marryat, *Mountains and Molehills: Or Recollections of a Burnt Journal* (London: Longman, Brown, Green and Longmans, 1855).

section of the image may allude to the fact that outsiders depended on the local knowledge and willingness of native inhabitants to successfully navigate the challenging terrain; this contradicts contemporary viewers' knee-jerk alignment of tropical inhabitants with the chaotic landscape. Two dogs accompany the travelers in the foreground; their partial domestication contrasts dramatically with the wildness of the scene. Indeed, although the illustration draws the eye right to left, it does not seem to "lead" the viewer through the image. Rather, *Crossing the Isthmus* disrupts the possibility of a clear reading and contrasts dramatically with the bird's-eye views that would come to define images of the canal zone and that offer long vistas and vast horizons for guidance, perspective and a colonial sense of dominion. "Up close," as Marryat's image suggests and art historian Darcy Grimaldo Grigsby later observed, "Panama was confusing" (Grigsby 2012, 122).

German-born artist Charles Nahl (1818–1878) also traveled through Panama on his way to California in 1850 and created a series of drawings, some of which he later transformed into paintings and/or illustrations for articles about emigrants seeking work in California (Manthorne 1989, 187). *Incident on the Chagres River,* 1867 (Figure 5.2), depicts one of the many hazards of the journey. Flat-bottomed boats were commonly used to ferry people down the river; native men with long poles for maneuvering stand at the front of the boat, while the back has a shade cloth to protect the travelers from the blistering sun. The boat has suddenly encountered rapids, causing one of the pole men to topple overboard; one of the passengers hangs precariously off the edge of the boat to size up the danger. A similar boat, wrecked and wedged under a flotsam-laden rocky outcropping in the river, stands as grim evidence of the journey's

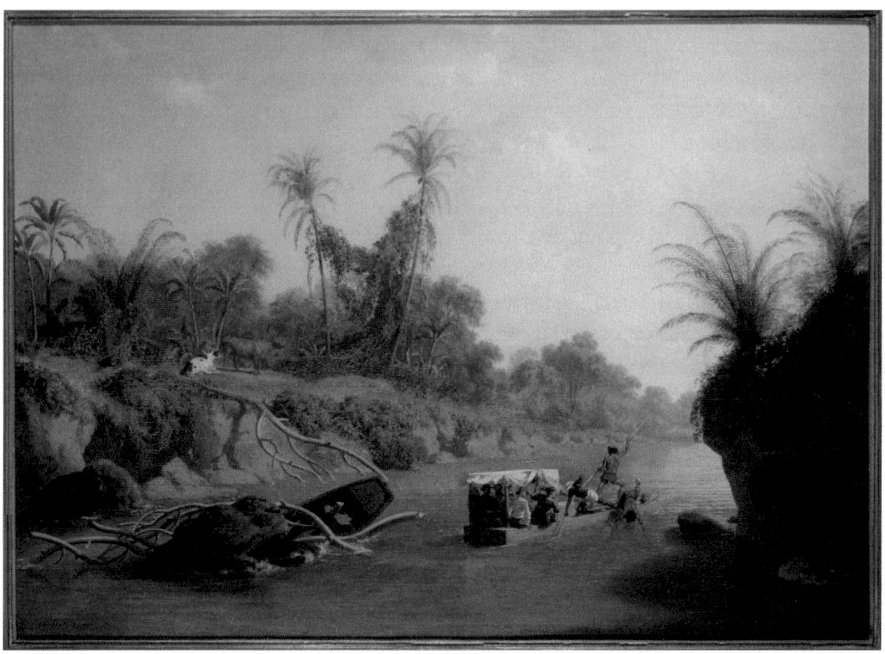

Figure 5.2 Charles Nahl, *Incident on the Chagres River,* 1867. Oil on canvas, University of California Berkeley, Bancroft Library.

risks. Dense, lush vegetation rises up on both sides of the riverbanks. Nahl positions the viewer at the river level in contrast to the far more common panoramic view from above; the latter implies a degree of colonial mastery over and safe distance from the scene depicted, while the former refuses a clear distinction between viewer and landscape. The jutting embankment on the right side impedes a glimpse into the distance, heightening the sense of danger and fear from which the viewer cannot retreat into the thrilling safety of the sublime. This tropical setting seems to exercise a particular ecological agency and reminds the viewer that navigating Panama is fraught.

Nahl studied with Horace Vernet in Paris before coming to the United States in the late 1840s. Such academic training can be seen in his treatment of the indigenous men whose naked-to-the-waist bodies allow the artist to show off his rendering skills; their three positions could be isolated as an academic exercise revealing the body from various viewpoints. In contrast with the native inhabitants in Marryat's image discussed earlier, whose unidealized bodies would have been understood as an embodiment of the chaotic Panamanian jungle, Nahl aestheticizes the indigenous bodies and as such renders them as amalgamations – hybrids, if you will – of the stable signifier of beauty and tradition – the academic nude – with the unruly indigenous body. Such an orientalist trope would have been familiar to Nahl given his academic training. Further evidence of the artist's strained deployment of academic strategies in the face of an unfamiliar and threatening landscape are the two cows on the hillside to the left, whose presence in the otherwise jungle setting suggests the pastoral tradition and a landscape, however delimited, that has graciously adapted itself to the forces of civilization. Indeed, Nahl's image creates an incongruous and contradictory scene: subject to the civilizing and benevolent aesthetic imperatives of the pastoral tradition – jungle becomes landscape – yet underscoring the inherent risks of navigating a wild and, by extension, unhealthful setting. The temporal congruity of the falling tree on the left, the boat-trapping deadfall in the middle of the river and the man falling from the boat on the right creates a tension between the prospect of accident and potential ecological agency.

The Chagres River in particular and the Panamanian jungle in general continued to inspire fear and repugnance throughout the pre-canal decades. One author defined the journey across the isthmus as "traversing the waste spaces" (Huntington 1924, 68). James Gilbert, who worked in the 1880s and 1890s for the Panama Railroad Company and later the United Fruit Company, was the self-declared poet of the Isthmus of Panama and transcribed his experiences there into a book of poetry, *Panama Patchwork*, published at the turn of the twentieth century. The poem "Beyond the Chagres River" etched in readers' minds the vile and deadly nature of his adopted home, and ironically, the place of his death from "yellow eyes," the descriptive term used for those suffering from malaria.

> Beyond the Chagres River
> Are paths that lead to death—
> To the fever's deadly breezes,
> To malaria's poisonous breath!
> Beyond the tropic foliage,
> Where the alligator waits,
> Are the mansions of the Devil—
> His original estates! (Gilbert 1901, 6)

Such words would ring stunningly familiar to those laborers, native and non-native alike, who later worked in the canal zone and set their sights on the eradication of disease.

Yankee Strip

The fraught experiences of Nahl and many others who crossed the isthmus before the building of the Panama Railroad could not stand in sharper contrast than with those who made the passage after its completion in 1855. Rather than a six-odd-week passage that was plagued by deep morasses, dense and intractable jungle, malaria- and yellow-fever-carrying mosquitoes, noxious reptiles, venomous insects, staggering heat and nearly incessant rain, the train journey – some 47 miles – across the "Yankee strip" of the continent took an average of two hours (Donoghue 2014, 12). One contemporary chronicler noted, "From the car windows we enjoyed rare glimpses of the virgin jungle, a tropical hortus of blooming trees with orchids and flowering vines . . . Cane huts, primitive as those pictured by the old chroniclers in the woodcuts of their first editions, basked in the shade of cocoa-nut [sic] palms." Indeed, the train compartments enabled the aestheticization of the journey, according to popular perceptions of lush Edenic abundance. Remarking on the speed with which the journey could now be made, he continued, "This is the only place upon the hemisphere where it is now possible to behold both oceans in a single day" (Peixotto 1913, 19, 22). The aesthetic of the technological sublime now defined the journey across the hybridized isthmus: thrilling in its passage through the jungle, safe in its distance from danger. Moreover, the journey on the train, framed by the passenger-car windows, could be understood as a corporeal parallel to the view from above that reassures the viewer with a sense of colonial mastery over the chaos below. The potential threats of the Panamanian jungle as well as the staggering price of human labor to build the railroad, due to disease and hazardous working conditions (Otis 1867, 34–36), receded from view as the speed of movement across the isthmus increased.

A number of texts and reports were published in the late 1850s and 1860s that recounted the "pioneering spirit" of the railroad's builders and proclaimed its completion as a technological triumph in the face of daunting prospects. As a contemporary historian of the Panama Railroad observed, "The character and geographical position of the country through which the line of the road had been carried was such as might well have made the hardiest projectors shrink from attempting its construction" (Otis 1867, 21). In an illustrated feature article in *Harper's New Monthly* Magazine, published in January 1859, the author was attentive to what his readers would understand as the stark contrasts between the modern railroad – the Yankee strip – and the chaotic Panamanian tropics in which it was located (Oran 1859, 145–169). The article included portraits of most of the key players, including the New York financiers who paid some $8 million for the project, and illustrations of the enterprise underway. *Running the Lines*, for example, offers a view of dense jungle foliage, a monkey hanging from the top of a coconut palm, and an indigenous boy, naked to the waist, holding a long spear. Embedded amid the foliage, the boy and monkey (in an unfortunate equation) are rendered as a part of the chaotic environment and well-adapted to their surroundings. The boy's youthfulness acts as a biological corollary to the assumed primitive nature of the Panamanian jungle. At the center of the composition is one of the United States' engineers on the project, fully dressed and wearing a wide-brimmed hat. He is shown in profile view and looking through the telescope of the theodolite, part of his geodetic survey equipment, mounted on a tripod. Depicted as qualitatively distinct from the natural world, the engineer, with his swagger, is an agent in the imperial project of making nature into a disciplined, hygienic and orderly space: mapping, measuring and instrumentalizing the environment while ostensibly immune to its disintegrating forces of excessive heat; humidity; and, of course, the bite of a disease-carrying mosquito. The cartographic information he produces will provide knowledge for the disciplining of the land through technological innovations: steam-powered machines, railroads and photography.

An ecocritical view, however, sees ecological agency at work in this pictorial record of the increasingly hybridized zone. The engineer is portrayed up to his thighs in a swamp and potentially in danger while the boy and monkey are marginalized, yes, but are also comparatively dry and safe. Indeed, the latter two are portrayed as a part of the ecosystem while the engineer seems profoundly out of place, even if he is at the center. We might propose that the engineer's corporeal integrity and separation from the environment is undermined – he is not the only actor at the party – and is reconstituted, using Alaimo's model of trans-corporeality, as invaded, pervaded and fundamentally interconnected to the environmental forces around him, seen and unseen. The Panamanian isthmus is no "mere empty space, an uncontested ground, for human development" (Alaimo 2010, 2, see also Glotfelty 2014, 221–238). Moreover, in spite of all his sophisticated equipment, the engineer can only "see" the landscape through his device, while the illustration's beholder can see boy, money and engineer performing their ideological roles. As such, we can read this image as a demonstration of mutually agentive work of humans, other-than-humans and indigenous inhabitants as well as the meaningful role of the viewer who participates in the production of knowledge. As culture (railroad) and nature (Panamanian isthmus) become a hybrid compound and, as such, deflate the certainties of closure offered by Cartesian dualisms, "it curves epistemologies into new orbits," and allows for new ways of thinking about the ecological agency and the environment (Cohen 2014, xi; Iovino and Oppermann 2014, 5).

The contrast between the unruly wilderness and native population, which this article defined as "too indolent and unaccustomed to labor to be depended on to any great extent" (Oran 1859, 148), and the technologies of progress – surveying tools, railroad lines, modern railway stations – and those who deploy them, was designed to underscore the heroic nature of the undertaking and the inevitability of the imposition of healthful forces of civilization on the jungle. The image, therefore, seems to easily conform to the underlying anthropocentric and colonialist assumptions that undergirded the enterprise of building the Panama Railroad and later the canal. And yet obscured from view was the staggering cost in human lives of the building of the railway; nearly 9,000 laborers died during the five-year construction period from malaria and abysmal working and environmental conditions. Might we argue that such costs, hiding in plain sight, demonstrated the region's agentic resistance to the enterprise (contemporary observers would not have used such language) and serve as a reminder that the Panamanian isthmus and its inhabitants pushed back as well as yielded? Such acknowledgment of the human cost of the venture, however, did not diminish the overwhelming consensus of the success of the venture. Indeed, until the completion of the trans-continental railroad in the United States in 1869, the Panama Railroad provided the safest and fastest travel from east to west coast. Moreover, it asserted the historical position of Panama as the keystone of global trade and imperial fantasies about a navigable passage between East and West. To the proselytizers of progress, the Panama Railroad served as a vital dress rehearsal for the future canal – overseas expansion as an extension of westward expansion – and articulated the coordinates of such progress and empire along political, technological and geographical lines (LaFeber 1989, 7). What remained unaccounted for in such a view, however, were the environmental implications of hybridization, the erasure of human labor and racial injustices. Each would become all the more defining with the building of the canal.

Landscapes of Human Disturbance

The complicated details of the building of the Panama Canal, as well as the colonial history of the Panamanian isthmus, remain outside of the purview of this essay. It should be noted, however, that the massive technological enterprise had two primary components: the structural building of the canal and the eradication of disease. It is the defining impact of the two together that begs an ecocritical reading of images of the canal and its construction. Indeed, one could argue that the hybridization of the Panamanian isthmus that resulted from the incursions of Euro-Americans with the building of the Panama Railway became exaggerated during the 11-year course of the canal-building enterprise and was the product of human and other-than-human forces – in this case disease and insects – working both with and against one another. With respect to the building of the canal within a colonial context, I would argue that the Panama Canal zone was an "ambiguous zone of empire" (Stoler 2006, 1) that assumed all of the entitlements of colonial privilege – access to international networks and vast capital, unchecked exploitation of resources and indigenous labor, implicit racism and bourgeois privilege – but did not attempt to engineer the disappearance of local inhabitants, as would be the case with settler colonialism (Verancini 2010, 3). Rather, the United States sought "to control a canal, not a colony" (Grigsby 2012, 15) and, in so doing, had a defining impact on local economies, ecologies and indigenous populations.

It is not that disease was a hidden insidious obstacle; by the time the United States took control of the Panama Canal zone in May 1904, disease-carrying mosquitoes were recognized as among the most formidable adversary the nation faced in its mission. Nor was it that the United States was lackluster in tackling disease. On the contrary, Army physician William C. Gorgas, who had cut his teeth fighting mosquitoes during the War of 1898, was appointed chief sanitation officer and swiftly implemented a wide range of sanitation measures to minimize, and ultimately eradicate, the spread of yellow fever and malaria. Where the enterprise went wrong was in the assumption that these contagions were an integral part of the local environment and could, just like the isthmus itself, be subjected and ultimately malleable to the rationalization of the environment. The truth proved otherwise.

Indeed, malaria was a trenchant adversary and its resistance to various sanitation measures raised the specter, as environmental historian Paul Sutter has argued, of the disease being a product of the environmental transformation of the Panamanian isthmus during the construction of the canal instead of being intrinsic to tropical nature. More broadly, Sutter proposes that sanitary engineering, in particular mosquito control for the eradication of disease, operated at the dynamic intersection of two competing forces: the "environmental ideology of tropical triumphalism," on the one hand, and the "ecology of malaria," on the other (Sutter 2007, 726). The former refers to the national metanarrative of anthropocentrism, triumphalism and progress at the turn of the twentieth century, when the tropics became both the actual and imagined place of American expansion within an extracontinental context, just as the western wilderness had functioned during continental expansion in the nineteenth century. The latter, by contrast, considers the ecosystem in the Panamanian isthmus as one of the material dimensions shaping American imperial expansion. In other words, nonhuman agents and a vast network of indigenous laborers who were forced to work within the institutionalized racial hierarchy that dates back to the building of the railroad (Green 2009, 62) resisted American hubris and the power of 90-ton Bucyrus shovels, and expanded the hybridized ecological zone begun in the 1850s.

As the tropics continued to be transformed from raw jungle to technological marvel, environmental conditions were created that promoted the spread of malaria vectors. Prominent entomologists were engaged by the Isthmian Canal Commission Sanitary Division to work on the eradication of malaria in the canal zone, and many found that the mosquitoes that carried the malaria virus had a particular affinity for "landscapes of human disturbance," such as those found on the isthmus (Sutter 2007, 743). The United States Department of Agriculture's August Busck, for example, visited Panama in 1907 and found that environmental changes caused by the construction of the canal were partially attributable to the successful breeding of mosquitoes. He noted, "The progress of each steam shovel or of each of the extensive dumps produces new problems to be solved" (Busck 1908, 53). Allan Jennings, Busck's former assistant, went even farther in implicating the role of canal construction – that is, the human-altered landscape – in the ubiquity of the particular mosquito species in question. He noted the close association of the malaria vector mosquito "with man and finds its most congenial surroundings about his habitations and in conditions he creates in the course of agricultural, engineering, and other work" (Jennings 1912, 133). *Mosquito Control in Panama,* published in 1916, drew on the work of several entomologists, sanitary engineers and physicians, and was the most conclusive to date in its assertion that malarial outbreaks in the Panamanian tropics were not inherent to the setting – an indigenous disease unearthed by the construction, or a product of an inherently unhealthful local environment – but instead was the result of the excavation and removal of earth (LePrince and Orenstein 1916, 87).

This brings us back to where we began, Jonas Lie's *The Conquerors* (Plate 2), painted during the artist's three-month sojourn to the canal zone in 1913. At first glance, the painting exemplifies the triumphalist rhetoric of the imperial enterprise as it was nearing completion (*Oh Panama!* 2016). Its subject, the famed Culebra Cut, cuts a slash from top to bottom of the large canvas – 60 × 50 inches – visually evoking the most trenchant obstacle to the entire canal project: the creation of the artificial canyon that cut through the continental divide in Panama which at its base was more than 300-feet wide and at its highest point, nearly 580 feet above sea level. Dozens of trains move along railway tracks at the base of the cut, spewing smoke and steam in long vertical trails; the tracks recede into the distance which is framed by a turquoise and lavender sky, suggesting the water that will soon fill the artificial canyon. Moreover, the tracks evoke the memory of the earlier project of the Panama Railway that prepared the ground for the later enterprise.

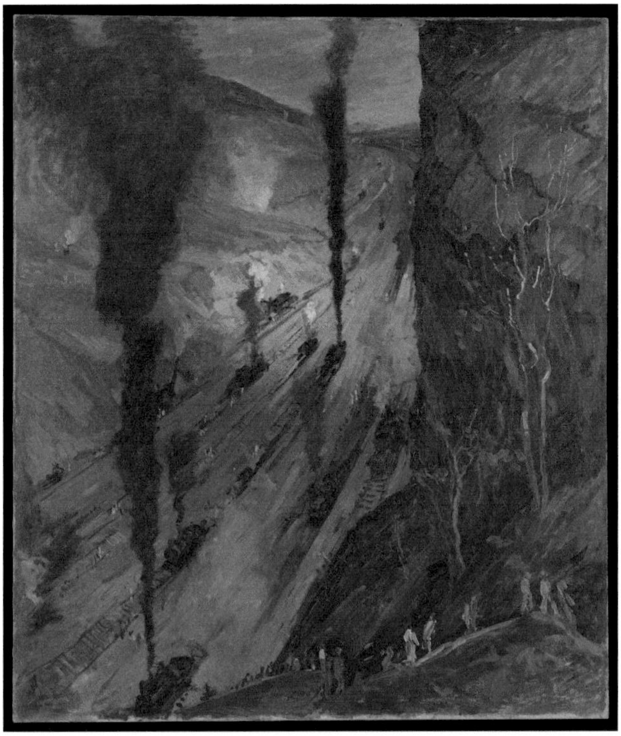

Plate 2 Jonas Lie, *The Conquerors (Culebra Cut, Panama Canal)*, 1913. Oil on canvas, Collection of The Metropolitan Museum of Art, New York George A. Hearn Fund, 1914, Image copyright @ The Metropolitan Museum of Art.
Image Source: Art Resource, NY.

Unlike the more common horizontal view from above used so widely in nineteenth-century landscape images, Lie's visual evocation of colonial domination and control over the land is drawn on vertical lines. Hovering precariously above in the top right corner, the viewer grasps the scene at a distance: the once intact mountains now split apart as if by the blow of a massive axe by a herculean god. Lie was a product of the early twentieth century and had turned his attention to the modern landscape of New York City before his trip to Panama. One could argue that he was primed by his interest in skyscrapers to approach the canal zone in Panama on a soaring vertical rather than horizontal axis. Moreover, the regularity of vertical lines evokes the logic of the engineer and the map, foregrounding technological predictability and precision over the erratic forces of nature.

And yet such verticality precludes a commanding view – one predicated on the hubris of the United States to subdue the tropics – and reveals instead the hybridized zone produced by the dynamic interchange between human and nonhuman actors. It shares with Charles Nahl's images from 1867 a jutting embankment on the right that prevents easy visual entry to the scene. The distinct landscape of Panama and the fact that the canal cut a winding path through the mountains made a panoramic view and its "illusion of a magisterial, complete view" impossible (Grigsby 2012, 147). The vast geography of the slender isthmus defied easy measurement and was difficult to visualize, thus compromising the rationality and productivity of the imperial venture (Allewaert 2013, 33). Two spindly and nearly leafless trees at the right top edge of the canyon along the embankment are dwarfed in scale by the technological enterprise below and seem to stand as mute reminders of the "nature" – the nonhuman actor – that was there before the construction of the canal began. Indeed, the colossal scale of the project erases the inordinate toil of the actual building of the canal, while the natural material that had once defined the region disappears into plumes of smoke. As Grigsby notes, "Emptiness is a strange achievement. Where did everything go? The Americans' reliance on dynamite made it seem that all the soil disappeared into thin air" (Grigsby 2012, 149). This historiography of hybridization and absence came full circle with the flooding of the canal at its completion, forever rendering unseen and hidden the forcible rationalization of space.

Even more diminutive in scale and sketched in very loosely, a single line of laborers emerge from the pit below and move toward the right edge of the canvas, effectively out of the canyon itself. Such details stray far from a heroic depiction of labor – in contrast with the colossal engineered machines, human beings become "their ant-like attendants" (Grigsby 2012, 6), and beg the question of Lie's premise in such an ostensibly bombastic work. Who, we might ask, is doing the conquering here? The colonial rhetoric of progress and triumph over nature is undermined, even if obliquely, by narrative details and aesthetic choices made by the artist. He leaves the viewer, instead, to consider this hybridized zone that refuses to become a landscape.[2]

Although Jonas Lie would never have used such language, we could propose that his images from the canal zone visualize a hybrid nature that the United States enterprise had helped to create and against which it struggled mightily, particularly in the eradication of disease (Sutter 2009, 319). Indeed, environmental changes to the Panamanian landscape – the scale of which was aptly described by James Bryce, then the British Ambassador to the United States, following a trip to the Canal Zone in 1913, as "the greatest liberty man has ever taken with nature" – promoted and perpetuated the spread of disease (Bryce 1913, 36; Miller and Reill 1996). Without the eradication, or at least containment, of yellow fever and malaria, the entire United States mission would have failed. To be sure, mosquito control was recognized by many as a defining factor in the transformation of tropical nature. However, triumphal and anthropocentric rhetoric dominated discussions and images of the massive canal-building enterprise, framing nature as something out there – inert, passive, awaiting the transformative touch of human agency and imperial desire – with which to struggle and ultimately overcome. Such views which construed Panama as a pristine and unalloyed nature to be conquered – to be turned into a landscape – fail to recognize the Panamanian environment as "an active historical force" whose very hybridity was, at least in part, the result of the anthropocentric endeavor as a whole and entangled with the disease the sanitation workers were sent to eradicate (Sutter 2007, 728).

An ecocritical reading of the enterprise, however, compromises such anthropocentric rhetoric and insists upon the centrality and agency of the environment. The Panamanian isthmus is a particularly rich material site for such an inquiry, as it does not constitute one object of analysis, a bounded geography or a monologic discourse. Instead, as environmental historian J. R. McNeill argues, history, as a rule, is a "co-evolutionary process involving society and nature," and the "links between human history and ecological history are robust, sometimes to the point where mosquitoes and viruses infringe on the fortunes of humankind in ways that seem unflattering to our species, making us seem mere playthings in dramas wrought (not directed) by tiny, mindless creatures." He adds, "Mosquitoes and pathogens could not make history on their own; human actions set the stage" (McNeill 2010, 15).

Indeed, the land in Panama became hybridized during the course of the construction of the railroad and canal, ironically increasing the breeding grounds for the mosquitoes that transmit malaria. Moreover, the insects themselves were never fully eradicated but merely controlled. Attention to the intra-actions between humans and more-than-humans, and the "centrifugal tendencies" which decenter and disrupt meanings and undermine the historical position of humans as reasoning agents having dominion over nature, encourages a rereading of the history of American empire building in the Panama Canal zone as one embedded within and defined by the local ecology as much as by international politics and imperial desires (Marafiote and Plec 2006, 60). Just as the water that passes through the Panama Canal – neither Atlantic nor Pacific but a hybrid of the two – so the Panamanian isthmus and its transformation with the building of the Panama Railroad and Canal insist upon a reading beyond the certainties of closure and of entanglements between desire, technology, anthropocentrism, economics, politics, colonialism, disease, sanitation, hygiene and the environment, with each as active players in its history.

Notes

1 For an expanded version of this essay, see Sarah J. Moore. 2017. "Mosquitoes, Malaria, and Cold Butter: Discourses of Hygiene and Health in the Panama Canal Zone in the Early Twentieth Century." *Panorama: Journal of the Association of Historians of American Art.* 3, 2 (Fall). http://journalpanorama.org/mosquitoes-malaria-and-cold-butter. The authors thank the editors of *Panorama* for their permission to use portions of this essay here.

2 *The Conquerors* was purchased by the Metropolitan Museum of Art in 1914 after its display, along with other canvases from Pie's Panama trip, at Knoedler Gallery in New York, December 29, 1913 – January 10, 1914 (Brinton 1915, 196–207).

Works Cited

Abbot, Willis J. 1913. *Panama and the Canal in Pictures and Prose.* New York: Syndicate Publishing Company.

Adams, Charles Francis. 1911. *The Panama Canal Zone: An Epochal Event in Sanitation.* Boston, MA: Historical Society.

Allewaert, Monique. 2013. *Ariel's Ecology: Plantations, Personhood, and Colonialism in the American Tropics.* Minneapolis: University of Minnesota Press.

Alaimo, Stacy. 2010. *Bodily Natures: Science, Environment, and the Material Self.* Bloomington: Indiana University Press.

Brinton, Christian. 1915. "Jonas Lie: A Study in Temperament." *The American-Scandinavian Review.* 3, 4:196–207.

Bryce, James. 1913. *South America: Observations and Impressions.* New York: Macmillan.

Busck, August. 1908. "Report on a Trip for the Purpose of Studying the Mosquito Fauna in Panama." *Smithsonian Miscellaneous Collections* 52: 53.

Cohen, Jeffrey Jerome. 2014. "Foreword: Storied Matter." In Serenella Iovino and Serpil Oppermann, eds., *Material Ecocriticism.* Bloomington: Indiana University Press, ix–xii.

Donoghue, Michael E. 2014. *Borderland on the Isthmus: Race, Culture, and the Struggle for the Canal Zone, American Encounters/Global Interactions*. Durham, NC: Duke University Press.

Frenkel, Stephen. 1996. "Jungle Stories: North American Representations of Tropical Panama." *Geographical Review* 86, 3.

Gilbert, J. S. 1901. *Panama Patchwork*. New York: The Burr Printing House.

Glotfelty, Cheryll. 2014. "Corporeal Fieldwork and Risky Art: Peter Goin and the Making of Nuclear Landscapes." In Serenella Iovino and Serpil Oppermann, eds., *Material Ecocriticism*. Bloomington: Indiana University Press, 221–238.

Green, Julie. 2009. *The Canal Builders: Making America's Empire at the Panama Canal*. New York: The Penguin Press.

Grigsby, Darcy Grimaldo. 2012. *Colossal: Engineering the Suez Canal, Statue of Liberty, Eiffel Tower, and Panama Canal*. Pittsburgh and New York: Periscope Publishing, Ltd.

Huntington, Ellsworth. 1924. *Civilization and Climate*. New Haven: Yale University Press.

Iovino, Sernella, and Serpil Oppermann. 2014. "Introduction: Stories Come to Matter." In Serenella Iovino and Serpil Oppermann, eds., *Material Ecocriticism*. Bloomington, IN: Indiana University Press, 97–113.

Jennings, Allan. 1912. "Some Problems of Mosquito Control in the Tropics." *Journal of Economic Entomology*. 5.

LaFeber, Walter. 1989. *The Panama Canal: The Crisis in Historical Perspective*. 2nd edition (1978 1st edition). New York: Oxford University Press.

LePrince, Joseph, and Alexander J. Orenstein. 1916. *Mosquito Control in Panama: The Eradication of Malaria and Yellow Fever in Cuba and Panama*. New York: Putnam's.

Manthorne, Katherine Emma. 1989. *Tropical Renaissance: North American Artists Exploring Latin America, 1839–1879, New Directions in American Art*. Washington, DC: Smithsonian Institution Press.

Marafiote, Tracy, and Emily Plec. 2006. "From Dualisms to Dialogism: Hybridity in Discourse about the Natural World." *The Environmental Communication Yearbook*. 3, 1: 49–75. London: Routledge.

Marryat, Frank. 1855. *Mountains and Molehills; or Recollections of a Burnt Journal*. London: Longman, Brown, Green, and Longmans.

McNeill, John R. 2010. *Mosquito Empire: Ecology and War in the Greater Caribbean, 1620–1914*. Cambridge: Cambridge University Press.

Miller, David Philip, and Peter Hanns Reill, eds. 1996. *Visions of Empire: Voyages, Botany, and Representations of Nature*. Cambridge: Cambridge University Press.

Moore, Sarah. 2018. "The Great American Desert Is No More." In Wendy Jean Katz, ed., *The Trans-Mississippi and International Expositions of 1898–1899: Art, Anthropology, and Popular Culture at the Fin de Siècle*. Lincoln: University of Nebraska Press.

Oh Panama! Jonas Lie Paints the Panama Canal. 2016. Yonkers, NY: Hudson River Museum and James A Michener Art Museum.

Oran. 1859. "Tropical Journeyings," *Harper's New Monthly Magazine*. 18, 104: 145–169.

Otis, Fessenden N. 1867. *History of the Panama Railroad*. New York: Harper's Brothers Publishers.

Peixotto, Ernest. 1913. *Pacific Shores from Panama*. New York: Charles Scribner's Sons.

Stoler, Ann Laura. 2006. *Haunted by Empire: Geographies of Intimacy in North American History*. Durham, NC: Duke University Press.

Sutter, Paul S. 2007. "Nature's Agents or Agents of Empire? Entomological Workers and Environmental Change during the Construction of the Panama Canal." *History of Science Society*. 98, 4: 724–754.

———. 2009. "Tropical Conquest and the Rise of the Environmental Management State: The Case of U.S. Sanitary Efforts in Panama." In Alfred W. McCoy and Francisco Scarano, eds., *Colonial Crucible: Empire in the Making of the Modern American State*. Madison: University of Wisconsin Press.

Verancini, Lorenzo. 2010. *Settler Colonialism: A Theoretical Overview*. Hampshire, England: Palgrave Macmillan.

Whitbeck, Ray H. 1921. "Geography and Man at Panama." *Bulletin of the Geographical Society of Philadelphia* 19: 8.

White, Richard. 2004. "From Wilderness to Hybrid Landscapes: The Cultural Turn in Environmental History." *The Historian*. 66, 3, 557–564.

Part 2

Material Ecologies

Climates and Extractive Logics

In this section, the four authors investigate the colors and textures, the plasticity and functionality of natural substances: the pure white Carrara marble that produced both gleaming sculpture and asphyxiating dust; the icy white of a frozen landscape that could arrest time and assuage fear of social change; the pure blue of the unpolluted British sky that contrasted with its contaminated 'plague-dark' counterpart; and the moral and material blackness of coal extracted to meet rising industrial demands. As Laura Turner Igoe points out, such new materialist methodologies challenge the often-lamented anthropocentrism of art history by shifting emphasis to the agency of the atmospheres, environments and nonhuman entities with which works of art are deeply engaged.

Political philosopher Jane Bennett's attention to material vibrancy proves useful in several of these essays, not merely in terms of highlighting matter's vitalized animacy, but also suggesting what Wai Chee Dimock has also described as its "diachronic historicism" or resonance (Dimock 1997, 1061; Bennett 2010). Igoe, George Philip Lebourdais, Corina Weidinger and Polly Gould track ripples spreading out from nineteenth-century events, evoking climatological 'teleconnection' – correlations between environmental phenomena that occur across great gulfs of both time and spatial distance (Taylor 2018). Exploring a host of concurrent discourses – from the ruinous quarrying of marble and mining of coal to the discoveries of physicists – these writers remind us that art history isn't a monolithic practice, but a process in which ecological interconnections are always at play. The materialisms within these texts exceed concern with the substances from which art is made to engage with aesthetics and criticism, philosophies and epistemologies: vital corollaries to the networks of imaginative visualization that objects set in motion. They encourage twenty-first-century viewers to think anew about the continued relevance of artistic products of a prior era to the concerns of the present day.

Igoe describes American artist Randolph Rogers's popular marble statue *Nydia*, considering how in both its medium and narrative motivation, it not only referenced the destruction of Pompeii but also sustained provocative comparisons with the toxic atmosphere produced by the famous Carrara marble quarry. Igoe reconnects *Nydia*'s marble to the human efforts and environmental transformation that enabled its procurement and determined its high cost and status as a luxury material. She emphasizes the function of materials within an ecological network, emerging as vital sites of production. The labor of material extraction and the processes of artistic fabrication don't exist apart from the philosophical, literary or anecdotal narratives that inspire works of art, but echo the observation made by literary scholars Iovino and Oppermann, that "stories come to matter" in more than one, purely semantic sense (2014).

Coining a term that reckons with frozen water's slippery ontological status, George Philip Lebourdais examines American artist George Henry Durrie's "cryoscapes": winter scenes that paradoxically both illustrate and curtail snow's agentive capacity. Tracking human-centered desire for reassurance and perfection in Durrie's willful distortions of time and place, he shows us that the stilled time of these paintings enacts a fantasy of control over weather's unstoppable forces. Durrie materially transforms environmental anxiety into a soothing vehicle for nostalgic escapism, attempting to forestall a cascade of mid-nineteenth-century social upheavals: industrialization, the generational trauma of the Civil War and the abolition of slavery. Both LeBourdais and Igoe also track the implicit bias that their objects reveal: the whiteness of snow and marble speaks to the implicit racial hierarchies that most nineteenth-century makers and viewers (presumed to be white themselves) accepted, seemingly without question. But when the materiality of objects and representations are brought into greater socio-economic and ecocritical focus, "whiteness" and its attendant racial prejudice are revealed, hiding in plain sight.

If Durrie's pristine snow-scapes suggest preternatural changelessness, Corina Weidinger and Polly Gould instead dwell upon the philosophical and aesthetic consequences of industrial development's attendant pollution. Weidinger's essay conjoins contemporary energy humanities to nineteenth-century utopian anti-industrial critiques, identifying nascent ecological thinking in the latter. She repurposes ecocritical scholar Stephanie Lemenager's inspired term "oil weather," applying it to the production of the nineteenth-century "coal weather," depicted in Neo-Impressionist paintings of Belgium's coal-mining region (Lemenager 2016). Although the work of Constantin Meunier and Maximilien Luce did not follow a consistent politic in representations of unchecked industrial destruction of the landscape, both artists brought socialist and anarchist concerns to bear not only on the ecological effects of coal mining but also on the careless exploitation of workers.

Whereas prior scholarship on such aesthetically radical late nineteenth-century artists has tended to 'sanitize' their images of rapacious industrial development by stressing the social value of their attunement to modernity (Rubin 2008), Weidinger reminds us that even amid upheavals in form and facture, artists were deeply aware of the consequences of representing industrialization. Departing from Nicholas Mirzoeff's essay "Visualizing the Anthropocene" (Mirzoeff 2014), Weidinger shows how art was complicit with capitalism, rendering environmental effects that were seen as the lamentable but inevitable consequences of economic growth and job creation. Her work thus intersects with those who favor the term *Capitolocene* in place of Anthropocene.

Particulate matter also concerns Polly Gould, who works to complicate the epistemological divisions that have traditionally distinguished nineteenth-century aesthetic philosophies from concurrent science. Gould employs Donna Haraway's non-hierarchical model of "diffraction" to highlight inter-disciplinary relationships, examining the mutable significance of nineteenth-century 'atmospheres.' Re-evaluating English critic John Ruskin's contributions to nineteenth- century aesthetic theories, Gould finds them thoroughly allied to the concurrent science he often decried. Exploring how the Cyanometer that physicist John Tyndall and Ruskin used to measure the blue of the sky coordinated color's ontological status with its 'purity,' Gould traces how atmospheric clarity was related to moral conditions: the purity of blue contrasted with the soul-compromising noxiousness of coal blackened storm clouds. Thus, although seemingly at odds with one another philosophically, both Tyndall and Ruskin shared a dedication to empirical observation of the natural world, and contributed early recognition of the aesthetic and material consequences of mounting air pollution.

Works Cited

Bennett, Jane. 2010. *Vibrant Matter: A Political Ecology of Things*. Durham, NC: Duke University Press.

Dimock, Wai Chee. 1997. "A Theory of Resonance." *PMLA*. 112: 1060–1071.

Haraway, Donna. 1997. *Modest_Witness@Second_Millenium. FemaleMan©_Meets_OncoMouse™*. New York and London: Routledge.

Iovino, Serenella and Serpil Oppermann, eds. 2014. *Material Ecocriticism*. Bloomington, IN: Indiana University Press.

LeMenager, Stephanie. 2016. *Living Oil: Petroleum Culture in the American Century*. Oxford: Oxford University Press.

Mirzoeff, Nicholas. 2014. "Visualizing the Anthropocene." *Public Culture*. 26, 2: 213–232.

Rubin, James H. 2008. *Impressionism and the Modern Landscape: Productivity, Technology, and Urbanization from Manet to Van Gogh*. Berkeley, CA: University of California Press.

Taylor, Jesse O. 2018. "Storm-Clouds on the Horizon: John Ruskin and the Emergence of Anthropogenic Climate Change." *19: Interdisciplinary Studies in the Long Nineteenth Century*. 26. Accessed 1 October 2018. doi:10.16995/ntn.802.

6 "A Gruesome Sight"

Randolph Rogers's Nydia in a Marble World

Laura Turner Igoe

In his 1846 travelogue *Pictures from Italy*, Charles Dickens remarked upon the massive amount of human and animal labor required to extract and transport large blocks of marble from the famous Carrara quarries in Tuscany's Apuan Alps.[1] He noted that many oxen and men perished while shepherding the stone down the mountains, prior to the introduction of railroads to the region. Dickens struggled to reconcile the seductive surfaces of finished marble statues with the crushing labor he witnessed: "Standing in one of the many studii of Carrara that afternoon ... full of beautifully finished copies in marble, of almost every figure, group, bust, we know – it seemed, at first, so strange to me that those exquisite shapes, replete with grace, and thought, and delicate repose, should grow out of all this toil, and sweat, and torture!" (Dickens 1846, 150) In the United States in the late nineteenth century, one of the more ubiquitous marble "exquisite shapes" was *Nydia, the Blind Flower Girl of Pompeii* (Figure 6.1), by the sculptor Randolph Rogers (1825–1892). A character from the 1834 novel *The Last Days of Pompeii* by Edward Bulwer-Lytton (1803–1873), Nydia, an enslaved, blind girl, helps the youth Glaucus and his lover Ione escape Pompeii during the eruption of Mount Vesuvius in 79 CE; she drowns herself after reaching safety because Glaucus does not return her love. Rogers portrayed Nydia as a symbol of feminine sacrifice and fidelity. She strains forward with one hand cupped around her ear, as if listening for Glaucus after she is separated from him when volcanic ash descends upon the city (Schiller 1993).

A popular success, more than 50 marble copies of *Nydia* were produced in two different sizes between 1867 and 1891. This essay concentrates on a life-sized version donated to the Princeton University Art Museum in 1896 from the estate of the physician and writer Abraham Coles. In his correspondence, Rogers noted that *Nydia*'s "striking" appearance made it "a very expensive statue to execute in marble...the flying drapery, deep cutting, and undercutting make it a very laborious undertaking" (Rogers 1971, 4–5). Working from a plaster model under the direction of the artist, carvers in Rome churned out marble copies of *Nydia* for the market as Rogers never learned to carve in marble himself. He instead modeled the figure out of clay, from which a plaster cast was made as a template for reproduction in marble (Rogers 1971, 4–5). A pointing machine allowed for the work to be duplicated, transferring the dimensions of the plaster version to gleaming white stone. The American artist David Maitland Armstrong (1836–1918) echoed Dickens's disturbed reaction to the dazzling products of anonymous human labor when he observed the seemingly mechanical reproduction of Rogers's masterpiece in his studio: "I once went to [Rogers's]

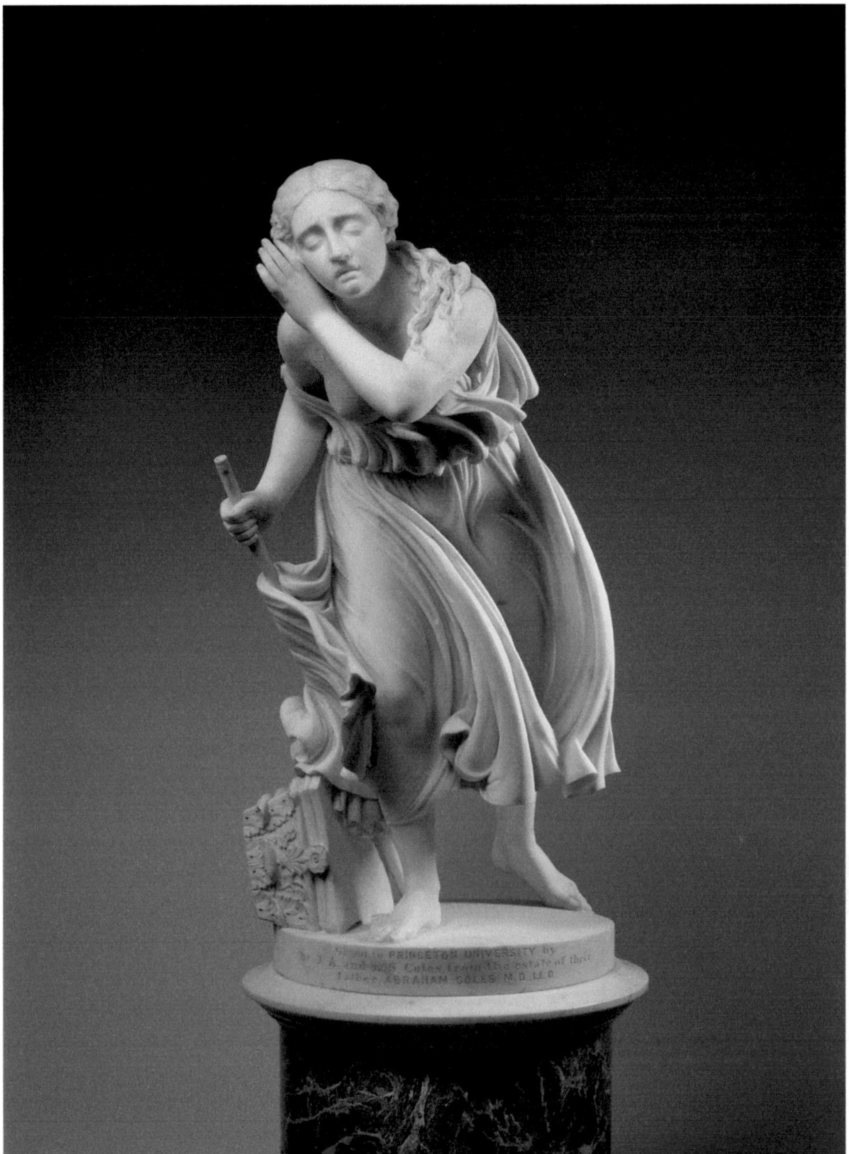

Figure 6.1 Randolph Rogers, *Nydia, the Blind Flower Girl of Pompeii*, after original model of 1855, marble, Princeton University Art Museum, Gift of Dr. J. Ackerman Coles and Emily Coles.

studio and saw seven *Nydias*, all in a row, all listening, all groping, and seven Italian marble-cutters at work cutting them out. It was a gruesome sight" (Armstrong 1920, 194–195). In 1892, *Harper's Weekly* estimated that Rogers earned $70,000 from sales of *Nydia* during his lifetime, the equivalent of nearly 2 million dollars today ("Randolph Rogers, the Sculptor" 1892).

Rogers's remarks and Armstrong's anecdote recast *Nydia* as a site of production and reorient our attention to the multiple hands and tools that transformed hard stone to articulated figure. The completed statue, reveling in the smooth, buttery characteristics of carved marble, similarly honors the provenance of its highly valued material, which was mentioned frequently in period accounts. The press, for example, championed *Nydia*'s material in its announcement of the statue's donation to Princeton by Coles's children in 1896. An anonymous reviewer for the *New York Examiner* praised "the magnificent life-size *marble* statue of 'Nydia,' made of the best *Carrara marble*" and "its *marble* pedestal," repeatedly underscoring the sculpture's esteemed medium.[2] Taking a cue from these repeated references to material and making, this essay reconnects *Nydia* with the genesis of her creation: the Italian marble quarry.

Art history frequently suffers from an environmental-historical amnesia about the physical stuff of artworks, a condition resulting from their abstraction as commodities divorced from their origins and transformed through processes of aesthetic production. Drawing attention to the socio-environmental history of art materials, however, opens up new perspectives on iconic works like *Nydia*. As art historian Robin Kelsey contends, "In the act of historical interpretation, we have a habit of separating our pictures from the material processes and economic desires that make them possible and give them form. We admire the art and forget the fuel" (Kelsey 2014, 13). Uncolored, Italian marble, the preferred medium of nineteenth-century neoclassical sculpture, initially thwarts deeper material investigation. Works like *Nydia* purposefully obscure any sign of the sculptor's hand and, as Charmaine Nelson has argued, uphold a cultural ideal of a racially pure and chaste whiteness divorced from "context, environment, and condition" (Nelson 2007, xiii). *Nydia*, however, is fundamentally a sculpture of dramatic peaks and valleys: a mountainous landscape of rippling cloth. Its captivating, undulating surface encourages viewers to walk around the entire statue as if surveying an unfamiliar terrain. Even the sculpture's narrative focal point, *Nydia*'s cupped ear, listening eagerly for the calls of her beloved, is a deep recess, echoing the yawning ravines of her billowing garments (Figure 6.2). Although the figure's flowing drapery and smooth skin obfuscate the solidity and weightiness of the marble stone, prominent blue-tinged veins snaking across the sculpture repeatedly break the illusion, reminding the viewer of its material origin. An ecocritical consideration of art materials helps us reconnect aesthetic objects like *Nydia* with their chains of production by recovering lost or neglected evidence of related environmental conditions that bear on politics, society and culture. This approach challenges the prevailing anthropocentrism of art history by recognizing the agency of the environments and nonhuman entities with which works of art engage (Braddock and Irmscher 2009).

The Last Days of Pompeii

On a narrative level, *Nydia* was inspired by an instance of cataclysmic environmental change: the violent eruption of Vesuvius. *Nydia* was a part of a growing body of popular imaginations of Pompeii's destruction, which ranged from dramatic paintings by the British artist John Martin to spectacular fireworks displays at London's Vauxhall Gardens. Bulwer-Lytton's novel even inspired a pyrodrama combining explosives and theater performed at Coney Island (Yablon 2007, 192–193). These artworks and events coincided with – and incorporated discoveries and evidence from – excavations at

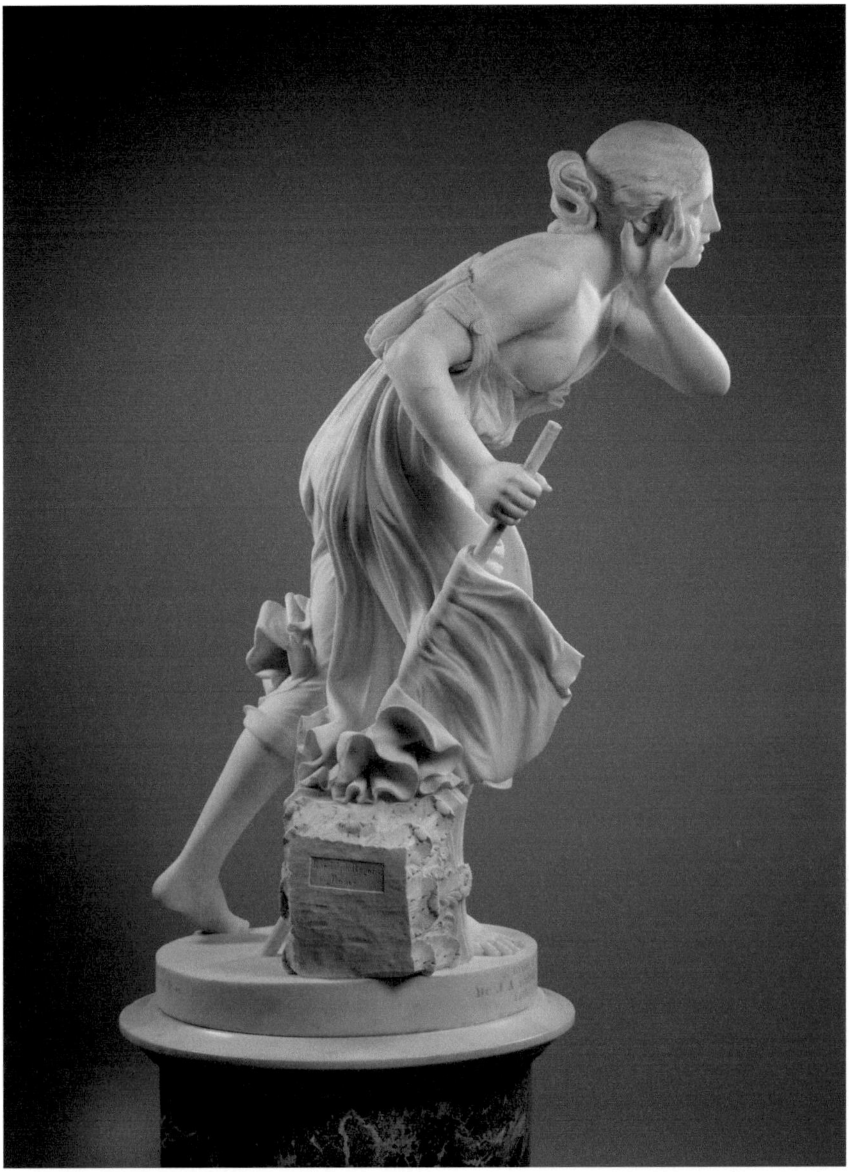

Figure 6.2 Randolph Rogers, *Nydia, the Blind Flower Girl of Pompeii*, after original model of 1855, marble, Princeton University Art Museum, Gift of Dr. J. Ackerman Coles and Emily Coles.

Pompeii and Herculaneum, important stops on the European Grand Tour, that began in the late eighteenth century and continued through the nineteenth century. *Nydia*'s crouching, twisting form recalls the plaster casts of Pompeiian victims introduced to the public in 1863 by Giuseppe Fiorelli, the newly appointed director of museums and archaeological excavations in Italy. These casts, made by pouring plaster of Paris into cavities in the earth, were displayed at the Pompeii Museum and lauded for their "artistic beauty" (Brunn 1863; Dwyer 2007, 178). The British archaeologist Austen Henry Layard described one body, commonly identified as "the pregnant woman," with language typically applied to sculpture, not human remains: "Her garments are gathered up on one side, leaving exposed a limb of beautiful shape. So perfect a mold of it has been formed by the soft and yielding mud, that the cast would seem to be taken from an exquisite work of Greek art" (Layard 1864, 172; Dwyer 2007, 182). Such a description could easily be applied to *Nydia* as well, visually linking the sculpture to actual bodies that once inhabited the ancient city.

In the nineteenth century, Pompeii became a moralized site through which to explore contemporary concerns of religion, slavery, urban industrialization and capitalist excess. Bulwer-Lytton's intimate rendering of life in Pompeii contrasted materialistic Romans with idealized Athenians and proto-Christians. This dichotomy appealed to Victorian American audiences during the Second Great Awakening, an era of heightened religiosity fueled by anxiety regarding increased consumption and moral corruption at a time of territorial and commercial expansion in the United States (Malamud 2001). These religious associations are undoubtedly what made *Nydia* so appealing to Abraham Coles, the original collector of the Princeton statue and a staunch evangelical Christian best known for his translation of Latin hymns. Marble statues of biblical subjects – including a bust of *Eve* by Hiram Powers (1805–1873) and a life-sized statue of Deborah by the Italian sculptor Giovanni Battista Lombardi (1823–1880) – filled Cole's Deerhurst estate in Scotch Plains, New Jersey (Coles 1892, xvi–xvii). *Nydia*'s inclusion in this ensemble would have contextualized the figure as virtuous and chaste.

Nydia, a blind slave, attracted attention at a moment when the country grew increasingly divided over slavery. Although Roger's *Nydia* is unshackled, her popularity as a literary figure would have broadcast her subjugated status to nineteenth-century viewers. Perhaps capitalizing upon the earlier success of Hiram Power's famous sculpture *Greek Slave*, Rogers represented his enslaved subject as a vulnerable, innocent, white female, a representation of slavery far removed – in appearance, as well as geographically and chronologically – from the realities of the black enslaved body in the American South. The predominant usage of white Carrara marble by nineteenth-century American sculptors underscored cultural connections between whiteness and purity, casting color as suspect and second rate (Nelson 2007; Droth and Hatt 2016). Rogers's own opinion regarding slavery is ambiguous; he completed several sculptural Civil War commissions for the Union, including many portraits of Lincoln, but he also worked on a major monument for Virginia during the Civil War and married into a slave-holding family (Coates, Lapatin and Seydl 2012, 201). *Nydia*'s vague, indirect reference to slavery meant that the statue could appeal to both Northern and Southern audiences during the antebellum period.

While *The Last Days of Pompeii* and *Nydia* tapped into nineteenth-century cultural anxieties about capitalism and slavery, they also explored the vulnerability of human civilization in the face of the awesome power of natural forces. For most of Bulwer-Lytton's novel, Vesuvius lurks ominously as a dormant, yet animated, presence in the background prior to its climatic eruption. Early on in the text, Ione and Glaucus note with trepidation a dark cloud gathering above the volcano that "seems to gaze with a steady face upon the city below, to point with one hand…over its glittering streets, and to raise the other towards the higher heaven" (Bulwer-Lytton 1880, 32). *Nydia* perhaps most effectively expresses the resulting violence of this natural disaster through the broken Corinthian capital – symbolic of ancient Rome and classical culture in general – which rests on its side by Nydia's right foot beneath a mass of bunched fabric (see Figure 6.2). This building fragment supplants the traditional role of the tree branch or stump, necessary structural elements that helped support the physical weight of marble figural sculpture since antiquity. In this context, the capital, however, is a symbol of a destroyed civilization, as it has, one presumes, toppled from a Pompeiian building during the volcanic eruption. "Huge fragments of a shattered column" bearing a bronze statue of the emperor Augustus crushed the villain Arbaces during a lightning strike in the final moments of Bulwer-Lytton's novel (Bulwer-Lytton 1880, 216). This material destruction is framed as divine retribution for the Egyptian murderer, who attempted to flee the city with his slaves hauling his accumulated wealth. Indeed, *Nydia*'s capital recalls another prominent Corinthian column in American art history: the solitary, standing column in *Desolation* (1836; Figure 6.3), the final work in the series *Course of Empire* by the British-born landscape painter Thomas Cole (1801–1848). Cole visualized his unease with what he perceived as a growing culture of materialism and excess in the United States – themes prevalent in *Last Days of Pompeii* – in five paintings representing the progression of a classical empire from the pastoral state through its corrupt and decadent consummation and culminating in its eventual ruin. As Angela Miller has argued, Cole witnessed the devastating, environmental impact of industrialization firsthand in his native Lancashire, England. *Course of Empire* expressed his anxiety about American expansion and development in the 1830s, revealing "the catastrophic results of falling away from nature" (Miller 2009, 92). Although the column still stands in Cole's *Desolation*, it is now a ruin, topped by a nesting bird. Both Cole and Rogers employ crumbling classical architecture to illuminate the ultimate triumph of the nonhuman over human civilization and its trappings of wealth and luxury.

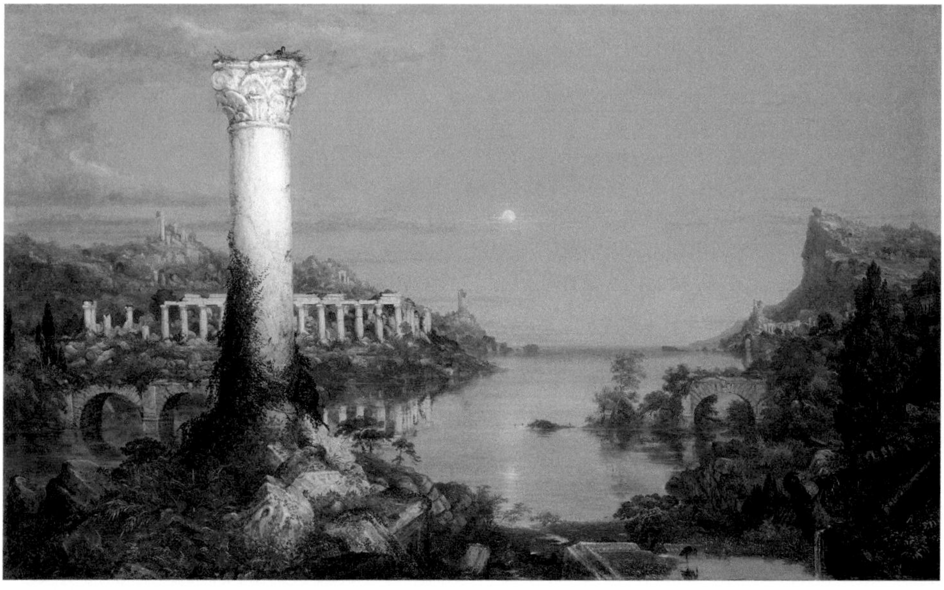

Figure 6.3 Thomas Cole, *Desolation,* from *Course of Empire*, 1836, oil on canvas, New-York Historical Society.

A Marble World

Nydia's ruined capital, a marble representation of a traditionally marble architectural ornament, additionally reminds the viewer of the sculpture's material origins and its act of creation. Despite its presumed violent demolition, the capital itself does not appear to have haphazardly broken from its column shaft; the underside facing the viewer instead displays ordered chisel marks, referencing the sculptor's own carving process (or, more specifically, that of his workshop). Indeed, the underside of the capital is prominently incised, "Randolph Rogers. Rome," drawing a parallel between the artist and classical antiquity, albeit one in ruin thanks to a natural catastrophe. This prominent inclusion of a precisely cut marble block encourages closer consideration of the sculpture's material and the context of its extraction from the Apuan Alps.

Carrara marble, most famously used by the Renaissance sculptor Michelangelo, has long been a desirable and revered stone for sculpture and architecture. Ancient Romans exploited Carrara marble in decorating Pompeii, complementing the narrative drama of *Nydia* with a certain degree of indexical realism (Mau 1899, 36; Cooley and Cooley 2014, 127). Increased international demand for Carrara marble after the mid-nineteenth century catalyzed modernization and restructuring of the local marble industry. An English entrepreneur, William Walton, helped finance Carrara's first modern port, which included a railway, mobile cranes and sawmills that used hydraulically driven machines to cut more than 2,000 tons of marble a year. Newly laid railroad tracks connected the quarries to the town of Carrara and the port, hauling up tons of sand used in the stonecutting process and shuttling excavated marble down the mountains. Thanks to the implementation of cutting-frames, sifting machines and other new technologies, sawmills that could previously cut and polish only two marble tables a week for the market could produce 250 tables in the same amount of time by the 1870s (Lorenzini 1994, 41–57). This increase in marble quarrying and production, however, came at a growing socioecological cost.

Although the transportation and milling of marble underwent significant modernization throughout the nineteenth century, quarrying itself did not change much until the end of the century, when the use of explosives became more widespread. Poor working conditions in the quarries were noted by both Dickens and the popular press. A *New York Times* headline in 1896 pithily proclaimed, "Carrara's Marble Quarries: Their Output is Large, Their Methods Most Primitive." Explosives used to dislodge large quantities of marble from the quarries produced prodigious amounts of waste, filling the landscape with pulverized debris (Lorenzini 1994, 64–65; Leivick 1999). A 1900 *Pearson's Magazine* article on the quarries entitled "A Marble World" noted, "The air is grey with marble dust" (Hart 1900, 631). Despite the growing use of explosives, quarries still predominantly relied on the labor of humans and animals to excavate and transport marble to local rail lines. To shepherd large blocks down the mountain, quarrymen used the *lizzatura* method, employed since ancient Roman times. Large blocks slid downhill on oak logs soaked in soap to reduce friction. As soon as a log rolled out, it was quickly moved to the front of the descending block creating a type of mobile conveyor belt. In the absence of roads, marble blocks were simply pushed off the mountainside, occasionally causing avalanches and extensive damage. These methods proved perilous for quarrymen. Between 1857 and 1879, an average of 25 deaths and 80 serious injuries in a workforce numbering approximately 4,000 were reported annually. Minor injuries that were less documented included broken arms, legs, hands, feet and ribs. Workers in sawmills labored 12-hour days, breathing a fine marble dust that irritated eyes, noses, throats and lungs, causing bronchitis, emphysema, pleurisy, silicosis and pneumoconiosis. Because of this high physical risk, quarry workers received a higher daily wage than other Italian industrial laborers, but they were also beholden to the variable weather conditions of the mountains. The unpredictable climate meant that quarrymen worked, on average, 230 days a year, resulting in a lower annual salary despite their elevated daily rate (Lorenzini 1994, 111–116). Even warm weather proved challenging as workers labored under a "hot sun whose brilliant rays were magnified into a blinding light by the reflective qualities of the white marble quarry walls and floors" (Lorenzini 1994, 114).

Due to these harsh, changeable conditions, the Apuan Alps loomed as animated beings in the minds of local quarry workers. Much like Bulwer-Lytton's Vesuvius, these mountains were capable of both immense cruelty and eternal replenishment. Until the nineteenth century, most quarrymen were predominantly farmers, supplementing their agricultural income with wages received from extracting marble. Theories of seasonal renewal therefore informed their views of marble production (Lorenzini 1994, 109). These beliefs date back to ancient Rome, as referenced by Pliny in his *Natural History*:

> Among the many marvels of Italy itself is one for which the acclaimed natural scientist Papirius Fabianus vouches, namely that marble actually grows in its quarries and the quarrymen moreover assert that the scars on the mountain side fill up of their own accord...If this is true then there is reason to hope that there will always be sufficient marble to satisfy luxury's demands. (Pliny 1855, 354–355)

In 1820, local geologist Emanuele Repetti published a full-scale study of the Apuan mountains, writing that quarry workers viewed the landscape as a constantly shifting, regenerating force. He contended that most laborers were "certain that motion resides within these mountains [and] that they produce secretions and excretions and are, like organic bodies, involved in perpetual destruction and reproduction" (Repetti 1820, 38; Scigliano 2005, 26). A network of veins – a word that refers to both blood and stone in Italian and English – circulated elements throughout the mountains. It was believed that marble also continued to transform, resting and settling after it was cut. This perception persisted into the twentieth century as demonstrated by a 1980s ethnographic study of workers who described the quarries as *agri marmiferi* or "marble fields" that naturally regenerate through quarrying (Leitch 1996, 238–239).

Despite this regional insistence on the cultivation and renewal of marble, quarrying radically altered the Carrara region. The quarries' architectural reshaping of the landscape, however, was occasionally commended as a triumph of human industry. According to "A Marble World," the mountains "have been so hewn and blasted for a thousand years, that parts of them have been cut and riven into peaked and pinnacled masses like vast cathedrals" (Hart 1900, 143). A recent *New York Times* photo-essay on the quarries similarly employed rapt, metaphor-laden language to romanticize the brutal landscape of the quarries as "their own isolated world: beautiful, bizarre, and severe. It is a self-contained universe of white, simultaneously industrial and natural, where men with finger-nubs stand on scenic cliffs conducting tractors like symphony orchestras" (Anderson 2017). Artist Edward

Burtynsky (b. 1955), who photographed marble quarries in Carrara, Vermont, India and China, described his interest in the subject as a desire to document "inverted cubed architecture on the side of a hill," which appeared like "inverted pyramids" or "skyscrapers" (Burtynsky and Mitchell 2007, 9). His *Carrara Marble Quarries #12, Carrara, Italy* (1993; Figure 6.4) seduces the eye with its depiction of quarry walls, reminiscent of precisely chiseled underside of *Nydia*'s broken capital. Burtynsky's image gives the impression that blocks of marble have been hewn and stacked within the quarry instead of blasted and extracted from it. One could also interpret Burtynsky's landscape as gaping wounds or gashes in the mountain, punctuated

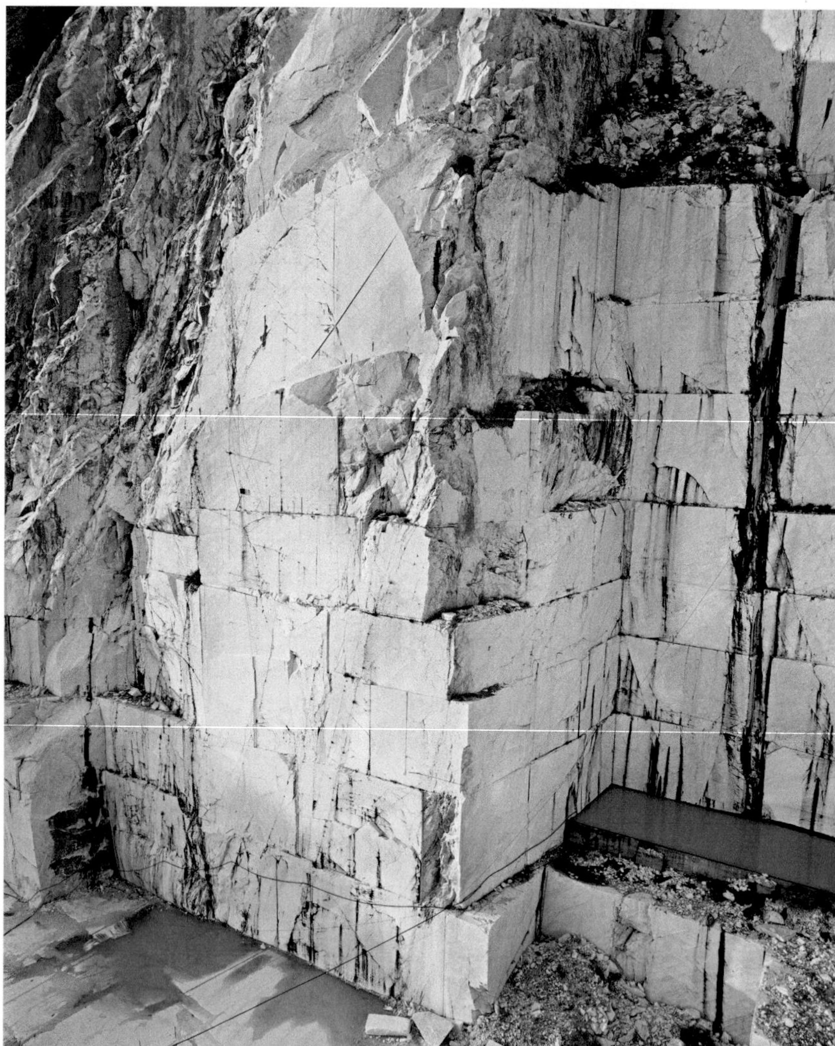

Figure 6.4 Edward Burtynsky, *Carrara Marble Quarries #12*, Carrara Italy, 1993, chromogenic color print, photo(s) © Edward Burtynsky, courtesy of Metivier Gallery, Toronto/Howard Greenberg Gallery and Bryce Wolkowitz Gallery, New York.

by brightly colored pools of contaminated water. The crevice-like drapery folds in Rogers's *Nydia*, including a particularly deep one in front of the figure's right leg, visually conjures these shafts. These undulating surfaces would have been initially built up through the accumulation of clay in Rogers's original model, but later carved out of a larger block by his workshop in the final marble version, echoing the ambiguity of addition and subtraction in Burtynsky's photographs and verbal description of the quarries.

In his 1967 *Tour of the Monuments of Passaic, New Jersey*, the land artist Robert Smithson (1938–1973) called similar man-made, cratered spaces, "ruins in reverse," harbingers of new construction that would rise from the industrial wasteland outside of Manhattan (Smithson 1996, 72; Coughlin 2009). The removed marble in Burtynsky's photographs may have resurfaced as a building, statue, architectural ornament or article of furniture elsewhere: a material accumulation that counterbalanced the void in the quarries. Most marble quarried from Carrara today, however, is pulverized into an industrial filler called ground calcium carbonate (GCC), used to whiten paper, stiffen plastics, prevent paint corrosion and thicken toothpaste. Carrara marble now infiltrates everyday objects and even our bodies, underscoring the porous interchange between human and nonhuman matter (Alaimo 2010). Although GCC produces less overall waste than block quarrying, it creates a fine powder that covers the quarries like snow, damming water flow down the mountains and causing landslides (Williams 2009, 171).

It is unlikely that Rogers consciously made any sort of visual and material connections between *Nydia* and the conditions of the Carrara quarries during the sculpture's conception. Indeed, Rogers probably never traveled to the region himself. His only record of purchased marble specified "statuary marble of the 1st quality" obtained from Livorno, a port near Carrara (Rogers 1971, 5). The social and environmental history of marble extraction in Carrara, however, is not merely tangential to Rogers's *Nydia*. The sculpture is layered with references to the impact of an eroding landscape on the human form. Nydia's hunched stance, straining forward with a cupped ear to hear the calls of her beloved Glaucus, evokes not only the crouched plaster casts of Pompeiian victims, but also the stooped postures of workers transporting heavy marble blocks down from the quarries, as photographed for "A Marble World" (Figure 6.5). When encountering *Nydia*, nineteenth-century viewers may have recalled Bulwer-Lytton's account of Pompeii as besieged by a "cloud, which had scattered so deep a murkiness over the day [settling] into a solid and impenetrable mass...Each hurried blindly and confusedly on" (Bulwer-Lytton 1880, 212). Falling ash transformed portions of the city with a "leprous and ghastly white" that similarly characterized Carrara quarries (Bulwer-Lytton 1880, 214). These imaginative descriptions conjure the dust-filled air of the marble quarry as described by period viewers and the press. A sculptural tour de force, *Nydia* persistently underscores its material transformation through its subject matter, undulating surface and material presence.

As this essay has demonstrated, marble sculpture cannot be disassociated from the labor and environmental transformation that enabled its procurement and determined its high cost and status as a luxury material. Through its depiction of a fictional Pompeiian, besieged by the wrath of an unstable, shifting mountainous landscape, Rogers's *Nydia* draws our attention to the conditions of the Carrara quarries that produced its celebrated material. Just as Dickens's new knowledge of marble

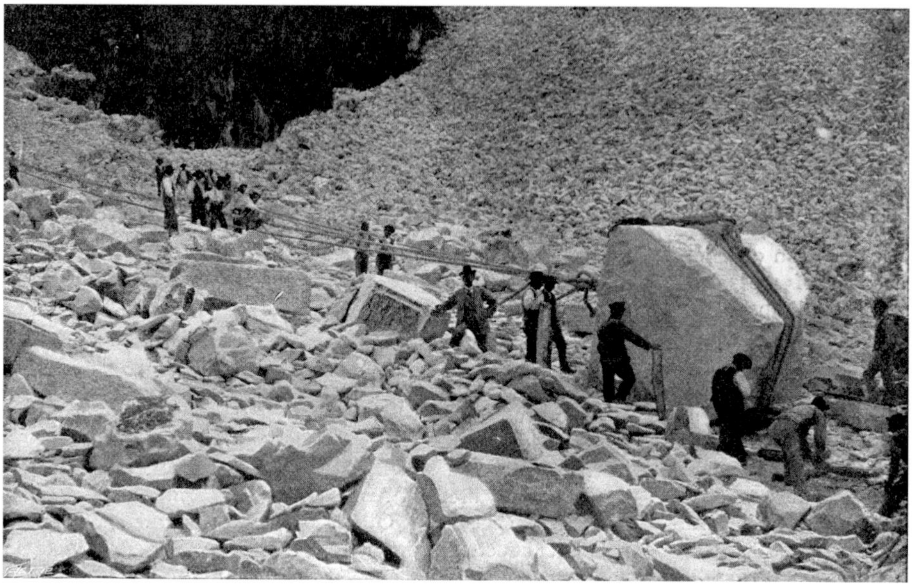

Figure 6.5 "Lowering a block from high to low level." In E. St. John Hart, "A Marble World," *Pearson's Magazine*. vol. 10, no. 60. December 1900.

quarries profoundly impacted his appreciation of marble statuary, so too should art historians seek to reconnect finished artworks with the socio-environmental agents that enabled their production and shaped their reception. Attending to the environmental history of *Nydia*'s medium ultimately brings together subject and material in a new interpretation of marble statuary.

Notes

1 This essays draws from research conducted while a Postdoctoral Research Associate at Princeton University Art Museum, a portion of which was published in Igoe (2018).

2 Italics added by author for emphasis. *New York Examiner* clipping in the object file for *Nydia*, Princeton University Art Museum. Thanks to the collaborative effort of Norman Muller, art conservator, Princeton University Art Museum, George Scherer, William L. Knapp '47 Professor of Civil Engineering at Princeton, Adam C. Maloof, associate professor of geology at Princeton, and Lorenzo Larrazini, professor at the Università IUAV di Venezia, isotopic analysis of a small sample of the statue's marble established that *Nydia* is, as advertised, made of Carrara marble. Maloof performed the isotopic analysis with a small sample of marble from the underside of the sculpture's base with the following results: $\delta^{13}C = 1.99$ and $\delta^{18}O = -1.46$ vPDB. This data was compared to Figure 5b of Gorgoni et al. (2002). It was determined that these values fall well within the range of Carrara marble and do not overlap with other fine-grained marbles that are very uniform in color.

Works Cited

Alaimo, Stacy. 2010. *Bodily Natures: Science, Environment, and the Material Self*. Bloomington: Indiana University Press.

Anderson, Sam. 2017. "The Majestic Marble Quarries of Northern Italy." *The New York Times*. July 26.

Armstrong, Maitland. 1920. *Day Before Yesterday: Reminiscences of a Varied Life*. Margaret Armstrong, ed. New York: Charles Scribner's Sons.

Braddock, Alan C., and Christoph Irmscher, eds. 2009. *A Keener Perception: Ecocritical Studies in American Art History*. Tuscaloosa: University of Alabama Press.

Brunn, Heinrich. 1863. "Scavi di Pompei, Cuma a Pesto." *Bullettino dell'Instituto di corrispondenza archeologica*. May–June: 86–105.

Bulwer-Lytton, Edward. 1880. "The Last Days of Pompeii." In *The Works of Edward Bulwer Lytton (Lord Lytton)*. Vol. 1. New York: P.F. Collier.

Burtynsky, Edward, and Michael Mitchell. 2007. *Edward Burtynsky, Quarries: The Quarry Photographs of Edward Burtynsky*. London: Thames & Hudson.

"Carrara's Marble Quarries: Their Output is Large, Their Methods Most Primitive." 1896. *The New York Times*. July 5.

Coates, Victoria C. Gardner, Kenneth D.S. Lapatin, and Jon L. Seydl. 2012. *The Last Days of Pompeii: Decadence, Apocalypse, Resurrection*. Los Angeles, CA: J. Paul Getty Museum.

Coles, Jonathan Ackerman. 1892. *Abraham Coles: Biographical Sketch, Memorial Tributes, Selections from His Works*. New York: D. Appleton & Company, 1892.

Cooley, Alison, and M.G.L. Cooley. 2014. *Pompeii and Herculaneum: A Sourcebook*. 2nd edition. New York: Routledge.

Coughlin, Maura. 2009. "'Inevitable Grottoes': Modern Paintings and Wasted Space." *Iowa Journal of Cultural Studies*. 10/11, Spring & Fall: 25–41.

Dickens, Charles. 1846. *Pictures from Italy*. London: Bradbury & Evans.

Droth, Martina, and Michael Hatt, eds. 2016. The Greek Slave by Hiram Powers: A Transatlantic Object. *Nineteenth-Century Art Worldwide Special Issue*. 15, 2.

Dwyer, Eugene. 2007. "Science or Morbid Curiosity? The Casts of Giuseppe Fiorelli and the Last Days of Romantic Pompeii." In Victoria C. Gardner Coates and Jon L. Seydl, eds. *Antiquity Recovered: The Legacy of Pompeii and Herculaneum*. Los Angeles: J. Paul Getty Museum: 171–188.

Gorgoni, Carlo, Lorenzo Lazzarini, Paolo Pallante, and Bruno Turi. 2002. "An Updated and Detailed Mineropetrographic and C-O Stable Isotopic Reference Database for the Main Mediterranean Marbles Used in Antiquity." *Proceedings of the Vth ASMOSIA Conference*. Boston, June 12–15, 1998: 1–25.

Hart, E. St. John. 1900. "A Marble World." *Pearson's Magazine*. 10, 60: 630–635.

Igoe, Laura Turner. 2018. "Creative Matter: Tracing the Environmental Context of Materials in American Art." In Alan C. Braddock and Karl Kusserow, eds. *Nature's Nation: American Art and Environment*. Princeton: Princeton University Art Museum: 140–169.

Kelsey, Robin. 2014. "Ecology, Sustainability, and Historical Interpretation." *American Art*. 28, 3: 8–13.

Layard, Austen Henry. 1864. "1. Pompeianarum Antiquitatum Historia, &c / 2. Giornale degli Scavi di Pompei / 3. Le. Case ed i Monumenti di Pompei disegnati e descritti da Fausto e Felice Niccolini…." *London Quarterly Review*. 115, 230: 161–180.

Leitch, Alison. 1996. "The Life of Marble: The Experience and Meaning of Work in the Marble Quarries of Carrara." *The Australian Journal of Anthropology*. 7, 3: 235–257.

Leivick, Joel. 1999. *Carrara: The Marble Quarries of Tuscany*. Stanford, CA: Stanford University Press.

Lorenzini, Pietro. 1994. "Tyranny of Stone: Economic Modernization and Political Radicalization in the Marble Industry of Massa-Carrara (1859–1914)." PhD dissertation. Chicago, IL: Loyola University.

Malamud, Margaret. 2011. "On the Edge of the Volcano: The Last Days of Pompeii in the Early American Republic." In Shelley Hales and Joanna Paul, eds. *Pompeii in the Public Imagination from Its Rediscovery to Today*. New York: Oxford University Press: 199–214.

Mau, August. 1899. *Pompeii, Its Life and Art*. New York: The Macmillan Company.
Miller, Angela L. 2009. "The Fate of Wilderness in Landscape Art: The Dilemmas of 'Nature's Nation'." In Alan C. Braddock and Christoph Irmscher, eds. *A Keener Perception: Ecocritical Studies in American Art History*. Tuscaloosa AL: University of Alabama Press: 85–109.
Nelson, Charmaine. 2007. *The Color of Stone: Sculpting the Black Female Subject in Nineteenth-Century America*. Minneapolis: University of Minnesota Press.
Pliny the Elder. 1855. *The Natural History*. Book XXXVI, Chap. 24. Translated by John Bostock and H.T. Riley. Vol. 6 of 6. London: Taylor and Francis.
"Randolph Rogers, the Sculptor." 1892. *Harper's Weekly*. February 6: 465.
Repetti, Emanuele. 1820. *Sopra l'alpe apuana ed i marmi di Carrara*. Fiesole: Badia Fiesolana.
Rogers, Millard F. 1971. *Randolph Rogers: American Sculptor in Rome*. Amherst: University of Massachusetts Press.
Schiller, Joyce K. 1993. "Nydia: A Forgotten Icon of the Nineteenth Century." *Bulletin of the Detroit Institute of Arts*. 67, 4: 36–45.
Scigliano, Eric. 2005. *Michelangelo's Mountain: The Quest for Perfection in the Marble Quarries of Carrara*. New York: Free Press.
Smithson, Robert. 1996. *Robert Smithson: The Collected Writings*. Jack Flam, ed. Berkeley: University of California Press.
Williams, David B. 2009. "The Trouble with Michelangelo's Favorite Stone—Carrara Marble." In *Stories in Stone: Travels Through Urban Geology*. 1st U.S. edition. New York: Walker: 152–176.
Yablon, Nick. 2007. "'A Picture Painted in Fire': Pain's Reenactments of The Last Days of Pompeii, 1879–1914." In Victoria C. Gardner Coates and Jon L. Seydl, eds. *Antiquity Recovered: The Legacy of Pompeii and Herculaneum*, 189–206. Los Angeles: J. Paul Getty Museum.

7 Cryoscapes

Snow and Fantasies of Freezing in the Art of George Henry Durrie[1]

George Philip LeBourdais

In the early 1840s, George Henry Durrie started using snow to stop time.[1] After scant training and competent work as a portraitist, Durrie began to paint quaint genre scenes and domestic landscapes, especially ones draped in winter snows. In and around his hometown of New Haven, Connecticut, the works earned praise for their sentimental portrayal of New England life under the bracing light of cold weather. A picturesque blanket of white is the common thread of these "snow scenes," as Durrie took to calling them, in which season, temperature and effects of weather conspire to catch fleeting, tender moments of rural life.[2] Yet while they purport to show human routines plowing on through docile drifts and mounds, Durrie's winter scenes emotively do the opposite. The paintings seem to celebrate the suspension of time, as if frozen in momentary perfection, by both embracing and repressing the material realities of snow. This essay undertakes an ecocritical reading that considers snow, a typically unquestioned precipitate of winter, within a deeper spectrum of materiality and meaning.

Like similar works by Thomas Birch and Joseph Morviller, Durrie's scenes frozen in snow, ice and wintry cold thwart the standard story arcs of mid-century landscape painting. Unlike the aestival promise on the horizon of Asher Durand's *Progress* (1853), or the transitional autumn tones of Thomas Cole's more ambivalent *The Hunter's Return* (1845), Durrie chose not to enhance reunion or plentitude with sympathetic seasonal allegories, but instead accentuated such themes through the contrast of winter. The artist molded snow into a passive foil for rising spirits and fortunes, making cold hardships all the sweeter to overcome. His snow scenes, therefore, register as a more subtle and ambitious subgenre of landscape painting than has previously been granted. In her book *Bodily Natures*, Stacy Alaimo defines a framework for interpreting scenes like Durrie's as "the emergent, ultimately unmappable landscapes of interacting biological, climatic, economic, and political forces" (Alaimo 2010, 2). With this injunction in mind, in the following pages, I reimagine Durrie's scenes as "cryoscapes," a term combining the art historical root "-scape" with the Greek prefix "cryo-," indicating a frozen state. Cryoscapes reflect both the repressed material agency of snow as well as the fantasy that representing it amounts to the control of natural time itself.

Winter in the Country, a canvas from 1857, epitomizes Durrie's reliable snow scene formula, in which life appears to soldier on through the year's harshest months (Figure 7.1). An archetypal farmstead sits under the clear light of a cold day. At left, the wide doors of a humble barn open to chickens and a cow that eat grain from the crunching earth outside. Even amidst the thinness of winter, animals persevere on this humble farmstead. Two sleighs shelter from snow in the adjacent carriage house,

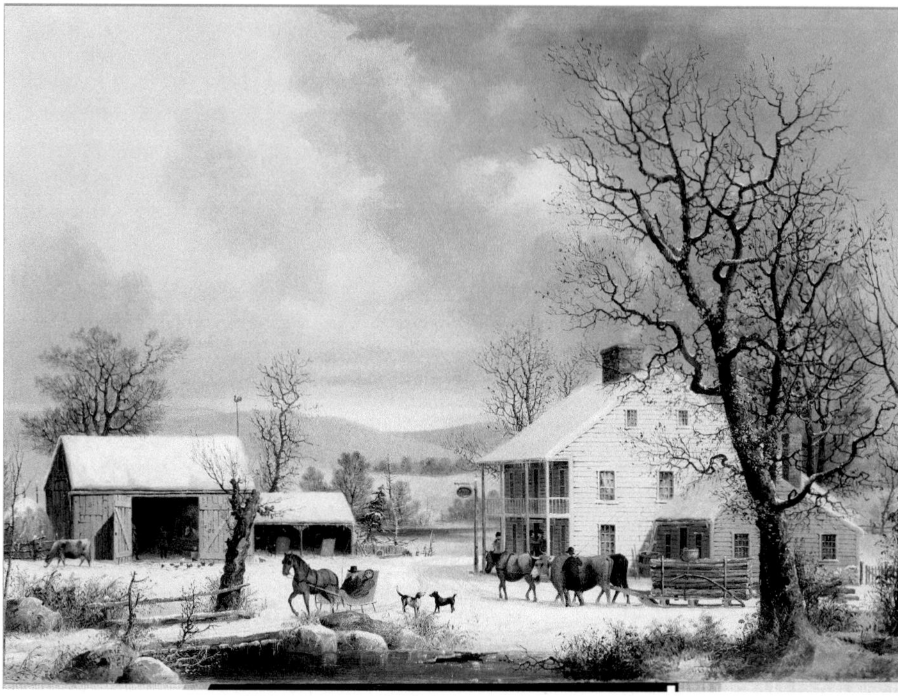

Figure 7.1 George Henry Durrie, *Winter in the Country*, 1857, oil on canvas, de Young, Fine Arts Museums of San Francisco.

while a horse draws a man and woman in third across a small bridge in the foreground. The open doors of the barn echo in opened arms at right; two men embrace on the porch of the two-and-a-half-story Jones Inn. Both emotional and physical warmth seem to be everywhere in evidence. Smoke curls from a stout brick chimney, and the building's yellow flank imbues a sense of spare, honest hospitality to the scene.

Pictorial gestures to movement – a departing sleigh and the approaching sledge at right for instance – allude to nineteenth-century America's self-anointed narrative of progress. Here, that story is spoken in part through the vernacular of architecture. While the inn's bright side dominates the scene, the small, brick-colored saltbox house huddled to its right points to humble beginnings. Though the Jones family may have begun life in that modest home, the conditions that lead to the construction of the current wayfarer's hotel fuel a larger arc of prosperity borne out by hard work and gifts from the land.

Yet the snow here, limp as a sheet, hardly invokes the severe storms typical of New England winters. Snow could fall lightly but also whip in violent winds. It could melt upon hitting the ground or pile inches by the hour, burying all things unsheltered. Americans were clearly aware of its dangers; the Donner Party's infamous winter captivity came to pass in 1846, and the Great Blizzard of 1888 crippled the Northeast, including Connecticut, for days on end. By reducing winter and its frozen precipitation from history-maker to decorative motif, Durrie undercut snow's material agency within its depicted environment. This is precisely the kind of pictorial move that ecocritical interpretations aim to explicate. Scraping past the veneer of Durrie's paintings reveals more of snow's time-bending agency in mid-nineteenth-century New England. Not only did it hinder human action; as we shall see, it also offered new metaphors and models for historical thought. In this sense, we can view snow according to what Jane Bennett calls *vibrant materialism*: "the capacity of things—edibles, commodities, storms, metals—not only to impede or block the will and designs of humans but also to act as quasi agents or forces with trajectories, propensities, and tendencies of their own" (Bennet 2010, viii). As a vein of the new materialism that animates much ecocritical thought, this perspective illuminates Durrie's snow scenes as complex depictions of human desire and environmental distortion.

By representing those rare happy moments plucked from winter's bone, Durrie catered to widespread public desires. Like a chord suspended over a change of key, his winter landscapes preserved and even sustained a passing moment of American history when a wholesome home-life in harmony with the land remained tenable. The immense notoriety that his paintings like *Home to Thanksgiving* achieved as prints by New York publisher Currier and Ives indicates that they fulfilled some deep-rooted need in American sentimental culture (Figure 7.2). Images like this may have become coveted not only for the seasonal family reunions they depict, but also for the impending cascade of upheavals they forestall: a generational migration from pious homesteads to morally questionable cities, the despoliation of picturesque rural lands through industrial greed, and, by the time of the print's publication in 1867, the tragic impossibility of reunion at all due to death in the Civil War.

Durrie's paintings thus register as icons of domestic stability and antebellum peace for a generation that saw the social fabric unraveling before its eyes. Such unraveling

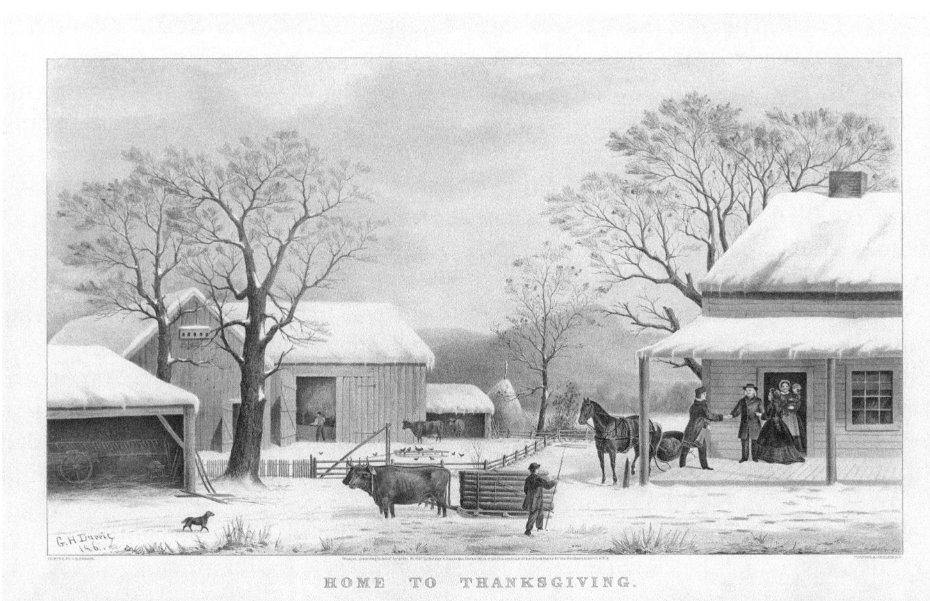

Figure 7.2 John Schutler, *Home to Thanksgiving*, 1867, lithograph for Currier and Ives after 1861 painting by George Henry Durrie, National Gallery of Art.

seems impossible for the plain, rigid figures in *Home to Thanksgiving*; both people and animals assume a stone-like stillness. As their New England iconography and deliberate brushwork was lent to mass-reproduction, these cryoscapes emerged not only as a cultural phenomenon but also as canonical representations of the American idyll. A passing glance at works by Thomas Kinkade suggests that his twentieth-century kitsch owes a great deal, in both content and ideology, to nineteenth-century precedents. Michael Clapper finds that Kinkade's work operates in much the same way as Durrie's did, with a nostalgia that "demands the virtual suspension of time, to the point where human activity is hushed and seen at a distance..." (Clapper 2006, 89). But rather than trapping people in their houses, snow operates as a seal that protects families from the threats multiplying beyond their walls. This nostalgic freezing suggests that Durrie's landscapes, ostensibly about human narratives read through the built environment, allude to weather's deferral of narrative more generally: the snow keeps everything where it should be. Indeed, the nostalgia – a word that etymologically refers to a beloved place as much as to past times – pervading these pictures and their reception suggests that we reconsider the environmental conditions that define them.

Recent art historical scholarship on the aesthetics of ice and freezing is instructive here. Maggie Cao has drawn attention to the material contradictions of ice, and how its blankness and volatility made it "difficult to assimilate into the language of American landscape painting" even for adventurous artists like Arctic enthusiast William Bradford (Cao 2017, 48). In that reading, Cao noted the "crystalline" aspects of ice, a term central to what Ellery Foutch has identified as a "perfectionist impulse" in late nineteenth-century American art more broadly. Varied works of visual culture, from glass flowers to hummingbirds painted close to the picture plane, expressed a pervasive "desire to stop time at a 'perfect moment,' pausing the cycle of growth, degeneration, and rebirth by arresting this perfect state, forestalling decay or death" (Foutch 2011, 8). While Durrie's scenes might portray movement, their tenor is convincingly frozen in this way. Fear of degeneration was harnessed to the way energy purportedly moved through natural systems. Winslow Homer's later paintings, to take one prominent example, were frequently described in terms of the language of thermodynamics, popularized through the influential writing of historian Henry Adams. Art historian Paul Staiti has called attention to these themes in Homer's works like *The Fox Hunt* (1909), a winter landscape in which the animal struggle to survive provides an allegory for the irreversible dissipation of energy in a world of winnowing resources (Staiti 2001, 12–14). For many American artists, the idea of freezing, in either time or temperature, emerged as a powerful pictorial trope in the mid-nineteenth century.

These ominous qualities of freezing make Durrie's positive spin on it all the more intriguing. The artist was a summer baby, born on June 6, 1820 in Hartford, Connecticut. His father was a bookseller, and his mother was a lineal descendant of William Bradford, Governor of the Plymouth Colony. By 19, he had received rudimentary training from Nathaniel Jocelyn, a local artist specializing in banknote engraving and miniature portraiture. Durrie began to earn a living painting signs and portrait commissions, which he showed in his father's shop, eventually moving to New Haven and establishing a studio there. An 1843 self-portrait in the National Portrait Gallery in Washington, DC shows him at 24, standing assuredly and dressed well. But the retreating curl of his left hand and placid expression convey the guarded character that defined his public life.[3] An entry from his journal in the frigid winter of 1845 rejoices in the quietude of domestic life: "I spent the evening at home, drawing, etc. I feel that it is happiness to spend my evenings at home with my wife and family around me. My little boy, the very picture of health, lies in bed sleeping. All is peace, contentment and enjoyment" (Cowdrey 1947, 4). Appreciation for the plain gifts of health and hearth – not to mention the vision of one's child in unmoving sleep – was intensified by a familiarity with death's stillness. In May and June of 1843, the year before his winter landscapes began to find favor, Durrie advertised his services in the *New Haven Palladium*: "G.H. Durrie, having taken the room No. 13 Phoenix Building, would be pleased to wait upon those who may be desirous of obtaining faithful and correct likenesses. Portraits taken after death, if application is made in due season" (Cowdrey 1947, 3). Durrie's emergent stature as a landscapist may have minimized these somber commissions, but so may have his medium; daguerreotypes became a popular means for death portraits in the 1850s. The frozen quality he gave to his landscapes certainly seems of a piece with the concurrent rise of photography in the same years.[4]

That the snows of Durrie's native landscape crystallized its vitality, rather than auguring death, reveals a certain fascination in local climate, one that seemed to permeate the air in New Haven. His adoptive hometown has the notable distinction of holding one of the longest continuous records of temperature in the world. It begins in June 1778, when Reverend Ezra Stiles became the president of Yale and began taking readings from his personal thermometer.[5] By the 1850s, when advances in thermometer technology made taking local temperature a common practice, regular people grew more and more curious about the observable effects of weather on their environs (Pollack 2014). Recording nature in this way, however, also seemed to solidify culture's dominion over it, by making sense of the world's unruly fluctuations. Such a view resonates in how Durrie's daughter described his painting as itself a process of unfolding measurements, aware of but largely unresponsive to natural variations. "I would go to him daily in his studio adjoining our home, for his greetings and kiss," she recalled:

> Sometimes I would linger, watch him put brush on canvas, starting a winter scene, and see what appeared: the roof of an old tavern, or the top of a farmhouse, or some barns, a cow enclosure; on my next visit a road, a sleigh, a load of wood going by, drawn by oxen, a little dog following — always a dog was feature of his winter scenes — perhaps the driver stopping to chat with a friend on the road. Such a hearty, whole-souled love of nature left an impress on his character, as a lover of Nature's God, which I profoundly feel has caused his works to live after him. (Durrie 1933, 3)

Proceeding from one domestic trapping to the next, Durrie carefully assembled a world for aesthetic comfort, enclosing weather in a predictable system like a delicate scene within a snow-globe.

Painting "as a lover of Nature's God," Durrie also had to weigh the ambivalent spiritual implications of his cryoscapes. A deeply devout man, he would often attend up to five different services on Sunday, repeatedly making the chilly pilgrimage he depicted in the 1853 canvas *Going to Church* (Figure 7.3). Here, the place of worship glows in contrast to the whiteness of surrounding snow, making prayer seem both natural and unaffected by nature. Throughout New England at that time, Congregationalist sermons swelled with a revivalist ardor that compelled sinners into the fold

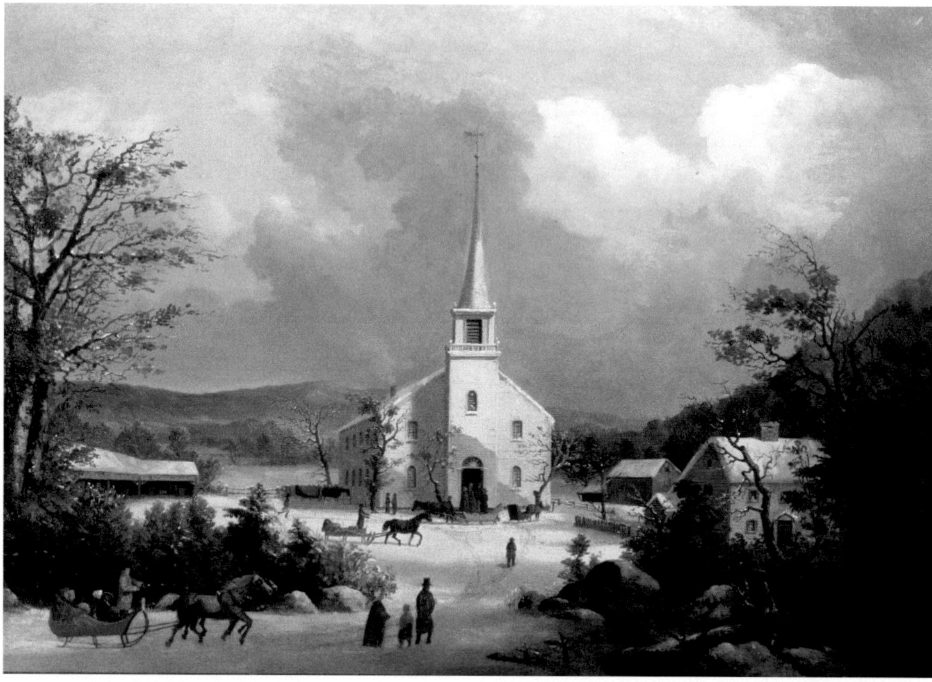

Figure 7.3 George Henry Durrie, *Going to Church*, 1853, oil on canvas, The White House Collection.

and galvanized the already faithful. Reformed Protestant theology, emanating from nearby Yale College, warned that sinners would burn in a fiery Miltonian Hell. The corollary was a Paradise braced in cooler climes. As theologian and preacher Jonathan Edwards had written in the 1738 sermon *Heaven is a World of Love*, "The paradise of love shall always be continued as in perpetual spring. There shall be no autumn or winter; every place there shall in be in perpetual bloom with the same undecaying pleasantness..." (Edwards 1999, 258). Yet as this citation makes plain, winter and freezing retained a deathly connotation among the pious. Emily Dickinson, Durrie's contemporary and another spiritual New Englander, consistently used snow to symbolize the end of life's journey. In a poem about snow, Dickinson paints a picture of the stuff's paradoxically shifting stillness: "It scatters like the Birds / Condenses like a Flock...It traverses yet halts / Disperses as it stays..." The concluding couplet references the zodiacal sign of the winter solstice, a word whose Latin roots mean "sun still": "It curls itself in Capricorn/Denying what it was."[6] As the image of stillness in Dickinson's poetic vocabulary, snow was a representation of eternity, a temporal state humanity could attain only through death (Anderson 1959, 410–411). Durrie must have recognized this conventional equation of winter with mortality.

These confusing matters of religious symbolism also led to more immediate questions of faith and morality. As an Episcopalian, Durrie heard sermons that were at the very least skeptical of the so-called "New England Theology" advanced by Nathaniel Taylor, another influential minister from Yale. Rejecting Calvinist Determinism, Taylor's liberal interpretations paved the way for progressive moral causes like the immediate abolition of slavery. And yet many of Durrie's more conservative peers resisted these rapid changes. While God- and neighbor-loving, their contentment with the status quo protected regressive social structures, such as the practice of "gradual abolition" that caused Connecticut to be one of the last states in New England to conclusively abolish slavery.[7]

In this context, another of snow's symbolic forces – its pure whiteness – leads back to broad questions about the veiling of history in Durrie's scenes. It is possible to read the pervasive whiteness of his cryoscapes as a code for possession, a way of racially claiming the landscape through color. And yet snow's materiality, both literally and metaphorically vibrant, complicates so neat a conclusion. As literary scholar Eduardo Cadava explored in his book *Emerson and Climates of History*, snow also carried a sense of uniting for the common weal in the nineteenth century, thereby advancing the path of history toward justice (Cadava 1997, 171–174). He notes that, in the two final stanzas of the poem "Boston Hymn," Emerson's God calls upon peoples of all races to cover America's flawed history with a bright mantle of freedom:

> Come, East and West and North,
> By races, as snow-flakes,
> And carry my purpose forth,
> Which neither halts nor shakes.
> My will fulfilled shall be,
> For, in daylight or in dark,
> My thunderbolt has eyes to see
> His way home to the mark. (Emerson 1904, 204)

Viewed through the prism of ecomaterialism, Emerson's dynamic imagery is a fitting counterpoint to Durrie's peaceful snows. Their opposed symbolic orders testify to snow's mercurial energy, its defiance of stable signification. Snow could preserve a nostalgic past, or unleash an avalanche of potentialities. The contingency of its meaning in the mid-nineteenth century serves as a reminder of how signification is always naturalized rather than inherently natural. That lesson is all the more valuable today as climate change alters the environment in unimagined ways. Indeed, what value will Durrie's paintings have for future generations who may never see New Haven blanketed in snow?

Until his death in 1863, the same year Emerson wrote his "Boston Hymn," George Henry Durrie tried to stop time with his paintings. His cryoscapes attempted to freeze history's flow, even as the winds of change stormed about.

Notes

1 An earlier version of this essay was presented at the 2018 C19 conference. I'm grateful to Nan Wolverton of the American Antiquarian Society for an invitation to present on her panel there, and for her thoughtful comments on the text.

2 The volume of record on Durrie and similar painters of American winter landscapes is Huston, 1977, signaling the need for new perspectives on the artist and his ilk.

3 Durrie's obituary in the *New Haven Daily Palladium* on October 17, 1863, two days after he died, acknowledged how meagerly he had connected with the community: "From his rare modesty, which has led him to shrink from society, and his devotion to home, many who have long been familiar with the genial face of the artist, have known little of the inner life and spirit of the man."

4 Though within this context I cannot engage this compelling connection between Durrie's still scenes and the concurrent rise of postmortem photography, which carried important social functions in memorialization and mourning, Jay Ruby's 1995 book *Secure the Shadow: Death and Photography in America* offers promising potential avenues of comparison.

5 This record extends from the beginning of Rev. Stiles's presidency until two days before his death on May 10, 1795. The record also notes an interruption from July 4, 1779 to January 22, 1780, caused by the invasion of British troops into New Haven, at which time the thermometer was broken. Joseph Milton Kirk, of the State Geological and Natural History Survey, published records from 1781 until 1935. See Waggoner, 1979.

6 The several variants of this poem (311), written between 1862 and 1864, further testify to the mercurial meanings of snow and its effects on human perception. See Dickinson, 1998, 231–234.

7 For an account of Taylor's effect on the Protestant dogma and the spread of liberalism in early nineteenth-century New England, see Sweeney, 2003.

Bibliography

Alaimo, Stacy. 2010. *Bodily Natures: Science, Environment, and the Material Self.* Bloomington: Indiana University Press.

Anderson, Charles R. 1959. "The Trap of Time in Emily Dickinson's Poetry." *ELH.* 26, 3: 402–424.

Bennet, Jane. 2010. *Vibrant Matter: A Political Ecology of Things.* Durham: Duke University Press.

Cadava, Eduardo. 1997. *Emerson and the Climates of History.* Stanford, CA: Stanford University Press.

Cao, Maggie. 2017. "Icescapes." *American Art.* 31, 2: 48–50.

Clapper, Michael. 2006. "Thomas Kinkade's Romantic Landscape." *American Art.* 20, 2: 76–99.

Cowdrey, Mary Bartlett. 1947. *George Henry Durrie, 1820–1863, Connecticut Painter of American Life.* Hartford, CT: Wadsworth Athenaeum.

Dickinson, Emily, and Thomas H. Johnson. 1998. *The Complete Poems of Emily Dickinson.* London: Little, Brown.

Durrie, Mary Clarissa. 1933. "George Henry Durrie: Artist." *Antiques.* 24, 1: 13–15.

Edwards, Jonathan, and Douglas A. Sweeny. 1999. *The Sermons of Jonathan Edwards: A Reader.* New Haven, CT: Yale University Press.

Emerson, Ralph Waldo. 1904. *The Complete Works of Ralph Waldo Emerson.* Vol. 9 of 12. Boston: Houghton, Mifflin & Co.

Foutch, Ellery. 2011. *Arresting Beauty: The Perfectionist Impulse of Peale's Butterflies, Heade's Hummingbirds, Blaschka's Flowers, and Sandow's Body.* University of Pennsylvania: Unpublished doctoral dissertation.

Hutson, Martha. 1975. "The American Winter Landscape." *American Art Review.* 2, 1: 60–78.

———. 1977. *The American Winter Landscape, 1830–1870.* Santa Barbara, CA: Santa Barbara Museum of Art and American Art Review Press.

Pollack, Henry. 2014. *A World without Ice.* New York: Avery.

Ruby, Jay. 1995. *Secure the Shadow: Death and Photography in America.* Cambridge: MIT Press.

Staiti, Paul. 2001. "Winslow Homer and the Drama of Thermodynamics." *American Art.* 15, 1: 11–33.

Sweeney, Douglas A. 2003. *Nathaniel Taylor, New Haven Theology, and the Legacy of Jonathan Edwards.* Oxford: Oxford University Press. http://public.eblib.com/choice/publicfullrecord.aspx?p=3053098.

Waggoner, Paul E. 1979. "Secular change in the variability of temperature." *Proceedings of the National Academy of Sciences of the United States of America.* 76, 4: 1547–1549.

8 Picturing Industrial Landscapes

Ecocriticism in Constantin Meunier's and Maximilien Luce's Paintings of Belgium's Black Country

Corina Weidinger

Belgian Realist sculptor and painter Constantin Meunier and French Neo-Impressionist painter Maximilien Luce repeatedly portrayed Belgium's Black Country, the southern industrialized part of Belgium, characterized by the prevalence of mines and factories. Meunier popularized the image of miners and ironworkers in paintings and sculptures that were successful not only in Belgium but also in France, prompting Luce to visit the industrial area around Charleroi to paint industrial landscapes and factory scenes. Both artists' industrial interests stemmed from their political beliefs. Meunier was closely associated with the Belgian Socialist Party and Luce was an anarchist. This essay will show that Meunier's and Luce's fin de siècle Black Country landscapes emphasized pollution of the soil and atmosphere, the most visible negative effects of industrial development on nonhuman nature.[1]

In so doing, these works diverge from industrial landscapes central to the canon of Western art, which, as Nicholas Mirzoeff has noted, depicted pollution in an appealing manner as a "sign of human superiority and the continuing conquest of nature"; in *Impression: Sunrise* (1874), Claude Monet rendered smoke emerging from the steamer smokestacks as visually stunning (Mirzoeff 2014, 222). This painting has since been regarded as an example of a scenic sunrise painted in Impressionist brushstrokes, while the environmental consequences of its industrial content have commonly been downplayed. At the same time, the environmentalist messages of Meunier's and Luce's paintings have not been thoroughly studied, although Meunier, at least, was a popular artist, whose works were exhibited throughout Europe and the United States during his lifetime. Luce enjoyed less fame than Meunier or his fellow Neo-Impressionists Georges Seurat and Paul Signac, and only a few art exhibitions have featured his industrial works during his lifetime or posthumously. Because Luce's industrial scenes have not been avidly collected or displayed in prominent art museums, scholars have generally ignored them. Moreover, most of Luce's paintings are held in private collections, making their study challenging. Although Meunier and Luce painted canvases that are part of a broader visual culture of Belgian industrial landscapes, these two artists were exceptionally dedicated to this subject (and were well acquainted with each other); their canvases mark a moment when evidence of pollution was harder to ignore or aestheticize.[2] Furthermore, Meunier's and Luce's works constitute a major change in the conceptualization of the relationship between industry and nature. If pre-1880s industrial landscapes suggest either that nature unproblematically provides raw materials for industry, or coexists peacefully with industrial activities, Meunier's and Luce's paintings imply that mining and industry have developed to such extent that they entirely remake their surroundings, contaminating them with solid and gaseous waste resulting from coal extraction and annihilating the vegetation as well as the previous agrarian uses of the land. This emphasis in their paintings creates a different kind of industrial aesthetic focused on dark colors, smoke and slag, not visually appealing as in Monet's paintings, but rather critical of the environmental changes caused by fin de siècle industrial development.

This essay applies concepts from energy humanities dealing with twentieth-century extraction and use of fossil fuels to the late nineteenth century. In their introduction to *Energy Humanities: An Anthology*, Imre Szeman and Dominic Boyer stress the importance of energy use since the modern period. As the cultural roles of fossil fuels have largely remained unstudied or little understood in the humanities, uncovering them contributes to envisioning a world powered by alternative sources of energy (Szeman and Boyer 2017, 1–10). The current essay emphasizes the importance of coal in creating modernity around 1900, about 100 years after the start of the Anthropocene, an era defined by humans' extreme impact on the environment through massive use of fossil fuels (Steffen, Crutzen and McNeill 2007, 614). Coal was so important in the nineteenth and early twentieth centuries that the most successful strikes started with miners and expanded to industries dependent on its transport, like docking, railways and shipping. Withdrawing coal from circulation could bring entire economies to a standstill (Szeman and Boyer 2017, 162).

On the one hand, coal consumption shares many commonalities with the use of oil in the twentieth century. Both coal and oil are fossil fuels, i.e. non-renewable energy derived from buried prehistoric vegetation and animals. While both create energy powering means of locomotion and industrial pursuits, both lead to pollution, destroy areas of extraction for the benefit of major cities and lead to the devaluation of the lives of those living and working in proximate environments. Stephanie LeMenager coined the phrase "oil weather" in her discussion of Los Angeles smog caused by car exhaust as well as by refineries and smelters (LeMenager 2014, 78). Atmospheric conditions in coal areas were even worse, as coal extraction and burning release sulfur dioxide, nitrogen oxides, particulate matter, mercury, cadmium, arsenic and carbon dioxide. Mining pollutes a region's water, changes the topography of its land and damages its ecosystems (Ebnesajjad 2017). As anarchist ecologist Elisée Reclus explained in 1866, coal pollution was inescapable in nineteenth-century European areas of extraction: "In Western Europe...there are a great many industrial valleys whose thick air is almost unbreathable to outsiders. The houses there are filled with smoke, and even the leaves on the trees are coated with soot. The sun almost always shows its yellowish face through a thick haze" (Reclus 2013, 109). Like oil, coal created its own kind of weather, which I will call "coal weather" in the rest of this essay. Since Reclus articulated the most cogent critique of pollution, this essay will rely heavily on his observations on pollution in analyzing Meunier's and Luce's works.

On the other hand, nineteenth-century coal mining and twentieth-century oil extraction differ in terms of their social and political contexts. Nineteenth-century coal mining took place mostly in the nation states that benefitted from it, and its associated pollution remained predominantly local. Southern Belgium went through a process similar to what Rob Nixon defined as the "slow violence" of the "resource curse" – unfair appropriation of resources in decolonized states by few power players working with multinational corporations (Nixon 2011, 70–71). Like the people from countries beset by the "resource curse," the inhabitants of the Black Country lived and worked in polluted areas, situated away from major cities and slowly damaged by industrial activities. Belgian mines and ironworks were generally owned by urban financial interests.

Most of the cast iron and steel as well as the girders, machines and locomotives manufactured there were transported away from areas of extraction to the capital or to regional centers to be used for constructing city buildings and laying railroad tracks. Industrial activities in the Black Country did not translate into a better life for its inhabitants. Instead, they improved and even beautified the lives of elites living in cities. Art Nouveau Belgian architects employed iron for structural purposes and as a medium of organic sculpture and decoration (Tschudi Madsen 2013). In Brussels, Victor Horta used iron in Art Nouveau buildings, like the Hôtel Solvay (1895–1898) and the Hôtel van Eetvelde (1895–1897), commissioned by a wealthy industrialist and the administrator of the Congo Free State, respectively, while Paul Saintenoy employed it in a highly visible manner in the Old England department store (1898–1899) (Kiefer 2002).[3]

The industrial landscapes discussed in this essay date from the 1890s to the early 1900s, just the period when Belgian iron and steel production peaked and when as many as 80,000 laborers worked in Charleroi (*L'industrie en Belgique* 1981, 25, 70–76). Their brutal working conditions fueled a series of strikes, which determined the government to create the *Commission du Travail* to investigate. The *Commission* endorsed freedom of association, legalization of trade unions and the exclusion of women and children under 12 from mines. As a result, Sundays were made non-working days in 1905 and the working hours were reduced to eight in 1909 (Witte and Craeybeckx 1987, 123–124; Levine 1996, 41–42). These changes were due to the Belgian Labor Party, a successful European left-wing party that promoted the concept of *art social,* which artists like Meunier embodied (*Art et société en Belgique* 1980, 23; Aron 1985, 23). According to Belgian Labor Party politician Edmond Picard, Meunier's works spoke so eloquently of laborers' suffering that "the measures that will bring these unfortunates out of the abyss of privations and suffering...are due to him more than they are due to discussions and books" (*Eugène Laermans* 1995, 55). Although Meunier's works helped bring about crucial changes in workers' lives, they did not have an immediate effect on legislation on the environment, as environmental laws only started emerging after World War II. Pollution was decried, yet no laws to contain it were passed in the first half of the twentieth century (Tarlock 2009, 42). Environmental damage was perceived as inevitable for economic growth and jobs.

Constantin Meunier's Black Country Paintings

In Meunier's Black Country paintings, pollution is shown in dark colors, denoting oppressive skies and dead soils. Although Meunier painted few mining landscapes, they had a powerful effect on his contemporaries, who noted that these works invited parallels between the industrial forces that caused the rapid expansion of mines and slag heaps, and the primeval forces that had modified the earth's crust. In *The Calvary*, the central panel of Meunier's monumental triptych *The Mine* (Figure 8.1; 1901), Meunier emphasized mining's pollution of the soil and atmosphere. The painting shows miners ascending a slag heap, a hill built by dumping unusable by-products of coal extraction. Slag heaps were by definition pollution and did not support life. They grew rapidly around mining galleries as underground sediments were dug up; sorted; and, if sterile, dumped. Meunier depicted slag heaps in dark colors with minimal vegetation. Beyond and above them, mine chimneys puff heavy smoke. Almost half of the canvas is dedicated to the sky, with smoke sluggishly dissipating into an impenetrable dark gray atmosphere. The color scheme suggests the foulness of the air, loaded with soot and coal particles.

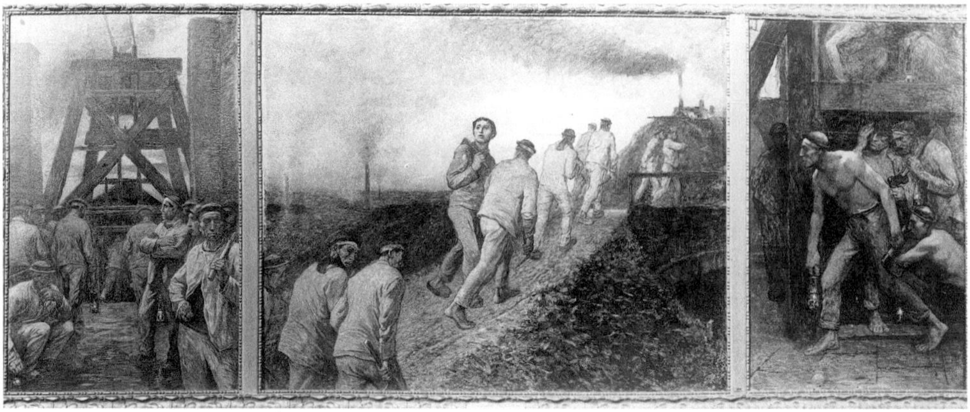

Figure 8.1 Constantin Meunier, *The Mine,* 1901, oil on canvas, Musée Constantin Meunier, Ixelles. Royal Museums of Fine Arts of Belgium, Brussels. © KIK-IRPA, Brussels.

Intensive mining engendered its own coal weather, in which the cloudiness typical of the area combined with pollution to create smog through which the sun barely shined. A viewer of the day would have made the association between Meunier's depiction of air quality and the many medical problems associated with mining, such as epistaxis, pneumonia, pleurisy, chronic bronchitis, tuberculosis, pulmonary phthisis and asthma (Boëns-Boissau 1862, 100–141). Art critics, such as Arnold Goffin, were stunned by such evocative depictions of smoke: "The central panel of a triptych: *The Miners* shows the panorama of an industrial landscape crushed under the tragic and colorless sky where chimneys rise, adorned with smoke and continuously vomiting the burning deleterious exhalation of furnaces and machines on the sterilized earth" (Goffin 1903, 90). Goffin's review emphasized a general lack of color and described the fumes acting like industrial "vomit" above a sterile land, devoid of plants and nonhuman life. The critic employed an anthropomorphic metaphor to explain the toxicity of industrial contamination. Art critic Francis Nautet wrote along similar lines: "the day is veiled by the vapors of the powerful industrial breath; the luminous exhalations never have a cheerful radiance here; in the atmosphere, black soot and exhalations of coaly and somber air circulate and veil the smiles of the sun" (Nautet 1891, 102). Nautet employed similes like "breath" and "exhalations," adopting words customarily applied to human or animal breathing to describe the fumes exiting industrial chimneys. These terms refer to exactly the upper respiratory processes that noxious vapors were capable of damaging.

Meunier also emphasized thick black smoke in *In the Black Country* (Figure 8.2; 1893), where it covers a third of the painting, creating a gloomy feeling. A railway separates two hills in the center, cutting a deep valley. Dark grays and browns are barely relieved by green highlights, suggesting feeble vegetation. Life seems to have been drawn out of the landscape. In 1892, when this painting was exhibited in France, anarchist journalist Octave Mirbeau described the pollution and dead vegetation in this painting as indicative of the suffering of miners underground and as an absolute image of despair:

> Under a sky dirtied by clouds that seem to ooze soot, under the infinite desolation of an unbreathable atmosphere, the valley is dug out...stagnating, deserted, all

black, and on each side rise stripped hillsides, bald slopes where here and there among the rocks grow rare and already dead weeds. No tree, no living being amongst this landscape of death. Beyond the hills one can see...factory chimneys raising their black trunks in the distance and giving off fumes....It is the country of coal. Its evocation is powerful and sinister. This work by M. C-tin Meunier is of an admirable character, of a painful pity. On can really hear erupting from the ground below the lament of the miners, the breath of chests that in the blackness give off the suffering cry of bodies wrecked by work. It all speaks of despair and death and revolt as well. (Mirbeau 1993, 469)

Mirbeau decried the air pollution annihilating all forms of life. Shocked by the lack of trees and people, evicted from the sinister landscape, he saw the painting as representing the effects of intensive capitalist development. The absence of human subjects did not prevent him from musing about the fate of miners and their inevitable revolt.

Critics attempting to come to terms with Meunier's landscapes compared the changes caused by mining in his paintings to cataclysms taking place in primeval times. In his Meunier monograph, Camille Lemonnier, Belgium's leading naturalist writer, remarked: "I have mentioned the wild landscapes, disheveled by structures and smokestacks, sooty horizons cut by bridges, barred by gallows ... the sinister and violent deformation of a land with ravaged and infected breasts ...

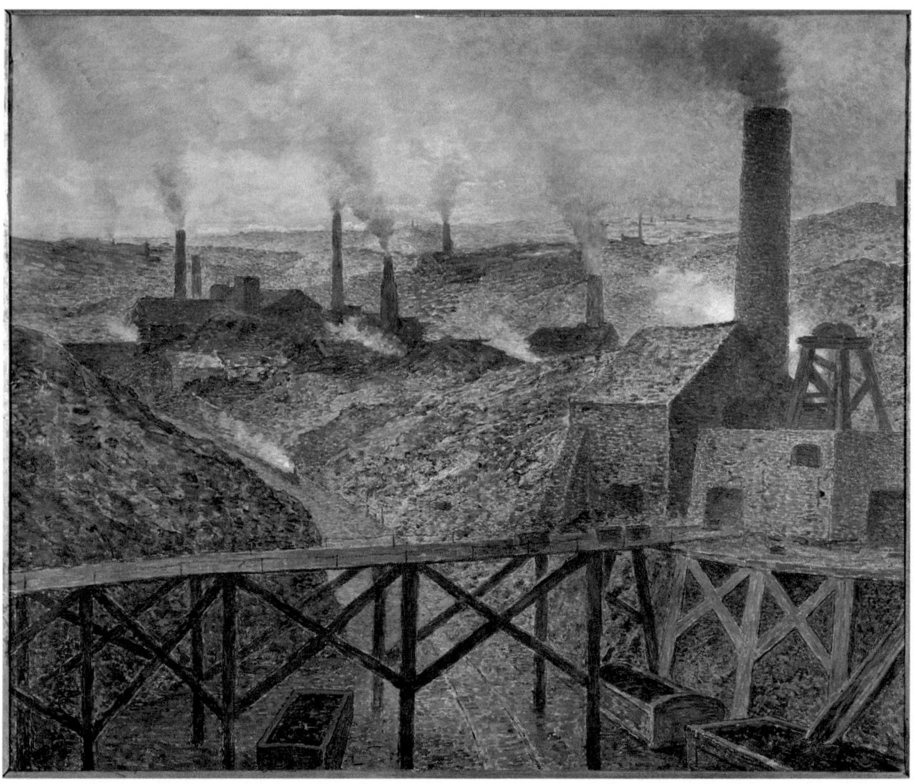

Figure 8.2 Constantin Meunier, *In the Black Country*, 1893, oil on canvas, Musée d'Orsay, Paris.

displaying its bare vertebrae, the terrifying scarred skeleton of lands resembling initial chaos" (Lemonnier 1904, 93). The critic compared the land to ravaged breasts to suggest that mining subverted its life-supporting role. Industrial activities removed layers of productive soil, exposing infertile deposits, whose hardness is invoked by the word "vertebrae." Moreover, Lemonnier likened the deformed mining land to the topmost layer of the earth during the chaos at the beginning of the world. Only primeval chaos on a geological scale could be compared to the man-made processes affecting the Belgian landscape. Novelist Eugène Demolder similarly noted that in Meunier's works, the impact of mining on the land had the violence of a cataclysm:

> The country, a century ago, sparkled idyllically, embossed by wooden hills, cut by fields, orchards, meadows...The sky seemed to smile, the leaves grew freshly. It really was a bucolic land, a country for Virgilian poets...But one day this bucolic landscape was shattered. That green poem of nature was burnt and darkened by the fire and smoke coming from *cataclysms*...Because under the fields and woods, under Walloon villages, one day, they started exploiting great layers of coal. They dug the soil, explored its entrails, violated its secrets, and tore out its secular treasures. The Hainaut carried on its flanks vast black wounds, gaping holes (Demolder 1901, 10–11).[4]

Demolder's passage evokes a landscape that has been destroyed through extreme pollution. Mining mutated Virgilian landscapes into ash and waste. The insides of the earth were exposed through deep wounds inflicted on the formerly verdant prairies. It seems hardly coincidental that terms like "initial chaos" and "cataclysm" had originated in the newly developed science of geology. In 1874, geologists founded the Société géologique de Belgique, and in 1881, they published *La Géologie de la Belgique* based on surveys of soils and their riches (Mourlon 1880, 11, 23). Geologists used terms like "cataclysm" to describe the colossal natural changes that took place during the Carboniferous period. Louis Simonin, a mining expert, used this term in his book *La Vie Souterraine: La mine et les mineurs* (1867) in a section explaining coal formation. Although coal deposits date from various eras, it was during the Carboniferous period that huge forests submerged under water formed bogs and later coal. The changes that brought down this luxuriant vegetation, forever altering the crust of the Earth, were cataclysmic events, never repeated again on the same scale (Simonin 1867, 7, 11). Nineteenth-century writers could only conceptualize the changes brought about by industrial development by comparing them with geological forces from the planet's distant past. They knew that humans were changing the land, atmosphere and vegetation in this area on a scale unseen since the Paleozoic. Their prescient wording anticipated the phrasing of Will Steffen, Paul J. Crutzen and John R. McNeill, who defined the term "anthropocene" in 2007, observing that "Human activities have become so pervasive and profound that they rival the great forces of Nature" (Steffen, Crutzen and McNeill 2007, 614).

Art critics also compared Meunier's mining landscape to apocalyptic scenes. In the nineteenth century, it was often believed that the end of the world would reiterate its formation and that future man-made catastrophes would reenact the upheavals that had formed the earth's crust. Verhaeren commented that Meunier depicted "smoky slag heaps ... panoramic views of roofs and factories, damned landscapes, sterile lands, skies of cataclysms and the end of the world" (Verhaeren 1997, 544). Demolder called the artist's factories apocalyptic and the changes caused by mining cataclysmic:

> Below skies tormented by chimneys spitting their whirls of smoke in sinister clouds...the rolling mills with their apocalyptic skeletons and the mines with their huge bizarre specters stand out. From the landscape set ablaze by furnaces the slag heaps rise, with their sterile schist, formed, one would think, by the lava of a volcano...and on the escarpment of this mined region, dented by cataclysms, a village hangs on tightly with its poor dirty houses. (Demolder 1889, 1)

The critic compared Meunier's slag heaps to volcanoes, another primeval geological force, and linked his smoke to that of primeval times, when volcanoes and lava forcefully altered the earth's crust.

For Meunier, the changes in the landscape echoed the suffering of miners. Intensive mining polluted not only the lands and skies of the region but also destroyed the people working in this environment. The interconnectedness of a terrible physical environment, horrible working conditions and lack of other economic opportunities caused the inhabitants of the Black Country to become physically and psychologically distraught. Meunier's use of the triptych format and the title *The Calvary* for the central panel associates the sacrifice of miners with that of Christ (Levine 1996, 143). The juxtaposition of the left and central panels showing miners before work with the right panel depicting miners after work suggests that mining turned clean, clothed and dignified people into fatigued denuded figures. In the bent-over miner leaving work in the right-hand panel, Meunier emphasized fatigue. Although the man's body is young, muscular and strong, he bends forward considerably, his arms hanging loosely by his body and his face betraying exhaustion. Meunier emphasized his mouth gasping for fresh air, his gaunt cheeks marked by dark shadows under the cheekbones as well as his deep-set eyes. Not surprisingly, the miner appears to have trouble breathing in the dark polluted atmosphere represented by Meunier (Levine 1996, 144). His pose contrasts with the poise of the miners in the left-hand panel, standing in classical contrapposto. Meunier's painting therefore implies that mining not only exhausts the landscape but also miners' bodies.

Pollution in Luce's Charleroi Landscapes

Unlike Meunier, Luce was mainly a landscape painter. While Meunier represented a generic Black Country landscape, Luce painted around the industrial center of Charleroi and often titled his paintings after their exact location.[5] Like Meunier, he highlighted pollution in his landscapes in the shape of slag heaps and toxic fumes. The perspective that Luce used in *The Slag Heap, Charleroi* (Figure 8.3; 1896) amplifies its scale, forcing viewers to take a position below its massive bulk. Composed of three sections, the heap dominates the painting, almost engulfing a small building on the right. Parts of the mine – the roofs of mine buildings and the rails on which a worker pushes a cart – can be seen on top of the heap. Luce suggested the close relation between mining and ironworks by including a factory in the distance on the left. He detailed the human community laboring on this toxic landscape: several women work at the bottom and on the heap. Some bend down to pick up small pieces of coal and others push wheelbarrows. They may be gleaners, who were allowed to take home small

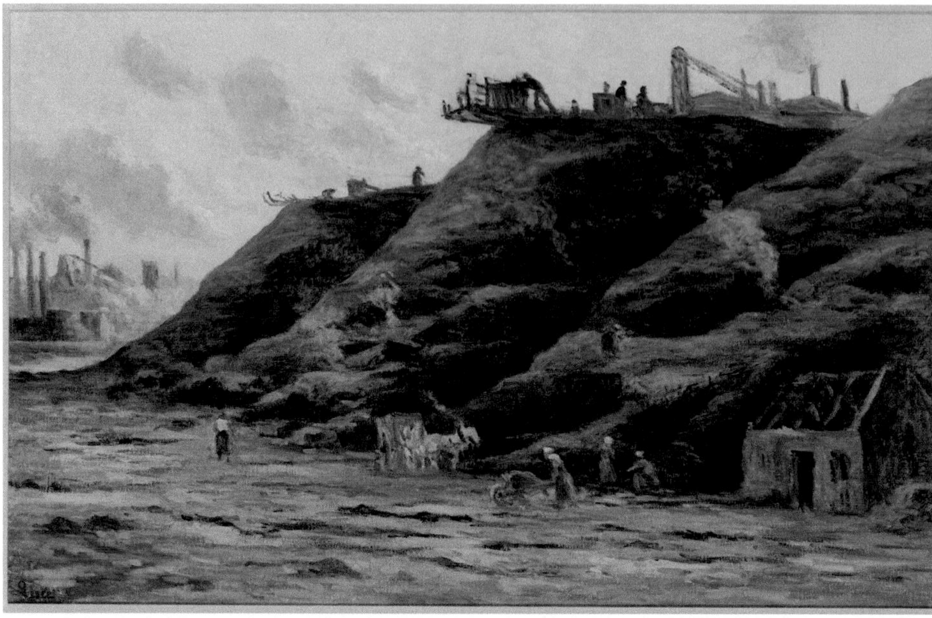

Figure 8.3 Maximilien Luce, *The Slag Heap, Charleroi*, 1896, oil on canvas, private collection. Photograph courtesy of Sotheby's, Inc. © 2012.

pieces of coal, or mine workers re-claiming small pieces of coal for the mine. Apart from the small patch of green in the bottom left corner, no vegetation can be seen on the heap, the soil in front of it, or in the factory surroundings. The colors in which the slag heaps are painted – browns, grays and blacks touched up with violet – suggest that contaminated mud and dirt have banished nature from the industrial area.[6]

In Luce's lithograph *The Factory of Briquettes on the Banks of the Sambre* (1896), he foregrounded the polluting smoke billowing from factory smokestacks and blown by the wind toward the other side of the river, stretching over half of the print and casting a dark shadow over the harbor on the right. The heavy fumes make the time of day uncertain; it could be daylight dimmed by pollutants and rainclouds, or it could be twilight. On the right, a heap of discarded materials leads the view toward factory buildings, giving the sooty waste of the sky a solid waste equivalent. The work recalls the words of Reclus, who complained that engineers "will do everything in their power to make their own work conspicuous and hide nature under piles of gravel and coal" (Reclus 2013, 158). What could have been a picturesque suburban view becomes an industrial site overpowered by piles of raw materials and a sooty haze. Reclus often decried the effects of fumes on landscapes, and so did Luce's critic Maurice Robin, who said of *The Factory of Briquettes*: "The mining centers inspired Luce to make stunning lithographs. Here are the banks of the Sambre in Charleroi, lugubrious buildings, black soils, chimneys, plumes of smoke: skies charged with soot" (Bouin-Luce and Bazetoux 1986, 174). An anonymous Parisian writer conferred: "Here is ... Couillet, the Borinage and the infernal atmosphere of this country of mist and soot" (C.H.S. 1904, 2–3).[7] Both of Luce's critics were based in Paris and described coal and iron weather on a scale unseen by those who never traveled to a Black Country. Lithographs' lack of color and ability to reproduce a variety of effects in gray tones made them well suited to depictions of smoke and pollution. Talking about Meunier, Lemonnier compared his paintings to engravings, as the gloom of the industrial landscape recalled the darkness of printing techniques (Lemonnier 1904, 27).[8] Paintings of industrial scenes seemed to reproduce the foulness of the landscape so fully that they ended up looking like images drawn directly on fragments of coal.

In the upper half of the painting *The River Sambre at Charleroi* (Plate 3; 1896), Luce focused on the transformation of industrial smoke into smog, displaying coal and iron weather. Hardly a healthy aquifer or ecosystem, the river is prominently included to allude to its role as a means of transportation of raw materials to factories and finished products to consumers as well as a water source for factories. To make the utility of the river even more apparent, a barge is shown moored on the left. Pollution has transformed the sides of the river, covering the left bank with large slag heaps and transforming the right bank into a muddy spot. Critics recognized that Luce highlighted the debris of modern industry. Art critic Gustave Geffroy described Luce's paintings as "industrial landscapes where factories activated by human labor vomit their fire and smoke into the skies" (Bouin-Luce and Bazetoux 1986, 172). Anarchist Victor Méric also noted that Luce exposed unbearable misery: "The painter fixed on

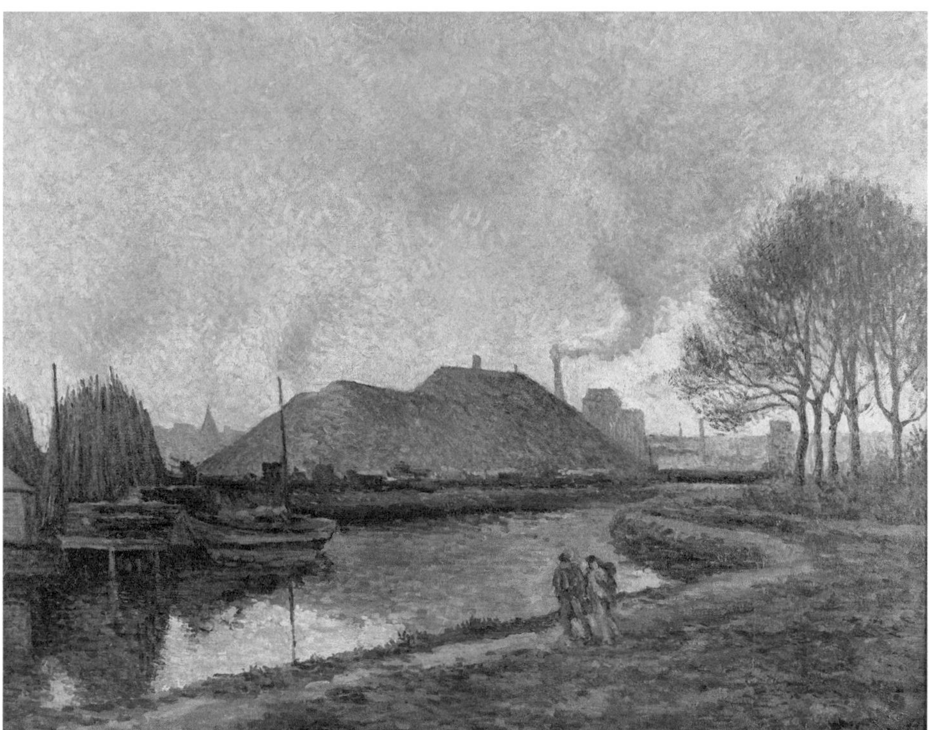

Plate 3 Maximilien Luce, *The River Sambre at Charleroi (La Sambre à Charleroi)*, 1896, oil on canvas, Chazen Museum of Art, University of Wisconsin-Madison, Elvehjem Museum of Art General Endowment Fund purchase.

his canvases all the misery, all the pain that hangs over tragically on this black land where man disappears among smoke, under the heavy sky dressed in mourning, in an infinite desolation" (Tabarant 1928, 48). Reviewers were well aware of what these kinds of landscapes entailed for workers, whether they were visible in the paintings or not.

A comparison between *The Sambre at Charleroi* and Claude Monet's *Argenteuil, the Riverbank in Flower* (1877, Pola Museum of Art) demonstrates to what degree Luce's industrial landscapes differ from canonical Impressionist landscapes. Instead of the flowers in Monet's painting, there is a dull, denuded foreground in Luce's work. Luce's thick dark smoke is a far cry from the frail puffs from the faraway chimneys in Monet's canvas. Luce's industrial river landscapes also differ from those of Camille Pissarro. In his discussion of Pissarro's industrial landscapes painted when the artist lived in Pontoise, Richard Bretell differentiated his views of small-scale *fabriques*, which are integrated effortlessly into the landscape, from those of larger *usines*, which sometimes look at odds with their environment (Brettell 1990, 77–97). Yet both kinds of factories depicted by Pissarro are much more integrated in their environments than their counterparts in Luce's works. Pissarro painted more elements of vegetation surrounding his factories, whether green fields, trees, bushes or even flowers. Smoke, when present, fails to darken the entire sky. In Luce's paintings, the "assimilationist aesthetic" articulated by Greg Thomas in regard to Impressionist landscapes like those of Pissarro gives way to a full-blown industrial aesthetic, in which nature holds a secondary position in an area defined as industrial rather than rural (2010, 49). To paint his industrial landscapes, Luce, a Parisian, traveled far from home, unlike Monet and Pissarro, who depicted the small factories in Argenteuil or Pontoise, where they lived. Most of Monet's and Pissarro's landscapes painted in these areas were not industrial; both of these Impressionist artists painted mostly non-industrial landscapes there. Unlike them, Luce traveled to Belgium's Black Country explicitly to paint and emphasize industrial scenes. Having discovered Charleroi while traveling from Brussels to the medieval-looking town of Thuin, Luce spent considerable time and financial effort to go paint in the Charleroi industrial area many more times despite his relative poverty (Luce to Cross, 1896; Rousseau 1966, n. p.; Luce 1995, 23; Ferretti Bocquillon 2010, 19). At the same time, he ignored the nearby picturesque tourist sites of Thuin and the Abbey of Aulne as subjects for his paintings.[9]

Luce depicted industrial landscapes in a less critical manner in a small number of Black Country paintings like *The Slag Heaps of Sacré-Madame* (1897, Musée du Petit Palais), where he rendered these vast piles of industrial waste in a mixture of blue, green, red and orange brushstrokes. The point of view, from across the river, the reflection of the heaps in the water and the relatively clear sky make the scene tranquil. Smokestacks are present only in the distance. The Sambre Canal, an artificial waterway connecting the Charleroi area to the capital, cuts diagonally in a straight line along the canvas. Several barges are moored on the left, waiting to be laden and moved. This is a mostly automated workplace with few slag heap workers present. Luce prominently represented an automated slag discharge system, an elevator that carried slag from underground to the top of the heap. The structure is much taller than the heap, anticipating its future growth. Luce may have depicted it because it was unique in the area and represented a technological advancement that relieved workers' exertion. The artist was interested in new technological machinery like this slag elevator or blast furnaces, which he portrayed repeatedly in his factory scenes. Like many other anarchists, Luce was ambivalent about technology and large-scale industry. He seems to have appreciated technological advancement and new kinds of machinery because they were a means to alleviate physical labor. Luce's interest in mechanization may be why he was less critical of pollution in this painting.

Anarchist intellectuals found it hard to conceptualize the environmental and labor issues brought about by coalmines and their closely connected ironworks. Pierre Kropotkin, who disliked large factories and preferred small decentralized industries scattered throughout a largely rural area, excluded the large ironworks of the kind that Luce painted from his vision of an ideal world made of fields and workshops. Yet he conceded their necessity in modern societies as well as the necessity of their agglomeration in specific locations, because he admitted that they had to be close to areas of extraction (Herbert 1961, 478; Capouya and Tompkins 1975, 243). Jean Grave, whose pamphlet *Le Machinisme* (1898) included a cover designed by Luce, argued in this very text that in a capitalist society, industrial machinery mostly injured the interests of workers, yet could lighten labor in an anarchist utopia (Grave 1898, 5). Although aware of the adverse effects of working with machinery, Luce probably agreed with Grave that technology was not intrinsically damaging. Yet he must have also been familiar with the writings of Reclus, whose words from 1866 anticipate the landscapes that Luce painted in Charleroi: "Vast regions which formerly were beautiful to behold and enjoyable to travel through are completely spoiled, and one actually experiences disgust upon seeing them...it matters little to the industrialist, operating his mine or factory in the middle of the countryside, whether he blackens the atmosphere with fumes from the coal or contaminates it with foul-smelling vapours" (Reclus 2013, 109). Although fully aware of industrial pollution's devastation of the environment, Reclus still believed that humans should control the environment, but should do it in harmony with nature (Reclus 2013, 20–22, 55).

Luce's letters suggest that, like other anarchists, he struggled to come to terms with the best way to conceptualize the Charleroi area. Although familiar with industrial landscapes around Paris, characterized by small-scale industries, he was shocked by the intensity of industrial development and pollution in the Black Country. In 1895, during his first visit, he wrote to his friend Henri-Edmond Cross:

> I am now in Charleroi...I find this country admirable, but what sadness! I have not worked much. The weather is atrocious, it rains in streams. I walk a lot and take notes. Oh, if I could manage to render what I feel here, I think I could do something interesting but very difficult. I do not know if you are familiar with this country, I for one had no idea. The environs of Paris, from an industrial point of view, are nothing. St. Denis is just a joke, what character. (Luce to Cross 1895)

Luce was so impressed by the Charleroi landscape that he disparaged the Parisian factories he previously painted. Charleroi seemed a much more interesting place because of the extreme intensity of its industrial development. He found it admirable and full of character, albeit sad and rainy. When he returned the following year, he also felt discouraged: "This country frightens me.... It is so terrible and beautiful that I doubt I can render what I see, and this I can do only with terrible difficulty" (Rousseau 1966, n.p.). Impressed by the area's "character" and terrible, sad beauty, Luce struggled to combine the gloomy parts of the landscape with its visual appeal.

His words reveal his efforts to determine his position regarding industrial development and pollution. He appears both impressed by intensive industrialization and appalled by it, split between a terror akin to Reclus's disgust and a hopeful musing similar to Grave's over the potential of new technologies. His letters include a mixture of repugnance and awe at human accomplishments, a paradoxical piece of praise amid condemnation.

Luce's critics agreed that his Charleroi landscapes emphasized pollution through the depiction of slag heaps; smoky skies; and muddy, lifeless industrial no-man's-lands close to mines and factories. Luce's works were, however, a little less bleak than Meunier's. Luce walked a more tenuous line between emphasizing the disastrous consequences of intensive industrialization and constructing an industrial aesthetic. Ultimately, both artists provided a potent critique of industrial development. Their works provoked strong reactions in the contemporary press because they visualized the massive changes brought about by large-scale mining and ironworking. Critics saw the damage done to this area as either a throwback to geological forces that created the earth's crust and its underground regions millions of years before, or as apocalyptic man-made forces turning a Virgilian landscape into a hellish, unbreathable, barren landscape. Although Meunier's and Luce's landscapes represented Southern Belgium, their reviewers had the remarkable foresight to understand the implications of these images: that environmental degradation would prove hard to contain, spreading far beyond that area and threatening to destroy the planet.

Notes

1 Although fatigue, the other main consequence of industrialization, was equally prevalent in their works, I will not discuss it at length here since I already examined it in Weidinger (2013).

2 Luce's and Meunier's works differ from the industrial sublime of paintings like Philip James de Loutherbourg's *Coalbrookdale by Night*. According to Daniels, De Loutherbourg's painting was seen as portraying a fallen world in a semi-religious perspective; see Daniels (1992, 195–230). Luce's and Meunier's paintings mostly eliminate the sublime from the industrial landscape.

3 While Ernest Solvay, the owner of the Hôtel Solvay, made his fortune mainly from the chemical industry, he also benefitted directly from coal. He co-owned an international company of coke ovens.

4 Wallonia is the southern, French-speaking part of the Belgium. The Hainaut is one of its mining regions.

5 Luce also painted scenes taking place inside factory buildings or within factory yards in the Charleroi area.

6 Robyn Roslak noted that Luce's usage of violet evoked sadness in accordance with aesthetician Charles Henry's theories. She also noted that socialist critics like Jules Christophe considered that Luce's uses of violet acted as "poignant signs of hardship and melancholy," see Roslak (2007, 76). Luce may have used violet in his Charleroi landscapes to the same end, or to alleviate the lack of color in the Charleroi area that he decried in his letters. See Rousseau (1966, n. p.).

7 Couillet is a town situated next to Charleroi. The Black Country was also called the Borinage.

8 Lemonnier was discussing an industrial painting by Meunier. His exact words were: "A suggestive and pathetic landscape engraved, it seems, on a page of shale."

9 Luce painted few canvases that do not portray industrial pursuits in the south of Belgium. He focused almost exclusively on industrial scenes in this area. Luce did paint numerous non-industrial landscapes throughout his career in regions that were not that industrially developed.

Bibliography

Aron, Paul. 1985. *Les Ecrivains belges et le socialisme, 1880–1913: L'expérience de l'art social d'Edmond Picard à Emile Verhaeren*. Brussels: Labor.

Art et société en Belgique. 1980. Palais des Beaux-Arts de Charleroi, 11 Octobre–23 Novembre.

Barrell, John, ed. 1992. *Painting and the Politics of Culture: New Essays in British Art 1700–1850*. Oxford and New York: Oxford University Press.

Boëns-Boissau, Hubert. 1862. *Traité pratique des maladies des accidents et des difformités des houilleurs*. Brussels: Tircher.

Bouin-Luce, Jean, and Denise Bazetoux. 1986. *Maximilien Luce: catalogue raisonné de l'œuvre peint*. Paris: Editions JBL.

Brettell, Richard. 1990. *Pissarro and Pontoise: The Painter in a Landscape*. New Haven and London: Yale University Press.

Capouya, Emile, and Keitha Tompkins. 1975. *The Essential Kropotkin*. New York: Liveright, 1975.

C.H.S. 1904. "M. Luce, exposition à la Galerie Druet." *Art et décoration, supplément* XV (Paris, May).

Cross to Luce. 1896. 2008. M.2, Getty Research Institute, Los Angeles.

Daniels, Stephen. 1992. "Loutherbourg's Chemical Theatre: Coalbrookdale by Night." In John Barrell, ed., *Painting and Politics of Culture: New Essays on British Art, 1700–1850*. Oxford and New York: Oxford University Press: 195–230.

Demolder, Eugène. 1889. *Impressions d'Art: Etudes, Critiques, Transpositions*. Brussels: Presses de Madame Veuve Monnom.

———. 1901. *Constantin Meunier*. Brussels: Deman.

Ebnesajjad, Sina. 2017. "Why Is Coal the Worst Fossil Fuel – Emissions, Climate, Health, Smog…?" *Elsevier Chemicals and Materials Now!* (May 15). Accessed January 12, 2018. https://chemical-materials.elsevier.com/chemical-rd/coal-worst-fossil-fuel-emissions-climate-health-smog/.

Eisenman, Stephen. ed. 2010. *From Corot to Monet: The Ecology of Impressionism*. New York: Skira.

Eugène Laermans, 1995. 1864–1940. Brussels: Credit Communal.

Ferretti Bocquillon, Marina. 2010. *Maximilien Luce néo-impressionniste: Rétrospective*, exh. cat. Giverny: Musées des Impressionismes.

Goffin, Arnold. 1903. "Constantin Meunier," *Revue générale*. 77: 81–94.

Grave, Jean. 1898. *Le Machinisme*. Paris: Aux Bureaux des. *Temps Nouveaux*, 1898.

Herbert, Eugenia W. 1961. *The Artist and Social Reform: France and Belgium, 1885–1898*. New Haven: Yale University Press.

Kiefer, David M. 2002. "Chemistry Chronicles: Soda Ash, Solvay Style," *Today's Chemist at Work* (February). Accessed January 7, 2018. https://pubs.acs.org/subscribe/archive/tcaw/11/i02/html/02chemchron.html.

LeMenager, Stephanie. 2014. *Living Oil: Petroleum Culture in the American Century*. Oxford: Oxford University Press.

Lemonnier, Camille. 1904. *Constantin Meunier, sculpteur et peintre*. Paris: H. Floury.

Levine, Sura. 1996. *Monumental Transformations: The Changing Status of Constantin Meunier's Monument to Labor*. PhD dissertation, University of Chicago.

L'industrie en Belgique: Deux siècles d'évolution, 1780–1980. 1981. Brussels: Crédit Communal de Belgique, Société Nationale de Crédit à l'Industrie.

Luce to Cross. 1895. ODO 2007.10.26. Archives Luce at the Documentation du Musée d'Orsay, Paris.

Luce to Cross. 1896. ODO 2007.10.26. Archives Luce at the Documentation du Musée d'Orsay, Paris.

Maximilien Luce, (1858–1941): peintre anarchiste: catalogue d'exposition, Charleroi, Musée des sciences de Parentville. 1995. Brussels: Université libre de Bruxelles, Institut de Sociologie.

Mirbeau, Octave. 1993. *Combats Esthétiques*, Eds. Pierre Michel, Jean-Francois Nivet. Paris: Seguier. Volume 1.

Mirzoeff, Nicholas. 2014. "Visualizing the Anthropocene," *Public Culture*. 26, 213–232.

Mourlon, Michel. 1880. *Géologie de la Belgique*. Paris: Savy; Brussels: F. Hayez, imprimeur de l'académie royale de Belgique; Berlin: R. Friedlander and fils.

Nautet, Francis. 1891. "Constantin Meunier," *La Fédération Artistique* (20 December).

Nixon, Rob. 2011. *Slow Violence and the Environmentalism of the Poor*. Cambridge, MA and London: Harvard University Press.

Reclus, Elisée. 2013. *Anarchy, Geography, Modernity: Selected Writings of Elisée Reclus*. Chicago, IL: Independent Publishers Group.

Robin, Maurice. 1910. "Peintre-graveur, M. Luce," *Le Livre et l'image* 1: 17–24.

Roslak, Robyn. 2007. *Neo-Impressionism and Anarchism in Fin-de-siècle France*. Aldershot: Ashgate.

Rousseau, Robert. 1966. *Maximilien Luce*. Charleroi: Palais des beaux-arts.

Simonin, Louis. 1867. *La vie souterraine; ou, Les mines et les mineurs*. Paris: Hachette.

Steffen, Will, Paul J. Crutzen, and John R. McNeill. 2007. "The Anthropocene: Are Humans Now Overwhelming the Great Forces of Nature?" *Ambio* 26, 8.

Szeman, Imre, and Dominic Boyer. 2017. *Energy Humanities: an Anthology*. Baltimore, MD: Johns Hopkins University Press.

Tabarant, Adolphe. 1928. *Maximilien Luce*. Paris: G. Crès.

Tarlock, A. Dan. 2009. *"Environmental Laws and Their Enforcement," History of Environmental Law vol. I*. Oxford: EOLSS Publishers/UNESCO.

Thomas, Greg. 2010. "From ecological vision to environmental immersion: Théodore Rousseau to Claude Monet" In Ed. Stephen Eisenman, *From Corot to Monet: the ecology of Impressionism*. Milano: Skira. pp. 47–57.

Tschudi Madsen, Stephan. 2013. *The Art Nouveau Style*. Mineola, NY: Dover Publications.

Verhaeren, Émile. 1997. *Écrits sur l'art*. Brussels: Labor.

Weidinger, Corina. 2013. "Fatigue, *Machinisme*, and Visual Spectacle." *Nineteenth-Century Art Worldwide*. 12, 2. www.19thc-artworldwide.org/autumn13/weidinger-on-fatigue-machinisme-and-visual-spectacle-in-maximilien-luce-s-l-acierie.

Witte, Els, and Jan Craeybeckx. 1987. *La Belgique Politique de 1830 à nos jours*. Brussels: Labor.

9 Ruskin's Storm-Cloud and Tyndall's Blue Sky

New Materialist Diffractions of Nineteenth-Century Atmospheres

Polly Gould

John Ruskin (1819–1900), the dominant British force in nineteenth-century thinking on art, and John Tyndall (1820–1893), one of that century's greatest scientists, both had things to say about the atmosphere. The topic of Ruskin's 1884 lectures "The Storm-Cloud of the Nineteenth Century" was the foggy gloom that he attributed simultaneously to coal-burning pollution and moral lassitude. Tyndall was also popular as a public lecturer. He gave his name to the effect that caused the sky to appear blue and devised a glass apparatus to demonstrate this during public talks. Tyndall and Ruskin, in their respective work, brought together previously discrete fields of knowledge into wide-ranging and associative inquiries. Both men shared practices that involved close observation of the natural world but they were frequently and sometimes publicly at odds with each other. They differed as to the significance of those observations and the worldview to which they should belong: on the one side was an attitude engaged with theological, mythical and vitalist cosmologies, and on the other the materialist scientific non-theological mechanistic, naturalistic and analytic approaches to explaining nature.

This essay, following the lead of Donna Haraway (1997), *diffracts* these two Victorian thinkers through a new materialist understanding in readings of feminist physicist and ecocritical philosopher Karen Barad's *Meeting the Universe Halfway* (2007) and political theorist Jane Bennett's *Vibrant Matter* (2010). "Diffraction," Haraway writes, "is an optical metaphor for the effort to make a difference in the world" (Haraway 1997, 16). "Reading diffractively" can develop a differential, boundary-crossing critique beyond adversarial self/other positioning and is in keeping with new materialist critique that displays "antipathy toward oppositional ways of thinking" (Coole and Frost 2010, 8). This essay applies a diffractive reading to the specific ways in which Ruskin and Tyndall, as well-known antagonists, visually illustrated the sky in their public lectures. Diffractions "are attuned to differences that our knowledge making practices make and the effects that they have on the world" (Barad 2007, 72). Tyndall and Ruskin, though adversaries within their own time, when read diffractively through twenty-first-century new materialism, are found not to align neatly on one side or another. Reading diffractively links the nineteenth-century's concerns to the present day, complicates the differences between then and now, and brings to light patterns of similarity.

Modern Painters, Ruskin's great five-volume work on landscape painting, was written as a critic's defense of J.W.M. Turner's work, and argued for the value of art derived from the observation of nature. Its focus of concern was with the quality of true seeing that goes beyond observational accuracy to reveal emotional truth

(Andrews 2008). Ruskin's own drawings and watercolor studies were themselves created in pursuit of a kind of empirical observation that served to connect both the maker and the viewer to a deeper meaning. New materialism resonates with Ruskin's nature admiration, and what Mark Frost describes in Ruskin's writing as a "dynamic materiality" that is approached through intensive passionate observation that leads into contemporary ecological thinking (Frost 2011, 368). In *Vibrant Matter: A Political Ecology of Things*, Jane Bennett theorizes a vital materiality and the necessity for politics to take account of the agency of things, combined with strategic anthropomorphism (Bennett 2010). *Vibrant Matter* does not construe an organicism of each part supporting the whole, as Ruskin would have it, but the globalized; interconnected; and, often conflictual, relations of "network, mesh, Empire" (Bennett 2010, 23). In describing her method, Bennett says that she "lavishes attention upon specific 'things,' noting the distinctive capacities of efficacious powers of particular material configurations" (Bennett 2010, ix). Bennett's ecology is "the study of story (*logos*) of the place where we live (*oikos*)" (Bennett 2004, 365), the narrative of organic and inorganic encounters and enmeshed agencies that "upset conventional distinction between matter and life, inorganic and organic, passive object and active subject" (Bennett 2004, 353). This approach attributes agency not only to the living nonhuman others such as animals but is also applicable to the non-living nonhuman earth others, such as atmosphere.

Karen Barad also engages with observation, stories and agency. Barad's new materialist philosophy of an ecocritical onto-epistemology has developed from her original disciplinary alignment with physics, specifically quantum physics, and this puts her in some sympathy with Tyndall, who was trained as a physicist and was engaged with atmosphere in physical ways. Barad's ground-breaking work in *Meeting the Universe Halfway* brings an understanding that puts the agency of things into a proper relation with discursive practices of human intentionality. Barad's agential realism, coming out of science studies, demonstrates a new synthesis of ontology and epistemology that shows that the apparatus of observation is a performed material-discursive mattering. Agential realism redistributes agency to show the boundary between object and subject to be moveable, and vitalizes non-living matter with agency. As she explains: "Objects are not already there; they emerge through specific practices" (Barad 2007, 157). It is important to note how Barad qualifies the social constructivist argument in her work: "the fact that scientific knowledge is socially constructed does not imply that science doesn't 'work,' and the fact that science 'works' does not mean that we have discovered human-independent facts about nature" (Barad 1996, 162). This emphasis has become especially critical in a post-truth era in which the validity of scientific evidence has come under attack, facts seem negotiable for the sake of political advantage, and relativism has been cynically exploited in the support of climate change denial.

Atmospheres

In defining atmosphere, one may encounter a struggle with the very vagueness that one hopes to describe: this is attributable to the double meaning of the atmosphere as both air and mood. In the seventeenth century, atmosphere was defined by science as the gaseous envelope around the earth, but later in the eighteenth century it also took on the figurative meaning of the influence, feeling or ambience of one's surroundings.

Ruskin's readers in the nineteenth century would have understood atmosphere in both senses, and it is partly this two-sided meaning that we might crudely apportion to Tyndall aligned with science on one side, and Ruskin with art on the other. Atmosphere is and *ought to be* interpreted in a much more entangled manner. Writing in the section 'Of Truth of Skies,' Ruskin contrasts the solid discrete objects of the suspended clouds in painting to the experience in nature of "looking not at but *through* the sky" (Ruskin 1912 v3, 347). 'Looking through' is a suitable description of what occurs in 'diffractive' ecocritical readings of cultural phenomena.

Science and art generate spheres of interpretation that have recursive relations, an example of which can be found in the study of clouds. Early in the nineteenth century, the chemist Luke Howard (1772–1864) developed nephology, the nomenclature system for clouds. It was first presented as a paper titled "On the Modifications of Clouds" in 1803 and then republished in Thomas Forster's second edition of *Researches about Atmospheric Phaenomena* in 1821 (Forster 1821; Thomes 1999, 52; Hamblyn 2001, 225). Howard and Forster are credited with inspiring artists in the scientific observation of clouds in nature and the depiction of particular rather than ideal clouds in art: John Constable (1776–1837) spent a summer painting cloudscapes in Hampstead Heath after reading their work. Yet Howard's choices for cloud-illustration in his articles and books were equally influenced by the picturesque in art: Howard's observational watercolor cloud studies were transposed into much harder-edged engraved illustrations that took liberties with the compositions and allowed unnaturally occurring cloud combinations to be assembled for aesthetic effect (Jardine 2014). In this case, what might have been considered as observational accuracy was sacrificed to artifice.

In literature, the blue of the sky has been associated with distance, openness and desire. It offers an exterior corollary for interior feelings of longing (Solnit 2005). Ruskin was also a keen observer of the blueness of skies. Notes in his biographical writing refer to a cyanometer, a device invented by the Swiss scientist and founder of alpinism, Horace Bénédict de Saussure (1740–1799) (Plate 4). Ruskin took one to Switzerland as an adolescent in 1835: "I shaded in cobalt a 'cyanometer' to measure the blue of the sky with," he wrote (Ruskin 1912, v35, 152); his diary entries are scattered with these readings. Howard too, in the introduction to the cloud-naming book, included a cyanometer in his list of measuring instruments applied to London skies, only to note that he had been thwarted in its use by the "continual recurrence of turbid skies" (Howard 1818, xxvii). Howard was writing in the early nineteenth century when this failure of the London sky to provide a measurable blue can be attributed to natural weather patterns, whereas at the end of the nineteenth century, the dull skies were increasingly the consequence of man-made smoke pollution.

The application of the cyanometer to blue skies and the project of cloud-naming share the common endeavor to identify the elusive properties of the sky. The study of clouds and blue skies challenges distinctions between what is objectively and measurably out there and what is subjectively experienced. Significantly, as Mary Jacobus proposes: "Clouds are confusing, not so much because they mix elements, or constantly change shape, but because they challenge the phenomenology of the visible" (Jacobus 2006, 221). Contemporary philosopher Gernot Böhme says that cloud-watching offers a type of beauty as feeling: an encounter with ephemerality, and "perception of the medium" that by "intensifying our existence" makes us "feel that we are there" (Böhme 2010, 30). Timothy Chandler names Böhme's approach

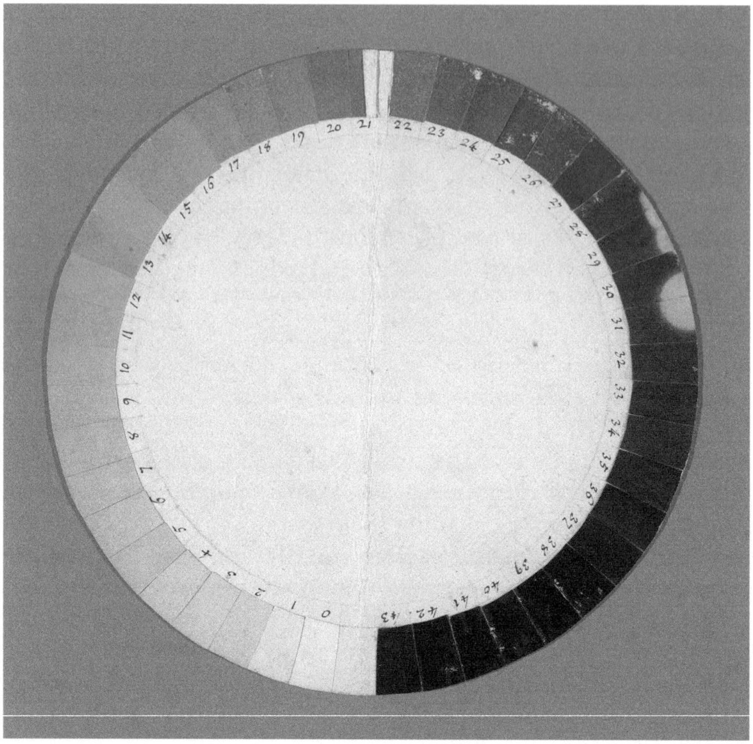

Plate 4 Cyanometer, invented by Horace-Bénédict de Saussure. Bibliothèque de Genève, Arch. de Saussure 66/7, pièce 7.

*eco*phenomenology (Chandler 2011, 554) and characterizes it as a "new aesthetics based on ecological embodiment" (Chandler 2011, 556). Chandler writes, "Böhme's aesthetic theory aims to eliminate the opposition between art and science, between what is thought of as subjective and objective approaches to nature, between concrete, sensuously given nature and prepared or cognized nature" (Chandler 2011, 556). Böhme's stated aim is to introduce "the aesthetic perspective into the science of ecology" (Böhme 2016, 1). According to Böhme that which "mediates objective factors of the environment with aesthetic feelings of a human being is what we call *atmosphere*" (Böhme 2016, 1).

Böhme, influenced by Naturphilosophie, adopts an organic and dynamic worldview against Cartesian dualism and the mechanistic and atomized understanding of nature. Diffractive reading allows the complexity of the patterns of genealogies of influence and interference down the centuries to our present time to be noted, so that, for example, the influence of eighteenth century thought upon Tyndall in the nineteenth century and Böhme in the twenty-first century can be tracked, as well as the similarities between Tyndall and Ruskin in their own time. Tyndall studied for his doctorate in Germany where he encountered Naturphilosophie in the work of Goethe (1749–1832), Fichte (1762–1814) and Schelling (1775–1854). Tyndall, in his first paragraph in *Essays of on the Use and Limit of the Imagination in Science* (and without intending any humor), refers to the books that accompanied him to the Alps, Goethe's *Theory of Colours* or *Farbenlehre* (the other being a treatise on logic by the philosopher Bain) as "two volumes of poetry" (Tyndall 1870, 13). Commenting upon Goethe's complaints against Newton, Tyndall writes, "We frequently hear protests made against the cold mechanical mode of dealing with aesthetic phenomena employed by scientific men" (Tyndall 1880, 320). Tyndall wrote here in a way that could be taken as a reply to Ruskin. He asserted that in all of humanity's history and experience, "poetry is an element just as much as science" (Tyndall 1880, 321). Tyndall, who himself wrote poetry, oscillates in his writing between scientific explanation and more poetic natural description.

Ruskin and Tyndall used observation as the method for considering the sky, but they had divergent intentions: Ruskin was always keen to affirm the moral God-infused implications of what nature had to show, whereas Tyndall, who felt strongly that things in nature were best explained without theological justifications, was emphatically materialistic and a proponent of the then unfashionable need to separate religion from science. Hewing to a humanistic tradition, Ruskin did apply scientific language in his efforts to better observe nature, but his ultimate aim was to see it as a theological cosmology with man at the center. Indeed, Ruskin made Tyndall the frequent focus of his attacks on science, peppering his writing with snide comments as to Tyndall's poetic efforts, criticizing Tyndall's inaccurate language when "endeavouring to write poetically of the sun" (Ruskin 1912 v34, 60) and chiding him for his lack of erudition in his use of certain metaphors.

Raymond Fitch, in his discussion of Ruskin, describes as 'a central impulse of nineteenth-century science: the tendency to explain natural phenomena in terms only of matter and energy and their laws' (Fitch 1982, 542). Fitch chooses as an example of this Tyndall's famous lecture that advocated this duality 'The Influence of Material Aggregation upon the Manifestations of Force':

> The system of the universe embraces two things, - an object acted upon, and an agent by which it is acted upon; - the object we call matter and the agent we call force; thus the luminiferous ether is the vehicle or medium by which the pulsations of the sun are transmitted to the organs of our vision. (Tyndall 1853, 254).

For Ruskin what is missing from this understanding is the aspect of form, which for him is the paramount concern, as it brings to the fore the agency of the maker's hand, whether that be God or the artist. Tyndall's explanation was also, for Ruskin, a reductive characterization of vision that misses the potential for revelation of deeper truth. Ruskin's observational methodology reaches beyond analytically corroborated facts to an encounter with what he considers to be the truth. If the object of this observation is a natural phenomenon such as a landscape or the sky, then this too could furnish the initiated observer with profound insights.

In another context, atmosphere became a question of epidemiology and environmental public health on two related fronts: disease transmission and pollution. First, atmosphere was discussed regarding the causes of disease in which 'bad air' or miasma, versus germ theory, offered rival theories of transmission. The miasmic theory, now obsolete, held that disease was caused by unhealthy air rather than by contagion through contact with disease agents such as bacteria: by the end of the nineteenth century, germ theory had prevailed. Second, commentators in the late nineteenth century noted the continuous twilight effect of atmospheric pollution caused by domestic and industrial coal burning. London and other UK cities were often shrouded in smog. In this context, the otherwise euphemistic 'change of air' meant the literal escape from the city in order to breathe. In 1891, the Public Health Act introduced legislation to mitigate the coal-fueled pollution and dense and poisonous fogs that industrialization had created in London and in other cities (Holdsworth 1891). The establishment of the Coal Smoke Abatement Society followed in 1898, later known as the National Society for Clean Air. During this second phase it was central to mobilizing effort toward the 1956 Clean Air Act after the Great London Smog of 1952. It subsequently became the independent charity Environmental Protection-UK, but closed in 2012, despite the continuing urgency in the UK with regard to the rise in invisible air pollution and the associated ill health and premature deaths today.

Tyndall's work as an empirical and experimental scientist ranged across many of these permutations of atmosphere, but it is his discussion of the optics of the sky's appearance that overlaps most obviously with Ruskin's interests. Tyndall's *Six Lectures on Light* (Tyndall 1885) explained light waves through experiment with descriptions and references to common experience in the natural world such as "The Blue of the Sky" (Tyndall 1885, 149–152). He explained that "the blue light of the firmament is reflected light" caused by the "rebound of the waves from the air itself, or from something suspended in the air" (Tyndall 1885, 149). Encountering an obstacle, these waves are not equally reflected since "the smaller waves are in excess, and, as a consequence, in the scattered light blue will be the predominant colour" (Tyndall 1885, 151). Now known as the Tyndall Effect – a special instance of diffraction or the bending and spreading of waves around obstacles – Tyndall demonstrated this to his contemporary audiences through the 'Artificial Sky' experiment (Tyndall 1885, 152–154).

Diffractions

During this century, developments in physics included bringing the 'atmospheric' phenomena of heat, light and magnetism under the scope of mechanistic views of natural laws of cause and effect, governed by a system of energy exchange. The term 'ether' – common at the time and derived from the Greek for the pure, upper air – that had initially signified the intangible air beyond the earth became the name given to the medium that was supposed to provide a vehicle for the transmission of heat and light and magnetic effects. This proposed *ethereal plenum* constituted a physical something rather than the nothingness of a vacuum. Ether fell out of use as a scientific term when phenomena that had been explained by it were later superseded by the understanding of the electro-magnetic field. Throughout this period, Tyndall was an active participant in many related experiments, and party to the many discussions upon others' efforts that took place at the Royal Institution. Toward the end of the century, the decades of experimental investigation into light, sound, electricity and magnetism and related work to develop a unified theory led to James Clerk Maxwell (1831–1879) proposing the electro-magnetic theory of light.

At the beginning of the century, one of the key early nineteenth-century experiments in this area of research was developed by the English physicist Thomas Young (1773–1829). He performed a two-slit experiment that supported the wave theory of

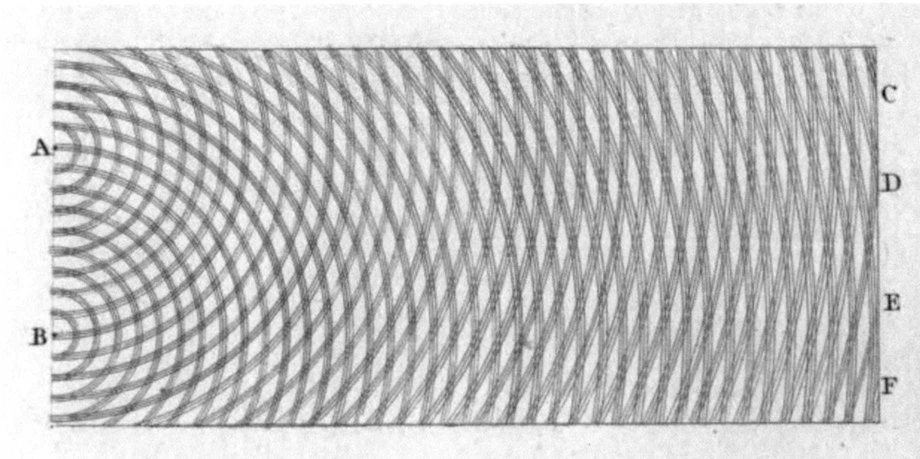

Figure 9.1 Diffraction Pattern Diagram, illustration in Thomas Young, *A Course of Lectures on Natural Philosophy and the Mechanical Arts.* (London, Joseph Johnson, 1807).

light, showing a diffraction pattern of interference similar to that observed in water ripples (Figure 9.1). Young describes his diagram of a diffraction pattern as follows:

> Two equal series of waves, diverging from the centres A and B, and crossing each other in such a manner, that in the lines tending towards C, D, E, and F, they counteract each other's effects, and the water remains nearly smooth, while in the intermediate spaces it is agitated. (Young 1807, note Fig. 267.)

Thomas Young's diffraction experiment was a classic mechanical physics experiment that Young believed proved the wave theory of light.

Barad's own onto-epistemology is derived from the twentieth-century double-slit experiment configured for a quantum mechanical measuring apparatus that confirmed the phenomena of wave-particle duality – that is, that light is both particle and wave – and the strange counter common-sense reality that, at any one time, light may show itself to be either one or the other depending upon the measuring apparatus applied in the observation. Young's two-slit diffraction pattern within the epistemological frame of classical physics is confident that the observer can recognize the objectively measurable fact of the wave-like behavior of light. Barad's two-slit diffraction pattern within the epistemological frame of quantum mechanics acknowledges that the observer along with their experimental equipment are part of the apparatus and are implicated in determining the possibility of what is observable.

Barad in a parodic thought-experiment proposes a lecture in which the audience according to their position in the auditorium would experience zones of amplification and silence within the diffraction pattern of the speaker's voice.[1] In reality, the acoustics of the spoken voice in an architectural space, amplified or otherwise, cannot be devised so as to create these effects.[2] It is nonetheless a powerful metaphor. The architectural design of the lecture theatre in which Tyndall presented many of his talks in the Royal Institution was built as a circular-domed lecture hall designed with concentric raked seating interrupted by stairs. Ruskin presented many lectures here too. "The Storm-Cloud of the Nineteenth Century" lectures took place at The London Institution which was an educational institute dedicated to science founded on the model of the Royal Institution with a lecture hall of similar design (Figure 9.2). A diffractive reading of Ruskin and Tyndall hears these two powerfully influential voices of the nineteenth century, as in Young's 1807 diffraction diagram, resounding out from podium A at the Royal Institution and podium B at the London Institution

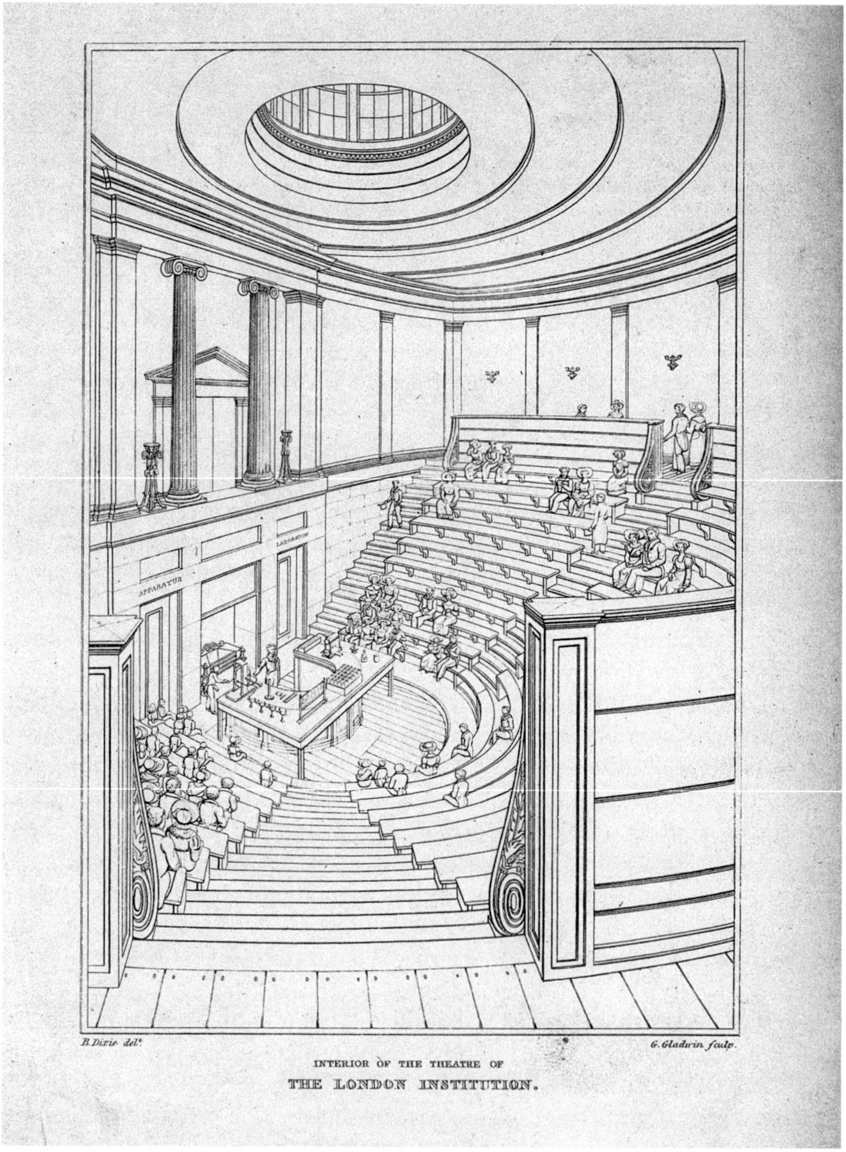

Figure 9.2 The London Institution, Moorfields: the interior of the lecture theatre, a demonstration in progress. Engraving by G. Gladwin after B Dixie.' by B. Dixie. With permission of Wellcome Collection.

into the ears and minds of their concentrically seated audience and out into the wider circles of cultural influence, creating patterns of interference down the years to our present time.

It is Barad's onto-epistemology based upon the second twentieth-century two-slit experiment that suggests diffraction as a praxis of analysis.

> Instead of imagining an unchanging wave-like form of cultural propagation in already fixed media, or stressing isolated encounters between cultural and technological objects, diffractive iteration demands a consideration of matter, medium, history and culture as dynamic, interactive and intra-active processes. (Edmond 2014, 246)

New materialist approaches affirm the openness of the past to reinterpretation and the entanglement of times – past, present and future. Sehgal writes that: "as a method, diffraction incorporates historicity and difference into the practice of theory itself" (Sehgal 2014, 188). The literary critic Jesse Oak Taylor, arguing for an ecocritical approach to Ruskin's 'Storm-Cloud,' discusses "the challenge of reading texts in their futurity" in which they acquire retrospective legibility and how this can invite the criticism of "teleological presentism" (Taylor 2018). This criticism is limited by a commitment to an overly restrictive sense of time moving in one direction. It loses validity when considered within the recursive temporality of a new materialist understanding. Taylor asserts that reading across times is needed for "life in the anthropocene" in which climate change connects consequences across vastly differing spatial and temporal scales. The blurring of the temporal distinctions between past, present and future is significant in terms of the role of atmosphere in climate change.

Lectures

As well as an interest in the sky and in practices of observation, these two men had in common the public demonstration of their discoveries and beliefs in 'illuminated' lectures. Tyndall was a talented and popular public speaker (J. Howard 2004) (Figure 9.3).

A reviewer in the *Guardian*, Sep. 21 1870, describing Tyndall, said that he had "pieces of paper in his hand but he rarely referred to them," and that "his thoughts seemed to flow forth with perfect ease," including the use of "apt metaphor" with "serious discourse" relieved by "pleasant banter" (Tyndall 1870, 4). The public lecture given by Tyndall on 15th January 1869, 'On Chemical Rays of the Light of the Sky,' to which he later made references in his 'Scientific use of the Imagination' (Tyndall 1879 v2), described the Tyndall Effect or Tyndall Scattering.

He devised an apparatus for use during public talks to demonstrate the blue sky effect made of copper alloy, iron, with wax sealant, and glass (Figure 9.4). As he described it, "The apparatus used to illustrate this consists of a glass tube about a yard in length, and from 2 ½ to 3 inches in diameter. The gas or vapour to be examined is introduced into this tube, and upon it the condensed beam of the electric lamp is permitted to act" (Tyndall 1885, 153). He further explained that by "sending through this compound and highly attenuated atmosphere the beam of the electric light, within the tube arises gradually a splendid azure" (Tyndall 1885, 154). Darkness was required during the experiment to present the spectacle of blueness to the audience, who viewed the apparatus from the side. Although not scientifically

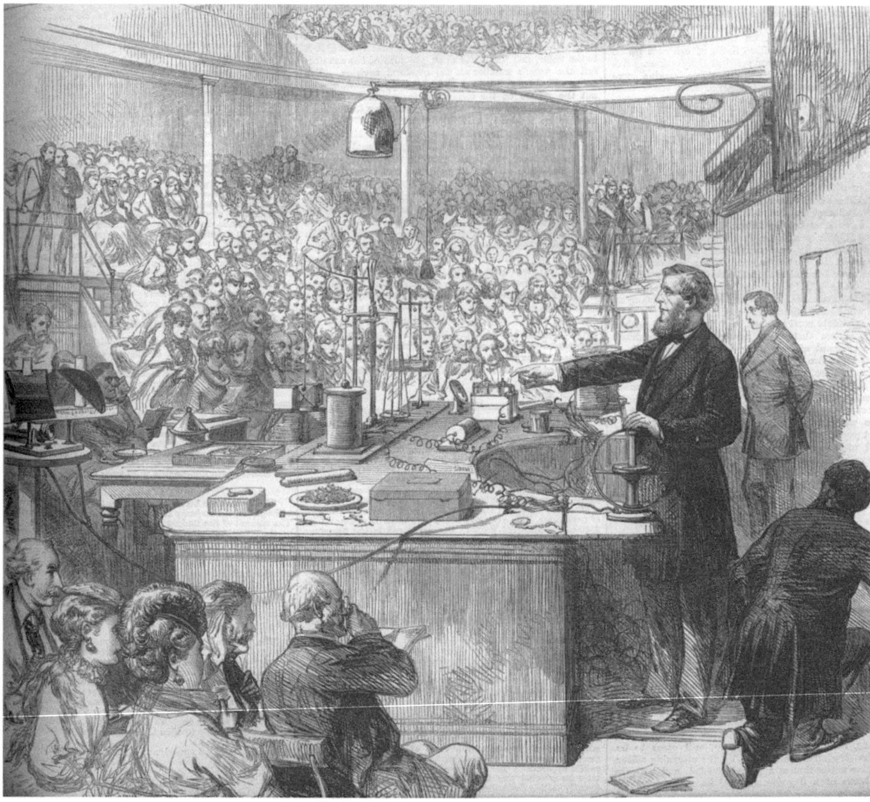

Figure 9.3 Tyndall Lecturing at the Royal Institution, *London Illustrated News*, 14 May, 1870.

necessary, Tyndall added theatre to the effect by also performing the emptying of the experimental tube between its gaseous illumination in complete darkness (Tyndall 1868–1869, 93). Clearly pleased with the results, and writing about these experiments in the entry titled "Production of Sky-blue by the decomposition of Nitrite of Amyl," Tyndall asserted, "for never, even in the skies of the Alps, have I seen a richer or purer blue than that attainable by suitable disposition of the light falling upon precipitated vapour. May not the aqueous vapour of our atmosphere act in a similar manner?" (Tyndall 1868–1869, 97). Then, Tyndall provided the quantities and instruction for a repeat experiment for any one inclined to follow his lead.

Tyndall's paper on the blue sky apparatus preceded Ruskin's 'Storm-Cloud' by 15 years but was presented the same year in which Ruskin published *Queen of the Air* (Ruskin 1912 v19), a collection in which he approached the significance of air, giving account of the myths and the human narratives of gods and goddesses that had some bearing upon the sky. Fitch identifies 'two ways of knowing' in Ruskin's writing, 'the passionate and organic, the dispassionate and dualistic,' (Fitch 1982, 577), and notes that these are set in conflict with one another in this book: an allegorical myth-making exercise. Ruskin wrote in his 1869 preface to it: "Even while I correct these sheets for press, a lecture by Professor Tyndall has been put into my hands, which I ought to have

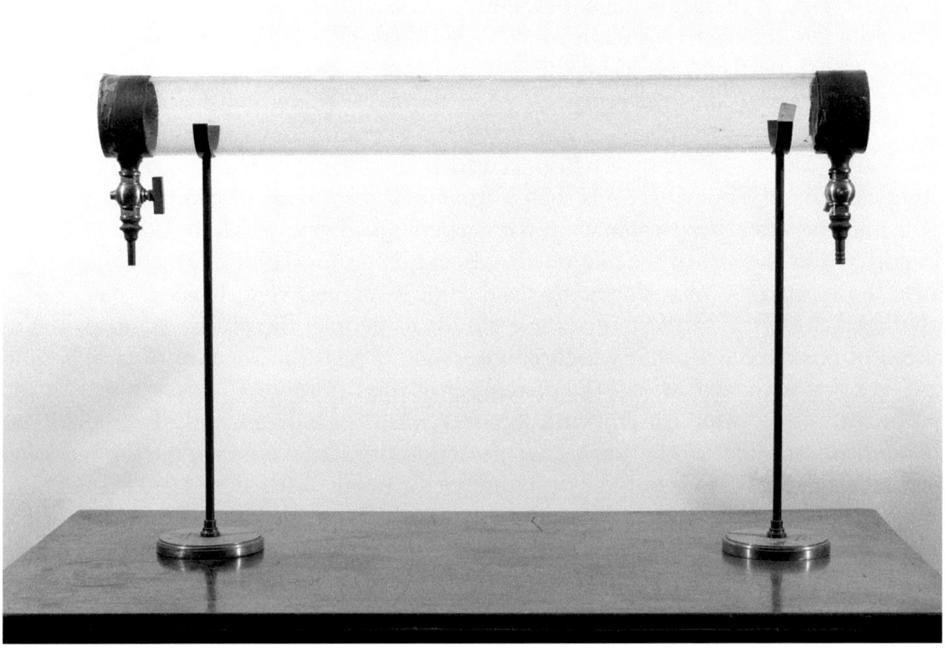

Figure 9.4 John Tyndall's apparatus for showing why the sky is blue, copper alloy, iron, glass and wax. The Royal Institution, UK/Bridgeman Images.

heard last 16th January, but was hindered by mischance" (Ruskin 1912 v19, 292). Ruskin, quoting Tyndall, perhaps envious of the spectacle that Tyndall had achieved, writes rather sarcastically: "Indeed, it would be difficult to imagine triumph more complete. To form, 'within an experimental tube, a bit of more perfect sky than the sky itself!'" (Ruskin 1912 v19, 292). Ruskin continued with the appeal: "Ah, masters of modern science, give me back my Athena out of your vials." Ruskin's asserted that the world is made to fit man "that the Air is given to him for his life" (Ruskin 1912 v19, 294). For Ruskin, it was most emphatically *not* man's place to fabricate an artificial sky to rival that in nature, and to do so was a victory of a Pyrrhic sort.

Ruskin's 'The Storm-Cloud of the Nineteenth-Century' was first presented as two lectures delivered at the London Institution on February 4 and 11, 1884. Ruskin's early attempts as public lecturer were an awkward combination of intoned passages read aloud, interspersed with improvised address (Birch 2015, 209), but his skill was to develop over his lifetime. Later, his delivery was improved, and a year after the 'Storm-Cloud' lectures, someone in Ruskin's audience described a "deep-toned, musical voice" (Ruskin 1912 v34, xxix). These lectures were a continuation if not, at points, a repetition of previous interlocutions with Tyndall and his work, and in them Ruskin frequently cited Tyndall's *The Glaciers of the Alps* (Tyndall 1860). Ruskin addressed the audience directly, offering to "bring to [their] notice a series of cloud phenomena" (Ruskin 1912 v34, 9). He continued, "I first noticed the definite character of this wind, and of the clouds it brings with it, in the year of 1871" (Ruskin 1912 v34, 31). Ruskin claimed that its range extends from the north of England to as far south as Sicily – this geographical reach constituting all of his known world. The language in 'Storm-Cloud' is apocalyptic. It identifies "plague-clouds" as a dismal mist: "For the sky is covered with grey cloud; - not rain cloud, but a dry black veil, which no ray of sunshine can pierce" (Ruskin 1912 v34, 32).

The preface identifies the storm cloud as damp air mixed with smoke (Ruskin 1912 v34, xxvi). Referring to his diary entry, Ruskin explains that it is comprised of "poisonous smoke" due to "at least two hundred furnace chimneys in a square of two miles on either side of me," but then adds that it appears to be "made of dead men's souls," thus replacing the scientific explanation for a theological one (Ruskin 1912 v34, 33). For Ruskin, the dark skies of the plague-clouds signified moral gloom, the blasphemy of science, the wrongness of scientific method. In a Romantic vein, Ruskin's 'Storm-Cloud' attributed the cause of these miserable clouds to people: the phrase suggests the implication of observer with that which is observed. "I pass the account of clouds that *are*, and – I say it with sorrow – of the *dis*temper of their observers" (Ruskin 1912 v34, 14).

The initial reception for Ruskin's lectures was dismissive, mostly from critics of his theological moralizing. He notes, "my assertion of radical change, during recent years, in weather aspect was scouted as imaginary, or insane" (Ruskin 1912 v34, 7) but the editor of the 1908 introduction to 'Storm-Cloud' affirmed the subsequent scientific corroboration of his observations. Ruskin himself claimed the authority of science when he asserted: "nor is there a single fact stated in the following pages which I have not verified with the chemist's analysis, and geometer's precision" (Ruskin 1912 v34, 8). Thus, while appealing to scientific fact, Ruskin also denigrated the one-sided limits of which he accused the approach of scientific method and warned that the listener must be wary of the explanations offered by men of science. Such caution notwithstanding, Ruskin, in contrast, promised no explanations, just the presentation of a few facts (Ruskin 1912 v34, 18). The visual support of those facts became part of his corroborating rhetoric.

For his public lectures, Ruskin commissioned an assistant to paint enlarged copies of his original small-scale watercolors, which were then held up at the appropriate moment (Figure 9.5): "Mr. Severn has beautifully enlarged my sketch of a July thunder-cloud of the year 1858, on the Alps of the Val d'Aosta" (Ruskin 1912 v34, 29). Notes to the published lecture inform the reader "The lecture itself had been illustrated by coloured enlargements from Ruskin's sketches, which were thrown on a screen by the lime-light" (Ruskin 1912 v34, xxvii).

Ruskin spent some time describing the watercolor painted illustration in which "pigments - brilliant to the height that pigments can be" were supported by "a white light as pure as that of the day" (Ruskin 1912 v34, 21). On showing another drawing of "An August Sky at Brantwood 1880," Ruskin asks of his audience: "But you must please here observe that while my first diagram did with some adequateness represent to you the colour facts there spoken of, the present diagram can only *explain*, not reproduce them" (Ruskin 1912 v34, 24). He then set out the limits of what his art was capable to represent in more detail and described the missing ruby and vermillion in the skies of the original view: "Only artificial and very high illumination would give the real effect of them – painting cannot" (Ruskin 1912 v34, 25).

Although his words conjured up impressions of the 'plague-cloud' – "the air one loath-some mass of sultry and foul fog, like smoke" (Ruskin: 37), Ruskin's illustrations did not show any such examples, but only demonstrated unpolluted atmospheres: his enlarged watercolors represented the clear skies that this storm cloud had put under threat: "The diagram is enlarged from my careful sketch of the sunset [...]

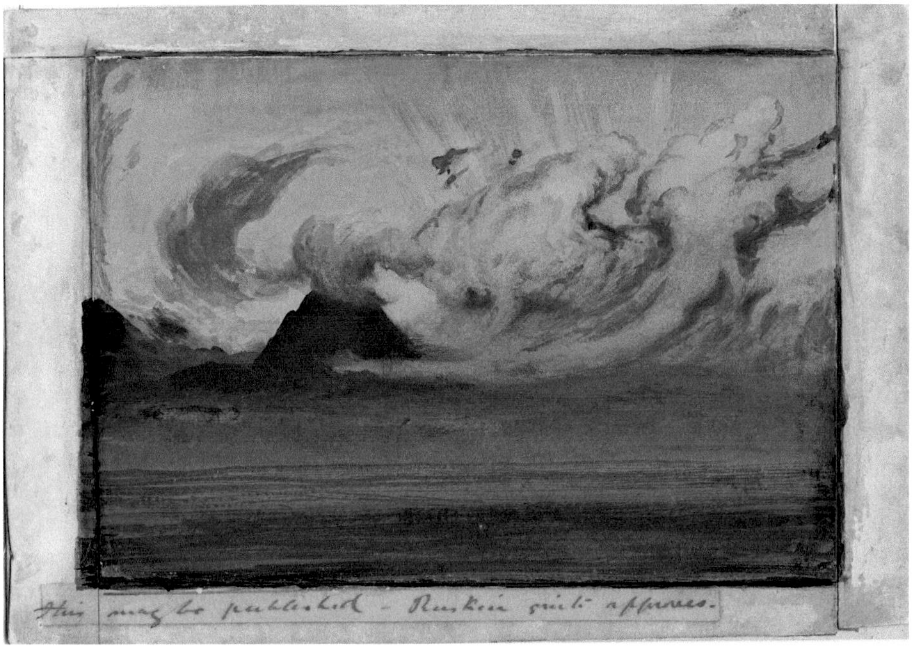

Figure 9.5 John Ruskin, *Thundercloud, Val d'Aosta*, 1858, original watercolor. Courtesy of Ruskin Foundation (Ruskin Library, Lancaster University).

unaffected by smoke" (Ruskin 1912 v34, 21). Yet while showing these illustrations, he simultaneously lamented their inadequacy to do justice to the skies to which they referred – "The brightest pigment we have would look dim beside the truth," he affirmed (Ruskin 1912 v34, 40), and reiterated: "No colours that can be fixed in earth can ever represent to you the lustre of these cloudy ones" (Ruskin 1912 v34, 21).

This sentiment is striking in contrast to the delighted tone and audacious confidence with which Tyndall spoke of his achievements – indeed, Tyndall's sky in the tube was not presented as a representation *of* the sky but as if it *were* the sky itself: "We can generate in air artificial skies, and prove their perfect identity with the natural one" (Tyndall 1885, 152–153). What did this scientific demonstration show, then, apart from the ostensible display of a scientific phenomenon, that is, the scattering of light? I propose that it may have manifested a rather hubristic enterprise: the human-power of the generation of 'Artificial Skies,' the same title denoting a sub-section in Tyndall's *Six Lectures on Light*. Perhaps Ruskin's emphasis on the inadequacy of his own representation was meant to rhetorically underline the irreproducibility of nature, and implicitly to chastise Tyndall's scientific presumptions. And as for the absence of illustrations of the storm cloud or polluted skies in his own lectures, Ruskin seems to suggest to his audience that all they need to see an example of the plague-cloud is to go outside and look for themselves (Ruskin 1912 v34, 40). Instead of illustrations, he offered a recipe with which his listener might make their own version:

> I should have liked to have blotted down for you a bit of plague-cloud to put beside this [an old-fashioned sunset illustrated]; but Heaven knows, you can see

> enough of it nowadays without any trouble of mine; and if you want, in a hurry, to see what the sun looks like through it, you've only to throw a bad half-crown into a basin of soap and water. (Ruskin 1912 v34, 40)

Ruskin frequently identified the combined effects of science and self-interested economic motives in the consequence of the ravaged natural environment. It is not an accident, then, that a coin is used in his example, which is also a kind of parody of Tyndall's experimental demonstrations. The efforts of Ruskin and Tyndall to put forward their respective points of view anticipate twenty-first-century ecocriticism. Like a diffraction pattern of pebble-provoked ripples in the water, current ecocriticism is the composite of previous waves, the first wave of nature writing in the contemplation of the Romanticized wilderness, and a second wave call to activism and environmental justice, the mixed heritages manifesting ambivalence as to the role of science, and contention as to the centrality or eccentricity of the human.

Ruskin's point of view on the imbrication of science with commerce often moved him to direct critical comment on Tyndall's empirical scientific observations. He writes: "the discoverers of modern science have, almost without exception, provoked new furies of avarice, and new tyrannies of individual interest" (Ruskin 1912 v26, 339). Ruskin had this attitude in common with many environmentalists and was aligned with others such as William Morris (1834–1896) and George Bernard Shaw (1856–1950) who argued for a romantic return to nature. This attitude can be traced in the tradition of conservation environmentalism resulting in the founding of The National Trust in the UK in 1895 (in which Ruskin was inspirational), as well as the London-based Coal Smoke Abatement Society in 1898. Ruskin held the men of science of Tyndall's ilk responsible for scientific naturalism and its associated misguided debasement of nature to human intentions and purposes and to laissez-faire political economy. Yet it is exactly through Tyndall's observations of the atmosphere that scientific arguments *against* the instrumentalization of nature have made progress: Tyndall's experimental proof for the theory of greenhouse gasses has been fundamental to the development of contemporary climate science, supporting the understanding of a history of changing climate and providing evidence for the "physical basis for anthropogenic global warming" (Hulme 2009, 121). Within our current perspective, Tyndall's work supports Ruskin's arguments against human hubris and the associated laissez-faire economic exploitation of natural resources.

New materialism, although mobilizing a strategic anthropomorphism, moves away from the classical humanist anthropocentric view and a human-centered spiritual world, toward a planetary-centered perspective that is both ethical and integrative, in which 'the milieu' or environment takes center place. This new materialism is negatively critical of human hubris derived from the mistaken anthropocentric view that all nonhuman matter is more or less inert and without agency. Bennett writes that: "the image of dead of thoroughly instrumentalised matter feeds human hubris and our earth-destroying fantasies of conquest and consumption" (Bennett 2010, ix). It is this, Bennett writes, that "may be one of the impediments to the emergence of more ecological and more materially sustainable modes of production and consumption" (Bennett 2010, ix).

Here, Ruskin's more traditional attitude to God resonates with the more spiritual strands of ecomaterialist thought. What would clearly have been contentious for Ruskin is, in new materialist thinking, the removal of a transcendental God from the equation. New materialism dispenses with theology as willingly as Tyndall had, so it may, in some ways, seem anachronistic to re-evaluate Ruskin through it, he who was so strongly critical of scientific naturalism in his own time. On the other hand, perhaps this is just our response in a conversation that Ruskin initiated: he addressed the time-to-come when he noted his own time as "a period which will assuredly be recognized in future meteorological history as one of phenomena hitherto unrecorded in the courses of nature" (Ruskin 1912 v34, 31).

Diffracting Tyndall's atmospheres through Barad's agential realism and Ruskin's through Bennett's vibrant matter reveals unexpected commonalities between these nineteenth-century antagonists. Tyndall's scientific materialism is associated with the development in physics of scientific study of the smallest scale and supports new materialism's arguments for the agency of non-living matter. Equally, new materialist thinking on the vitalism of matter resonates with the sensibility for a kind of dynamic materialism or animism that can be found in Ruskin's view of the world, but without God and Man at the center. Every interpretation is an intra-action that according to Barad's terminology creates patterns of interference.

Ruskin and Tyndall's voices, while speaking of atmosphere, resonated as concentric circles of sound waves into the ears of the audiences assembled in the radiating levels of raked seating in the architectures of the lecture theatres and created semiospheres of influence through the decades to our current moment, the significance of their persuasive words and associated visual demonstrations at points canceling out and at others amplifying one another. Every 'storied matter' in its retelling creates these interference patterns.

For Ruskin, the 'plague-wind' that he writes of in his 'Storm-Cloud' lecture seems to be a direct consequence of the scientific investigations that had made it possible for Tyndall in his lecture to demonstrate the generation of artificial sky in a bottle: this same scientific naturalism had treated the natural world as an object to exploit, thereby polluting the environment and robbing all of our common and divine right to the blue sky above. Ruskin's lecture argued for an attitude that we should bring when we observe the sky, one that in his worldview included an anthropocentric and theological understanding, but which was also vitalist and replete with the atoms of a dynamic materialism. He criticized Tyndall, not entirely fairly, for an instrumentalizing attitude toward nature. Yet although Tyndall's demonstration of 'sky in a bottle' might be interpreted as hubristic, in profound ways, his empirical and scientific materialism has led to the foundation of climate science and to an ecological understanding about the atmosphere with which Ruskin might well sympathize.

Taylor, alluding to J. Hoffmeyer's *Signs of Meaning in the Universe*, critiques the over-emphasis on ecology as a set of physical relations that thereby underestimates the role of signification (Hoffmeyer 1996, vii). He argues for the role of linguistic mattering, so citing Haraway's 'storying,' and Hoffmeyer's 'semiosphere' (Hoffmeyer 1996). How Ruskin and Tyndall told their contrasting versions of the storm cloud and blue sky constitutes the semiospheres of their atmospheres: their choice of visual illustrations and demonstrations, the architectural and institutional platforms, the vocal performance and their rhetoric.

Tyndall's enthusiasm for combining a poetic approach with the analytical does resonate with current ecocritical thinking. Tyndall frequently engaged in thoughtful excursions in his writing that pondered the connectedness of all matter-energy in ways that could be seen as 'vitalist.' Yet he was equally engaged, for example, in thorough experimental efforts to disprove the theory of spontaneous generation – the erroneous idea that life could spring out of non-living matter, rather than by descent from similar organisms – still current despite the work of Louis Pasteur (1822–1895) on germ theory. My argument here is that both men can – in part – be figured as presciently ecomaterialist, and that what divides them are obsolete epistemological boundaries that were the consequences of epistemological practices of their own time.

Ruskin had claimed during his 'Storm-Cloud' lecture that he only wanted to present "a few facts," but as Barad has shown, the facts exist as part of practices of material-discursive matterings: "Making knowledge is not simply about making facts but about making worlds [...] in the sense of materially engaging as part of the world in giving it a specific material form" (Barad 2007, 91). Phenomena, Barad shows, must be understood within the expanded notion of apparatus in which observer and observed are implicated. 'Apparatus' for Barad is the material-discursive boundary-making practice that is formative of matter and meaning. Barad's apparatus is more than the apparatus of the instruments of experimentation set up in some laboratory scenario, but by considering apparatus of this limited kind, as the technical equipment to support some action, and in its old-fashioned and outmoded examples, we can de-naturalize our current knowledge-making practices. By paying attention to the Tyndall's nineteenth-century science demonstration using wax, glass and vapor, or Ruskin's public lecture using limelight and enlarged watercolors held up by assistants, we can become more attuned to how our own knowledge-practices are constituted.

By bringing new materialist diffractions of the nineteenth-century visual illustration of the sky in these two men's respective lectures, we can find patterns of interference that are useful for coping with understanding of anthropogenic climate change. First, the necessary understanding that the strict division between the objective-scientific and the poetic-subjective is untenable (Iovino and Oppermann 2014), since the matter-meaning composites of new materialist thinking can be related to the re-entanglement of subjective poetics and objective science. Atmosphere and semiosphere are ecologically entangled as spheres of matter and spheres of meaning. Second, the necessary understanding of the timeliness of anachronism with regard to anthropogenic climate change in the anthropocene. We must not be overly fearful of making an anachronistic error when looking back in light of what we now know, as we need to actively seek out and recognize the forces at play in the present through the patterns of diffraction across legacies and futurities. An ecocritical approach to the old-fashioned visual illustrations and dated rhetoric of Tyndall and Ruskin's lectures can challenge our current apprehension of our atmosphere and provoke new understandings of our ecological entanglement.

The atmosphere is an exemplary entity that proves the complexity to which new materialist approaches are suited. Attention to atmosphere brings us into a felt encounter with our ecological entanglement, more so than with some other natural entities. Atmosphere is doubled, associated both with the internal subjective feeling of mood (Jacobus 2006, 220) and air as external objective environment. For all of Ruskin's criticism of Tyndall's sky-in-a-bottle reductionism, the two shared a sensibility of ecological entanglement. In 'Climbing in Search of the Sky,' Tyndall wrote, 'We live *in* the sky, not *under* it' (Tyndall 1870, 13).

Notes

1 Karen Barad: Re-membering the Future, Re(con)figuring the Past: Temporality, Materiality, and Justice-to-Come, Feminist Theory Workshop Keynote - Duke Gender, Sexuality & Feminist Studies. Published on May 19, 2014. Available at www.youtube.com/watch?v=cS&szDFwXyg. 1:05:52, 31:41-34:43.

2 Thanks to my friend the acoustic engineer Pedro Novo for dicussing this with me.

Works Cited

Andrews, Malcolm. 2008. "The Emotional Truth of Mountains: Ruskin and J.M.W Turner." *Caliban: French Journal of English Studies*. 23, Le Montaigne: 21–28.

Barad, Karen. 1996. "Meeting the Universe Halfway: Realism and Social Constructivism without Contradiction." In Lynn Hankinson Nelson and Jack Nelson, eds., *Feminism, Science, and the Philosophy of Science*, Synthese Library (Studies in Epistemology, Logic, Methodology and Philosophy of Science) 256. Springer, Dordrecht: 161–194.

———. 2007. *Meeting the Universe Halfway: Quantum Physics and the Entanglement of Matter and Meaning*. Durham, NC: Duke University Press.

———. 2014. "Re-membering the Future, Re(con)figuring the Past: Temporality, Materiality, and Justice-to-Come." *Feminist Theory Workshop Keynote - Duke Gender, Sexuality & Feminist Studies*. Published on May 19. Available at www.youtube.com/watch?v=cS&szDFwXyg. 1:05:52, 31: 41–34:43.

Bennett, Jane. 2004. "The Force of Things: Steps toward an Ecology of Matter." *Political Theory*. 32, 3: 342–372.

———. 2010. *Vibrant Matter: A Political Ecology of Things*. Durham, NC: Duke University Press.

Birch, Dinah. 2015. "Lecturing and Public Voice.'" In Francis O'Gorman, ed., *The Cambridge Companion to John Ruskin*. Cambridge: Cambridge University Press: 202–216.

Bōhme, Gernot. 2010. "On Beauty." *The Nordic Journal of Aesthetics*. 39: 22–33.

———. 2016. *The Aesthetics of Atmospheres*. Abingdon: Routledge.

Chandler, Timothy. 2011. "Reading Atmospheres: The Ecocritical Potential of Gernot Bōhme's Aesthetic Theory of Nature." *ISLE Interdisciplinary Studies in Literature and Environment*. 18, 3: 553–568. doi:10.1093/isle/isr079.

Coole, Diana, and Samantha Frost, eds. 2010. *New Materialisms: Ontology, Agency, and Politics*. Durham, NC: Duke University Press.

Edmond, Jacob. 2014. "Diffracted Waves in World Literature." *Parallax*. 20, 3: 188–201.

Fitch, Raymond E. 1982. *The Poison Sky: Myth and Apocalypse in Ruskin*. Athens and London: Ohio University Press.

Forster, Thomas. 1821. *Researches about Atmospheric Phaenomena*. 2nd Ed. London: Baldwin, Craddock, and Joy.

Frost, Mark. 2011. '"The Circles of Vitality": Ruskin, Science and Dynamic Materiality." *Victorian Literature and Culture*. 39: 367–383.

Hamblyn, Richard. 2001. *The Invention of Clouds: How an Amateur Meteorologist Forged the Language of the Skies*. New York: Picador.

Haraway, Donna. 1997. *Modest_Witness@Second_Millenium.FemaleMan©_Meets_OncoMouse™*. New York and London: Routledge.

Hoffmeyer, Jesper. 1996. *Signs of Meaning in the Universe*, translated by Barbara J. Haveland, Bloomington: Indiana University Press.

Holdsworth, William Andrews. 1891. *The Public Health (London) Act, 1891: With an Introduction, Notes and Index*. London: George Routledge and Sons.

Howard, Jill. 2004. "'Physics and Fashion': John Tyndall and His Audiences in Mid-Victorian Britain." *Studies in History and Philosophy of Science Part A*. 35, 4: 729–758.

Howard, Luke. 1803. "On the Modifications of Clouds, and On the Principles of Their Production, Suspension, and Destruction; Being the Substance of an Essay Read before the Askesian Society in the Session 1802–3, Part 1." *Philosophical Magazine*. 16, 62: 97–107.

———. 1818. *The Climate of London Deduced from Meteorological Observations Made at Different Places in the Neighbourhood of the Metropolis*, 2 Vols. London: Philips.

Hulme, Mike. 2009. "On the Origin of "the Greenhouse Effect": John Tyndalls' 1859 Interrogation of Nature." *Weather*. 64, 5: 121–123.

Iovino, Serenella, and Serpil Oppermann. 2014. *Material Ecocriticism*. Bloomington: Indiana University Press.

Jacobus, Mary. 2006. "Cloud Studies: The Visible Invisible." *Gramma*. 14: 219–247.

Jardine, Boris. 2014. "Made Real: Artifice and Accuracy in Nineteenth-century Science Illustration." *Science Museum Group Journal*. Accessed October 2, 2018. doi:10.15180/140208.

Ruskin, John. 1903–1912. *The Library Edition of the Complete Works of John Ruskin*. Edward Tyas Cook and Alexander Wedderburn, eds., 39 Vols. London: George Allen.

Sehgal, Melanie. 2014. "Diffractive Propositions: Reading Alfred North Whitehead with Donna Haraway and Karen Barad." *Parallax*. 20, 3: 188–201.

Solnit, Rebecca. 2005. *A Field Guide to Getting Lost*. New York: Viking.

Taylor, Jesse Oak. 2018. "Storm-Clouds on the Horizon: John Ruskin and the Emergence of Anthropogenic Climate Change." 19: *Interdisciplinary Studies in the Long Nineteenth Century*. 26. Accessed October 1, 2018. doi:10.16995/ntn.802.

Thomes, John E. 1999. *John Constable's Skies: A Fusion of Art and Science*. Birmingham: The University of Birmingham Press.

Tyndall, John. 1853 "On the Influence of Material Aggregation Upon the Manifestations of Force." *Proceedings of the Royal Institution*. 1: 254–259.

———. 1860. *The Glaciers of the Alps*. London: John Murray.

———. 1868–1869. "On a New Series of Chemical Reactions produced by Light." *Proceedings of the Royal Society of London*. 17: 92–102.

———. 1870b. *Essays of on the Use and Limit of the Imagination in Science*. London: Longman's, Green and Co.

———. 1879. *Fragments of Science: A Series of Detached Essays, Addresses, and Reviews by John Tyndall, F.R.S.* Vol. 1 and 2, 6th ed. London: Longmans. Green, and Co.

———. 1880. "Goethe's Farbenlehre: Theory of Colors II." *The Popular Science Monthly*. 17: 312–321.

———. 1885. *Six Lectures on Light delivered in the United States in 1872–1873*. 4th ed., London: Longmans, Green, and Co.

Young, Thomas. 1807. *A Course of Lectures on Natural Philosophy and the Mechanical Arts*. London: Joseph Johnson.

Part 3

Depletion and Conservation of Natural Resources

Describing the alarming loss of biodiversity in our current era of the impending sixth mass extinction (Kolbert 2014), historian Dipesh Chakrabarty notes that "our footprint was not always that large. Humans began to acquire this agency only since the Industrial Revolution" (Chakrabarty 2009, 207). Although ecological thinking took root in the nineteenth century, it was a time when species loss resulting from hunting, fishing, habitat decimation and the unintended consequences of introduced species (Columbian exchange) began to accelerate. Nineteenth-century extinctions caused by hunting were compounded by the trophic cascades that followed the removal of key species from interdependent ecosystems. The species loss of the nineteenth century includes a list of animals whose names are exotically distant from us: the bluebuck, the mysterious starling, the Mauritius blue pigeon, the Tonga ground skink, the quagga, the great auk, string tree and sea mink; the broad-faced potoroo, the Falkland Islands wolf, Labrador duck, Atlas bear, eastern elk, Hokkaido wolf and eastern hare-wallaby.

The three authors in this section consider wild animal populations, long-conceived as abundant and inexhaustible "natural resources" and the human communities that fished or hunted them, at just the moment that new ecological understandings arose of the crisis of depletion and the need for conservation and stewardship. Naomi Slipp's close examination of a shad caught in a gilded net, depicted on the American Presidential official State Dinner service, discusses the latent eco-anxieties of Gilded Age Americans about species abundance, artificial propagation and wildlife management. As she explains, when the service was used at state events, it produced a culinary spectacle of the abundance of American flora and fauna. Images on plates reminded diners of the vitality of their meal and the permanence of such fish – and by extension game, grain and vegetation – on American dining tables thanks to the work of federally supported bodies such as the U.S. Commission of Fish and Fisheries. However, a deeper perusal of the imagery suggests period tensions regarding formerly plentiful species and their decline, and builds an agential narrative that far exceeds the role of ornament.

Using "shifting baselines" – a term that describes subjective perceptions of marine species in fisheries science – as an apt descriptor of human coastline change, Maura Coughlin examines the tragedy of the commons in Atlantic cod fishing from the shores of France, reading paintings of fishing communities alongside historical and contextual displays in maritime exhibitions. In contrast to those who celebrated tourism's commodification of the shoreline, coastal artists Charles Cottet and Francis Tattegrain developed, in Tim Ingold's words, a "dwelling perspective" (2000) on the communities they inhabited. In paintings and in displays of material culture, Coughlin engages with complex ecological relationships such as the flows of fish and animals, seaweed, people, sand, stones, boats and other actors that moved across and through the strand, to construct a body of imagery that permits us to visualize new modes of sustainability.

As Jessica Landau writes, wilderness photographer George Shiras proclaimed a conservation-aesthetic in his proposal of the trap camera as an alternative to hunting "game animals," such as deer, with a rifle. Yet, while endeavoring to protect endangered deer populations, Shiras unfortunately campaigned for wolf slaughter. Eliminating an apex predator is today understood (not without controversy) to trigger a trophic cascade within the wolves' ecosystem; thus, Landau examines the paradox that Shiras's conservation efforts ironically served to destroy fragile ecosystems. While Landau reads these equally 'captured' animals in terms of early American wildlife photography's "optical unconscious" (Benjamin in Smith 2013), she also implicates the viewer of images of animal distress. Using the notion that the viewer must acknowledge responsibility for images of suffering, promoted by photo theorist Ariella Azoulay (2008), Laudau reads the un-articulated violence underpinning Shiras's photographs of animals that have recorded their own likenesses – whether mortally or metaphorically trapped.

Works Cited

Azoulay, Ariella. 2008. *The Civil Contract of Photography*. New York: Zone Books.

Chakrabarty, Dipesh. 2009. "The Climate of History: Four Theses." *Critical Inquiry*. 35, 2: 197–222.

Ingold, Timothy. 2000. *The Perception of the Environment: Essays on Livelihood, Dwelling and Skill*. London: Routledge.

Kolbert, Elizabeth. 2014. *The Sixth Extinction: An Unnatural History*. New York: Henry Holt.

Smith, Shawn Michelle. 2013. *At the Edge of Sight: Photography and the Unseen*. Durham, NC: Duke University Press.

10 Gilded Age Dining

Eco-Anxiety, Fisheries Management and the Presidential China of Rutherford B. Hayes

Naomi Slipp

Introduction: Reading Eco-Anxiety in Cultural Objects

In a 2008 article in *The Independent*, Michael Hewitt described "eco-anxiety," a new psychological condition stemming from severe anxiety about the environment, including panic over global warming, rising sea levels, genetically modified crops and fear of mass extinction. Examples of contemporary cultural products that reflect eco-anxiety include eco-disaster films that pit man against nature, such as *The Day after Tomorrow* (2004), *2012* (2009) and *Geostorm* (2017) (Hewitt 2008). While "eco-anxiety" describes the widespread apprehension felt by twenty-first-century citizens, it could just as easily describe the ecological anxieties of the Gilded Age.

This essay argues that similar concerns occupied the minds of late nineteenth-century Americans, specifically when it came to natural resources, wildlife conservation, species abundance and fisheries management. Indeed, a recognition of how environmental concerns create social anxieties and are reflected in cultural productions indicates that we might likewise identify references to environmental anxieties embedded in the visual form and iconography of works of fine art and material culture. Eco-critical analyses of cultural artifacts challenge anthropocentrism by adopting interdisciplinary approaches that look for connections across disciplines. In this essay, I consider what an eco-materialist approach to reading the decorative arts might teach us about the environmental anxieties of Gilded Age Americans and their relationships to populations of natural species.

By drawing upon scholarship in environmental history, material culture studies and the history of marine fisheries, we can begin to understand how even domestic objects like fine china reflect American worries about decreasing species, national dominance and environmental decline. Specifically, in 1879, President Rutherford B. Hayes commissioned a State Dinner and Dessert Service designed by Theodore Russell Davis. The elaborate hand-painted 562-piece service captured native North American flora and fauna in dramatic detail across 130 unique designs. Subsequently mass-reproduced by Haviland and Co., Limoges, France for a limited public market, the service was highly popular, if somewhat idiosyncratic. Later presidents, including Hayes's successor Grover Cleveland, a forerunner in forestry conservation, also used the design (Detweiler 2008, 60). What interests me most in considering this service are the ways in which concurrent issues regarding species abundance, the legislation of conservation at a national level and Gilded Age dining practices coalesce around these distinctive examples of elite material culture. I consider how these utilitarian objects might have participated in assuaging concerns, while also visually and functionally supporting conservation efforts in wildlife management and fisheries science.

Painting Native Flora and Fauna

Following Rutherford B. Hayes's election to the presidency (served 1877–1881), First Lady Lucy Hayes sought to replace the 587-piece Presidential china set of Ulysses S. Grant with a service that represented the diversity of North American flora and fauna. She selected *Harper's Weekly* correspondent and illustrator Theodore Russell Davis (1840–1894) to provide the decoration and Haviland & Co., Limoges, France as the manufacturer. Davis's illustrations served as models for etchings transferred onto dishes at the Haviland & Co., factory via the relatively modern printing processes of chromolithography and decalcomania. The images were then shaded, enameled and gilded by hand by 16 artisans. The service was comprised of 130 distinct designs across 562 pieces. Plates featured native game, such as pheasant, quail, wild turkey and deer, and seafood, including shad, pompano and smelt, along with blue claw crabs, oysters and lobster. Soup bowls and fruit plates pictured indigenous fruits and vegetables, including pecans, tomatoes, persimmons and wild apples, while ice cream plates whimsically took the form of snowshoes. The Hayes service was delivered to the White House on June 30, 1879 and first used for an official state dinner in December of 1880 with 40 attendees, held to celebrate the return of Mr. and Mrs. Ulysses S. Grant from an around-the-world tour. Finally, the Presidential china service was mass-reproduced by Haviland and Co., for a highly popular limited release. It appealed to middle-class American buyers, who purchased individual pieces in order to emulate their upper-class counterparts.

While the whole set features meticulously rendered flora and fauna, the Fish Platter (Plate 5), depicting a large Shad caught in a gilt gill net, serves as a useful case study. An analysis of the platter demonstrates a distinctive yet interconnected narrative about the species of the United States and their scientific management at the moment of the service's execution. The platter is shallow and rectangular, measuring 24½ inches long by nine inches wide, with widely curled corners that roll forward like rounded ears. This device creates a basin-like effect, heightening the dimensionality of what would otherwise be a flattened surface. Instead, the upturned corners operate as a three-dimensional frame and establish depth. In this way, the scene depicted exists behind or beyond the edges of the platter in another pictorial space. The illusion is further exaggerated by the fact that the gilding applied to the lip of the platter runs over onto the white porcelain back and blooms into painterly strokes with softened edges. This is visible on the backs of the curled corners, so that the back of the platter is apprehended at the same time as the front. This heightens the sense of artifice and contrasts with the image presented within, which is highly realistic. In other words, while the outer edges and corners of the platter appear to be artificially constructed by a brush and exist in the same space as the viewer, the scene within is presented as perspectivally and stylistically removed.

On the platter, a large shad is shown in a foreshortened three-quarter view. Its head faces to the left, while the caudal fin is bent behind it mid-swish, indicating the fish's power as it propels itself toward the left side of the scene. Its mottled body is rendered in varied tones of blue, purple, green and brown. The soft color application gives the appearance of blooms of aqueous watercolor paint, approximating the visual effects of an underwater scene and offering a subtle counterpoint to the mechanical nature of the chromolithographed porcelain decoration process. The background is rendered in a wash of mauve, while the environs are vaguely indicated by dark indistinct vertical

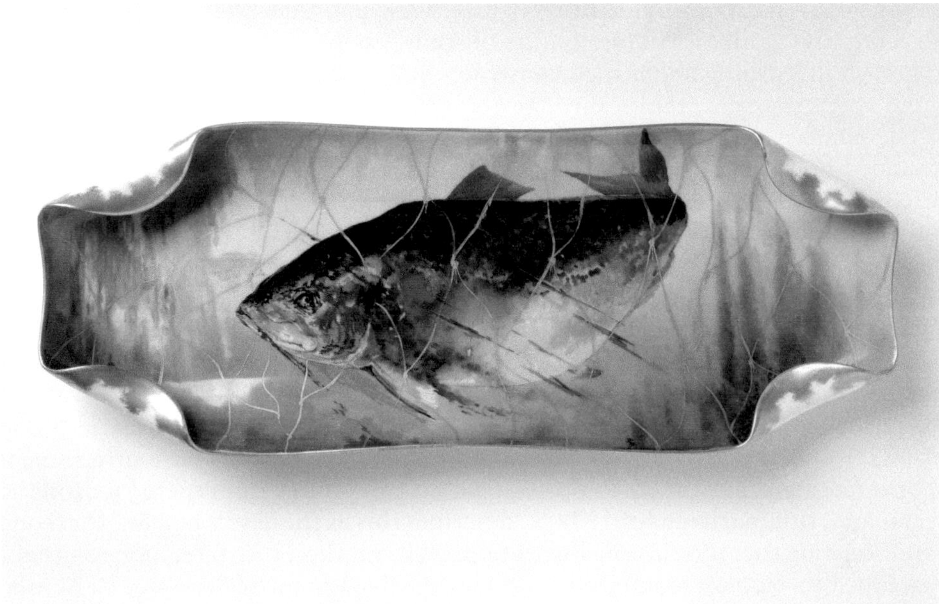

Plate 5 Designed by Theodore Russell Davis, made by Haviland & Co., Limoges, France, "Fish Service Platter – The Shad," c. 1880–1887. Porcelain with chromolithograph, enamel and gilt decoration, Philadelphia Museum of Art, Gift of the McNeil Americana Collection, 2006.

forms framing the fish's fat body. The gills of the fish have been caught by a delicately hand-knotted golden gill net, which covers the entire surface of the dish, connecting at the platter's rim to the gilded edge decoration. Deep red blood streams from the wounded gill and the fish's mouth. The fish eyes the viewer as if in appeal. However, since the net covers the entire face of the platter, its fate appears sealed and plight certain. There is no escape.

The fish platter is beautiful in its painterly application of color and gilding, poignant in its depiction of animal suffering, and moving in its matter-of-fact representation of the powerful fish netted and facing certain death. It is one of the more violent images in the service and, although no fisherman is present, the scene actively represents man as the agent of death via the net. This allows the diner to imagine themselves as the successful fisherman.

Fisheries Science, Population Decline and the Atlantic Shad

The iconographic details of the fish platter, and indeed its pathos, indicate a fundamental shift in the ways that Americans defined "nature" and their relationship to it in the 1870s. The idea of nature is culturally constructed through vastly different cultural, social, scientific and individual value systems, which vary across societies and eras. As historian Elizabeth Ann R. Bird writes, "scientific knowledge should not be regarded as a representation of nature, but rather as a socially constructed interpretation with an already socially constructed natural technical object of inquiry"

(Bird 1987, 255). In other words, scientific knowledge is itself based upon the ways in which social groups agree to recognize, view and understand contingent natural objects and organize them into bodies of study, such as "forestry" or "fisheries." The approach to learning about such constructs and managing them, likewise, represents those very epistemic frames through the reproduction of these cultural constructs. While scholars identify the Conservation movement, John Muir and the Sierra Club (founded in 1892) as integral to encouraging Americans to believe that conserving resources, like land, water and wildlife, was of national importance, beginning in the 1870s, "nature" itself was being reshaped by scientists, local communities and national government into manageable categories.

During the Gilded Age, the "hunter-conservationist movement" or "Progressivism" model emerged, which approached nature as an abundant scientifically managed resource, or as nature defined in relation to man's needs. Natural resources, like fish and trees, were increasingly legislated through national conservation and management efforts, including the U.S. Commission of Fish and Fisheries (founded in 1871) and the Division of Forestry (founded in 1881). These governmentally sponsored divisions aimed to protect such commodities in perpetuity. Roderick P. Neumann, summarizing Karl Marx, describes this as the development of "second nature," writing that it is "... the institutions – the market, the state, money – that have developed to regulate commodity exchange" (Neumann 1998, 26–27). While this may seem counter-intuitive, national policy toward nature was driven by capitalism and focused on natural resource management, with the specific aim of maximizing commodity yield over time in order to increase profits. Therefore, the protection and management of species in order to facilitate their harvest and consumption reflected upon a collective national identity based upon the financial success of natural resources to stimulate economic growth. By implication, the abundance of key species reflected collective ideals about natural and national superiority, while suggestions of population decline created anxieties about national decline that extended well beyond a single animal.

In the case of American fisheries, radical technological changes in fisheries during the mid-nineteenth century, including the adoption of hand-lining, tub trawling, expanded hook footprints, steel boats and cod traps, dramatically increased offshore and inshore catches. However, while the industrial mechanization of fisheries enabled larger landings, it also contributed to steady population declines and, although it produced maximized corporate profits, independent fishermen suffered. During the 1860s, fish landings of major stocks, long regarded as abundant – including Atlantic cod, herring, halibut and shad – markedly decreased. In this light, Gilded Age anxieties about fisheries stock abundance and decline reflected concerns about national and economic integrity.

As relationships between the United States, Canada, Great Britain and Newfoundland (which did not confederate until 1949) strained over declining fisheries, the United States Government adopted efforts to reverse depleting freshwater and saltwater fisheries. In 1871, the U.S. Congress created the National Fish Hatchery System and appointed ichthyologist and Smithsonian Secretary Spencer Fullerton Baird as the U.S. Commissioner of Fish and Fisheries. Significantly, this was the first attempt at national environmental management. Baird's job was to promote "the Investigation of the causes of the decrease in the supply of useful fishes of the United States, and of the various factors entering into the problem, and the employment of such active measures as may seem best calculated to stock or restock the waters of the rivers, lakes, and the sea" (*Report of the Commissioner* 1887, xix). In short, the government wanted to know: where were the fish going and could they be replenished?

The Commission founded permanent research stations at Woods Hole, MA and Halifax, Nova Scotia and spearheaded the first U.S. government-sponsored artificial propagation efforts and marine fisheries studies, which sought to secure threatened populations in inland and coastal waters. The following year, Baird began studying migration patterns of staple fish, like herring, alewives and shad, eventually recognizing that the creation of dams prevented these anadromous fish, who returned to the same localities every year, from spawning. This caused declines in Atlantic cod, which fed on the young. Baird recognized that fisheries populations fluctuated regionally and man-made and natural factors impacted their abundance, establishing causal relationships between environmental conditions, overfishing and reductions in catches. Attempting to counteract widespread belief in oceanic abundance, Baird described overfishing as unlike "dipping water out of a bucket, where the vacancy is immediately filled from the surrounding body," and "more like taking lard out of a keg, where there is a space left that does not become occupied by anything else" (Baird 1873, xxix). In order to combat the "empty bucket" effect, Baird established hatcheries to boost and replenish struggling populations. He set the standard for a balanced approach to fisheries, which sought to establish "how many fish may be taken and leave enough for seed" (McKenzie 2010, 136). In the process, Baird modernized fisheries management and proved its financial value to the U.S. government and the American people.

One of his key achievements was the successful introduction of Atlantic shad on the Pacific coast. The largest member of the herring family, *Alosa sapidissima*, is native to the Atlantic coast and is enjoyed for their meat and roe. In late spring, they enter freshwater rivers to spawn and are easily netted in the shallow waterways. In 1873, the U.S. Fish Commission sent Atlantic shad fry on a much-publicized trans-continental railroad journey to California, where they were introduced into the Sacramento River. Federal and State artificial shad planting continued through 1881, creating a non-native population of Atlantic shad on the West coast. The shad success assured the public that artificial propagation and introduction of species to non-native areas was possible. By 1884, the Commission was rearing 21 varieties of fishes, plus oysters, clams and lobsters in 13 hatching stations throughout the U.S. Hugh M. Smith wrote in 1893 in the popular journal *Science* that:

> If these far-reaching ... results attend the planting [of shad] on few occasions, of small numbers of fry in waters to which the fish are not indigenous, is it not permissible to assume that much more striking consequences must follow the planting of enormous quantities of fry, year after year, in native waters? (Smith 1893, 88–89)

As Smith's words indicate, the overwhelmingly positive results of the shad experiments suggested that the U.S. Fish and Fisheries Commission might have similar successes in future, thereby turning the tide on population declines of other fisheries, and – in the process – spurring national optimism about artificial propagation and species management in general.

An Iconography of Dining: Natural or Artificial?

In this context, the Shad Fish Platter for the Presidential service of Rutherford B. Hayes communicated circulating concerns surrounding marine abundance and declining fisheries stocks. Simultaneously, it visually presented a positive message regarding American advances in artificial propagation and national success in the introduction of non-native species. The shad, caught in the gilded gill net on the surface of the platter, was an illustration of man mastering nature and, at the same time, of the tensions implicit in such a task. The use of the service directly referenced man's power over nature, as chefs in the kitchen transformed raw materials into an edible delicacy fit for consumption. Ironically, however, while fisheries scientists were hard at work experimenting on ways to *artificially* promote abundance in declining fisheries, their efforts were undertaken expressly to counteract the increasing yields created by the industrialization of fisheries technology of *natural* populations, in order to support the consumption of such fish by the American consumer. Materially, the service telegraphed an increasing ambivalence about the tensions between the natural and the artificial. These were accentuated by the format of the State Dining Room, which conflated indoors and outdoors, the correspondence between the service's individual iconography and menu items, wherein one dined on the animals in the decoration, and through the techniques utilized to create the service, which eschewed handwork in favor of industrial reproduction.

Under the Hayes presidency, the State Dining Room – where visitors engaged with the Hayes service – troublingly operated as an artificial extension of the natural environment. Socialite Mrs. John A. Logan wrote, "It was at her [Mrs. Hayes] suggestion that the billiard-room, which was formerly between the conservatory and the state dining room, was made an extension of the conservatory, and by this means guests ... enjoy a beautiful vista of arching palms and blooming flowers while sitting at a state dinner or luncheon" (Logan 1901, 679). While indoors, visitors had the unnatural sensation of dining inside a garden or forest. In this way, the State Dining Room blurred the lines between the reproduction of a wild but artificial outdoors and an indoor, formal, social space, thereby heightening the ideological effects of the Hayes service's iconography.

Gilded Age diners also distanced themselves from the once living animal by engaging in civilized and highly circumscribed dining rituals, which, as Susan Williams explains, "were a constant, visible reassurance of an established and secure social position in a complex and changing world" (Williams 1996, ix). Specifically, the material culture and etiquette of dining affirmed social position, while dining rooms established spatial parameters for the performance of gentility. In the White House, such rituals also illustrated political power and natural dominion. The ceremonial and ritual functions of dining were amplified when the President of the United States dined on a service depicting a pantheon of America's edible species. As when an English landowner ate a stag from his own land, this activity emphasized the preeminence, wealth and power of the President and located him at the top of a hierarchy of species. The flora and fauna pictured on the services' surfaces and served upon them literally telegraphed the physical consumption of native delicacies in a lengthy parade of courses illustrating American natural bounty. However, the service and its use also created an uneasy conflation between the painted representation of the living animal and its transmutation as a consumable commodity.

For example, at a state dinner for 26 guests hosted by the Hayes Administration on February 3, 1881, 24 different dishes were offered, including at least two that match the dinner service's iconography: rotis of wild turkey and chestnuts, and canvas back ducks with redcurrant jelly ("Party Politics" nd). Canvas back ducks appear on a game plate flying above a cranberry bog, while the wild turkey is pictured on a large platter, strolling through a snowy landscape with small chestnuts scattered around his feet (Figure 10.1). Dinners

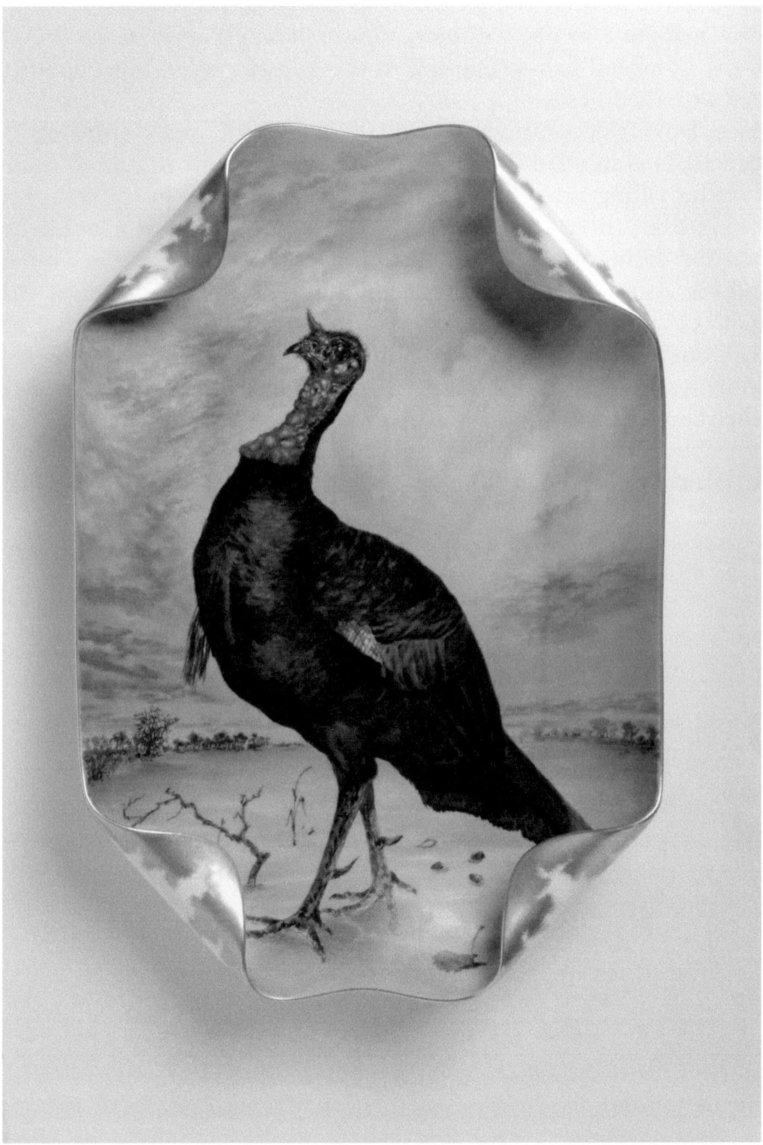

Figure 10.1 Designed by Theodore Russell Davis, made by Haviland & Co., Limoges, France, "Dinner Service Platter," c. 1882. Porcelain with chromolithograph, enamel and gilt decoration, Philadelphia Museum of Art, Gift of the McNeil Americana Collection, 2006.

at the White House were served *a la Française*, meaning that all of the items of each course were arranged on the table and then served. Imagine, then, the performative nature of the Hayes service, as ducks in red sauce and turkeys with chestnuts were consumed by diners to reveal their vivified, imperishable selves painted beneath.

This visual confusion extends to the Fish Platter as well. The average adult shad ranges from 20 to 24 inches, matching the length of the fish painted on the platter's surface. As one dined on the large cooked shad laid out whole on the platter, the watercolor analogue of the animal being consumed slowly appeared as a reminder of the previously enlivened state of one's dinner. Viewers could ruminate on the visual representation of the life-sized fish struggling in the gilt gill net, while enjoying the taste of its poached or baked flesh.

Across the service, Davis's decorations are equally whimsical and disturbing. "Seafood plate" (Figure 10.2) is molded into the shape of the interior of a crab shell and pictures a large lobster riding upon a wave toward a small crab in the foreground. The bottom of the plate has three "feet" that take the shape of molded gilt crab claws. Expressly designed for eating seafood, one encounters the very creatures one consumes, reanimated for visual delectation. Indeed, White House guest Clover Adams complained that she could not eat her "soup calmly with a coyote springing at her from a pine tree" (Gould 2001, 149–150). The lifelike quality of the painted subjects frightened some diners and made it difficult to ascertain the differences between artifice and nature, the representation and the real.

Finally, the confusion between the natural and the artificial extended to the creation of the service itself, which was oriented more towards modernized machine reproduction

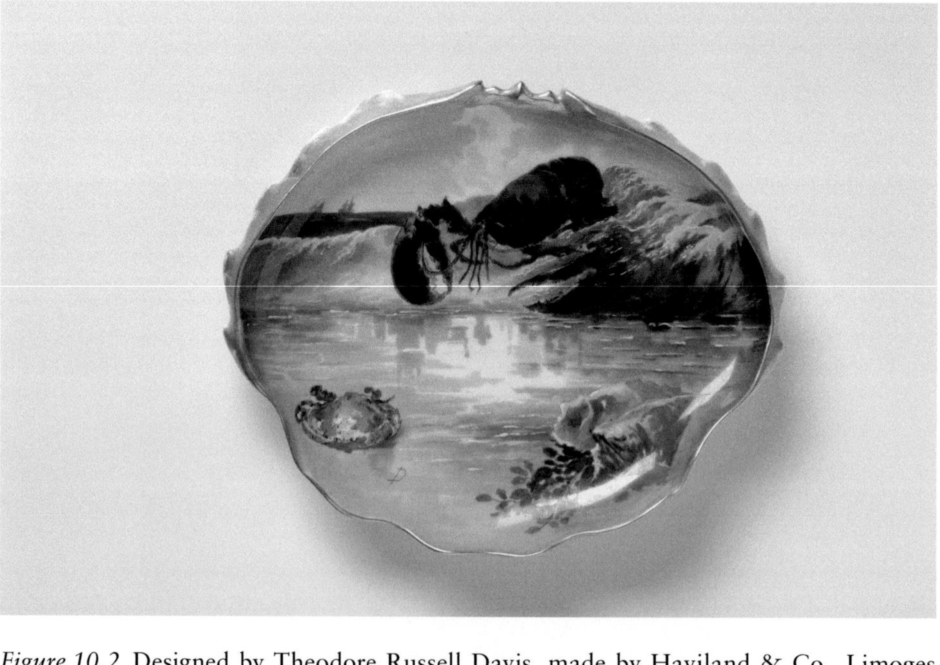

Figure 10.2 Designed by Theodore Russell Davis, made by Haviland & Co., Limoges, France, "Seafood Salad Plate," c. 1880. Porcelain with chromolithograph, enamel and gilt decoration, Philadelphia Museum of Art, Gift of the McNeil American Collection, 2006.

than traditional methods of porcelain decoration, such as hand-painting. Although Davis provided the designs, the use of chromolithography and decalcomania – new ceramic printing processes perfected in the 1870s – to transfer the designs automated the production process. The service also shows the stylistic influence of *Japonisme*, a fact that was negatively commented upon in the press as a foreign influence, when it was later marketed to the public. Perhaps to combat negative connotations associated with mechanical reproduction and "foreign" influences (the *Japonisme* style and French manufacturer) and to increase marketability, later editions featured several inscriptions, including a polychrome seal of the President of the United States, the artist's signature and monogram, and two Haviland & Co., marks. These indicate a concern for assigning American, "native" authorship, perhaps in the face of its seemingly "unnatural" and foreign method of production.

In addition, Haviland produced an 88-page booklet, which introduced the artisans involved, discussed the pieces and subjects, and included pen-and-ink drawings of the designs. The text which accompanied the drawings emphasizes the American-ness of each native, wild species and links that species to national preeminence. For example, the wild turkey is described as "so distinctively American ... that Benjamin Franklin urged its adoption as the national bird of the United States. The tame turkey differs considerably from its wild progenitor" (Klapthor 1975, 222). Likewise, the explanation of the shad platter proclaims: "The American shad is highly prized as food. ...It has been for years the custom to send to the President of the United States the first shad taken from the waters of the Potomoc river. ...Species of shad, differing from and inferior to the American shad, are found in some of the rivers of Europe" (Klapthor 1975, 208). While such species may have been artificially managed or "tamed" by 1879, the text makes clear that the public regarded native game and fish as distinctly American and only wild-caught examples were fit for the table of the President of the United States.

In spite of concerns about the nationality of the service, Lucy Hayes defended their artistry, writing to Davis on August 2, 1880: "It is a delight to study the beautiful forms and paintings. One almost feels as if such Ceramic art should be used for no other purpose except to gratify the eye" (Klapthor 1975, 113). Here, Hayes highlights the difficulty of admiring the decoration while eating off the surface. Should one use the object as intended, which obscures the decoration, or simply admire it as a fine work of art, thereby delimiting its function? Indeed, material culture studies must balance a recognition of the use value of an object against attention to its material and iconographic program and the larger historical contexts that inform their production and reception. Specifically, a consideration of the Shad Platter reveals a larger narrative about the early efforts of fisheries scientists to manage shad populations in concert with the U.S. government. Created as new methods for industrial ceramics decoration and the scientific management of species concurrently developed, both the use and design of the Hayes service point to contemporary concerns regarding the role of science in managing fisheries and an increasing ambivalence over the distinctions between the artificial and the natural.

Conclusion: A Narrative of Abundance

In conclusion, an eco-critical reading of the Hayes presidential china service illuminates the "eco-anxieties" of Gilded Age Americans, who were caught up in concerns over the scientific management of natural resources during a key period in environmental history. The Hayes Presidential Service was used at Official State Dinners, extremely formal affairs to recognize significant citizens and dignitaries, and then mass-produced and publicly sold. The abundance of American flora and fauna and its culinary delights were laid out as a spectacle for diners to visually appreciate and consume. When making use of the Shad Fish Platter, diners were reminded of the vitality of their dinner. They were also assured of the permanence of such fish – and by extension game, grain and vegetation – on American dining tables thanks to the work of scientists, the United States President, and federally supported bodies like the U.S. Commission of Fish and Fisheries. At the same time, however, such displays implicitly suggested a tension between species abundance and decline through consumption, the painted in contrast to the real, and an increasing ambivalence regarding natural versus artificial animal propagation. Equally significant, perhaps, is that these objects of material culture signal a new understanding of the expanding role of the government to manage and conserve native species, transforming Americans' ideas about nature, conservation and their dinner.

Works Cited

Baird, Spencer F. 1873. *Report of the Commissioner – United States Commission of Fish and Fisheries*, Part I. Washington, DC: Government Printing Office.

Bird, Elizabeth Ann R. 1987. "The Social Construction of Nature: Theoretical Approaches to the History of Environmental Problems." *Environmental Review: ER*. Special Issue: Theories of Environmental History. 11, 4: 255–264.

Detweiler, Susan Gray. 2008. *American Presidential China: The Robert L. McNeil, Jr., Collection at the Philadelphia Museum of Art*. Philadelphia: Philadelphia Museum of Art.

Gould, Lewis, ed. 2001. *American First Ladies: Their Lives and Their Legacy*. New York: Routledge.

Hewitt, Michael. 2008. "Agonising over the ICECAP or Frantic About Floods? You May be Suffering from 'Eco-anxiety." *The Independent* (20 March).

Klapthor, Margaret Brown. 1975. *Official White House China: 1789 to the Present*. Washington, DC: Smithsonian Institution Press.

Logan, Mrs. John A. 1901. *Thirty Years in Washington; Or, Life and Scenes in Our National Capital*. Washington, DC: A.D. Worthington & Co.

McKenzie, Matthew. 2010. *Clearing the Coastline: The Nineteenth-Century Ecological and Cultural Transformations of Cape Cod*. Hanover and London: University Press of New England.

Neumann, Roderick P. 1998. *Imposing Wilderness: Struggles Over Livelihood and Nature Preservation in Africa*. Berkeley: University of California Press.

"Party Politics: Entertaining at the White House." Nd. The National First Ladies' Library. Accessed December 1, 2017. www.firstladies.org/documents/partypoliticsbrochure-webcompress.pdf.

Smith, H.M. 1893. "Fish Acclimatization on the Pacific Coast." *Science*. 22, 550: 88–89.

Unattributed. 1887. *United States Commission of Fish and Fisheries: Report of the Commissioner for 1885*, Part XIII. Washington, DC: Government Printing Office.

Williams, Susan. 1996. *Savory Suppers and Fashionable Feasts: Dining in Victorian America*. Knoxville: University of Tennessee Press.

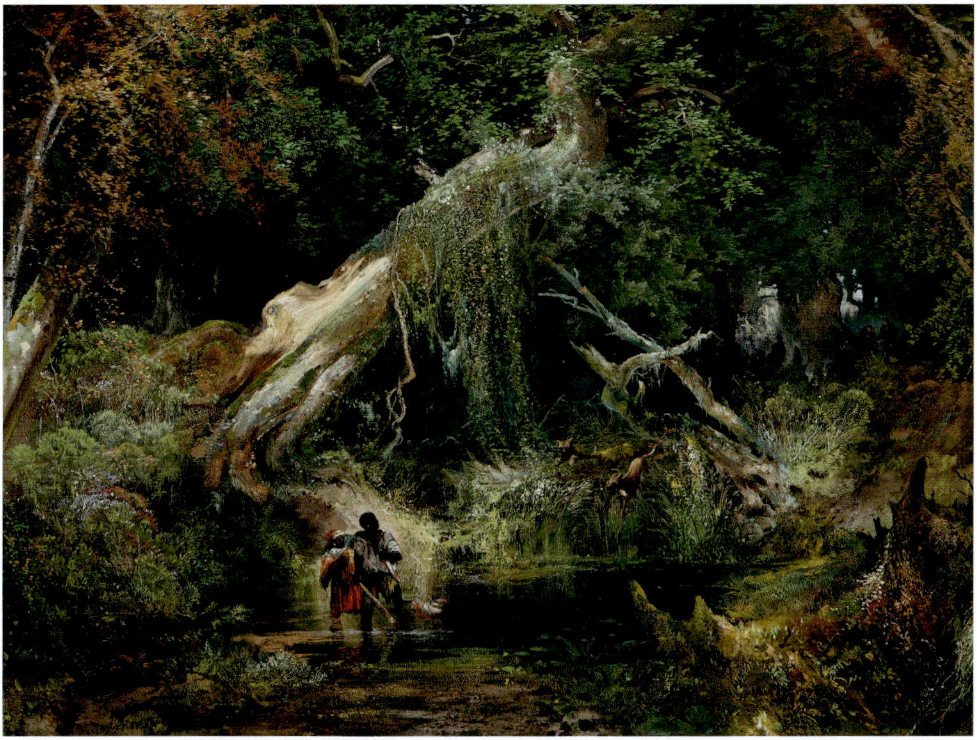

Plate 1 Thomas Moran, *Slave Hunt, Dismal Swamp, Virginia*, 1862, oil on canvas, Philbrook Museum, Tulsa.

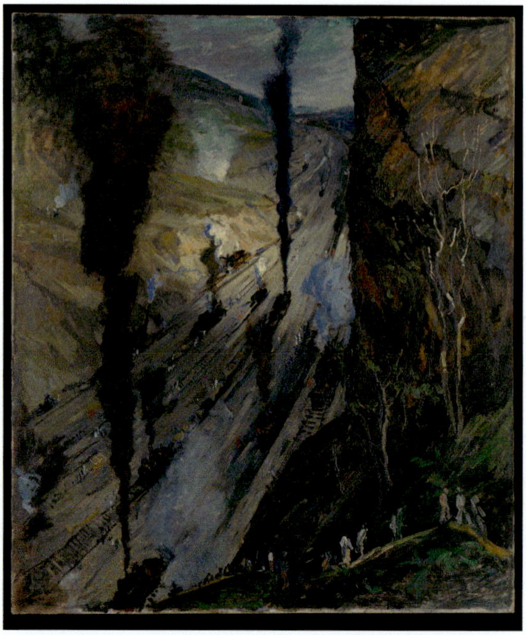

Plate 2 Jonas Lie, *The Conquerors (Culebra Cut, Panama Canal)*, 1913. Oil on canvas, Collection of The Metropolitan Museum of Art, New York George A. Hearn Fund, 1914, Image copyright @ The Metropolitan Museum of Art.

Image Source: Art Resource, NY.

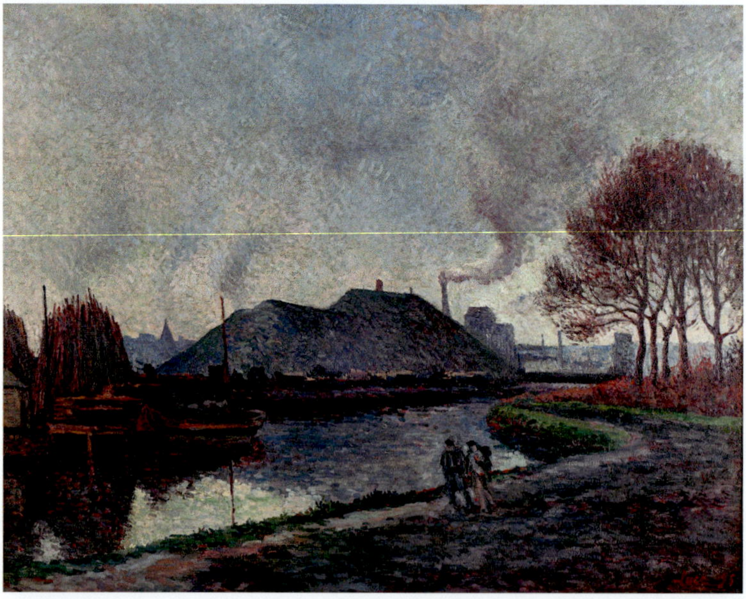

Plate 3 Maximilien Luce, *The River Sambre at Charleroi (La Sambre à Charleroi)*, 1896, oil on canvas, Chazen Museum of Art, University of Wisconsin-Madison, Elvehjem Museum of Art General Endowment Fund purchase.

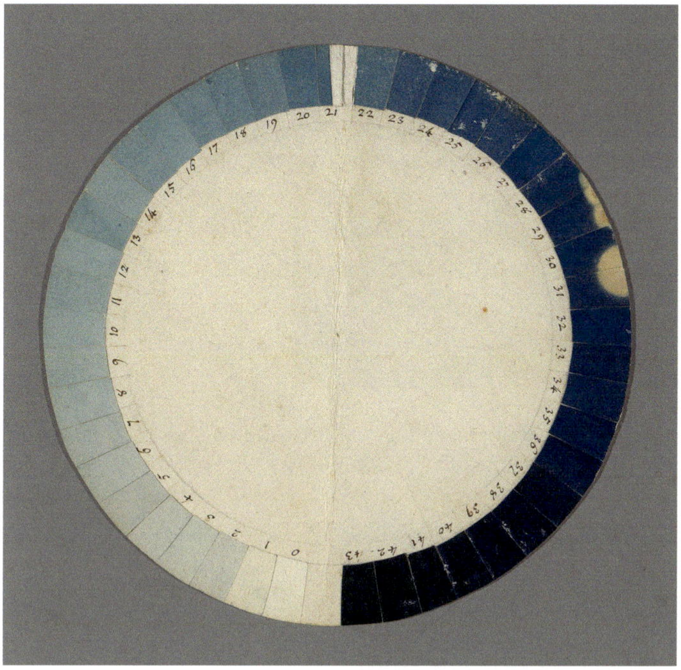

Plate 4 Cyanometer, invented by Horace-Bénédict de Saussure. Bibliothèque de Genève, Arch. de Saussure 66/7, pièce 7.

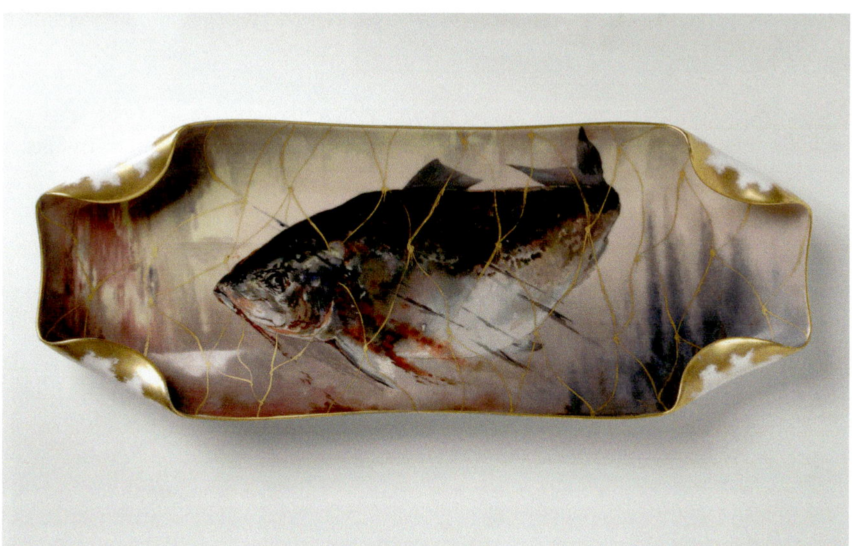

Plate 5 Designed by Theodore Russell Davis, made by Haviland & Co., Limoges, France, "Fish Service Platter – The Shad," c. 1880–1887. Porcelain with chromolithograph, enamel and gilt decoration, Philadelphia Museum of Art, Gift of the McNeil Americana Collection, 2006.

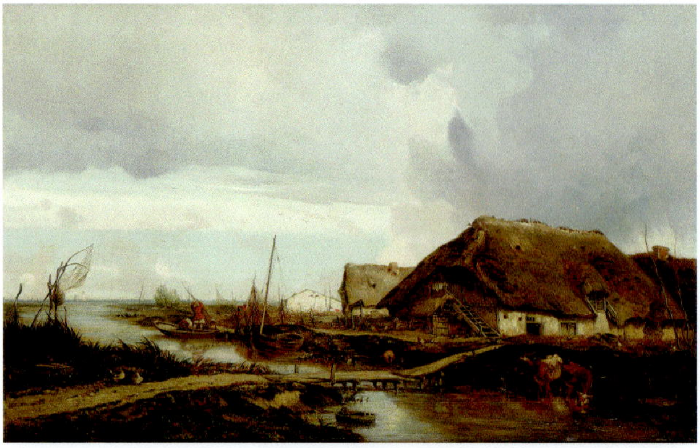

Plate 6 Camille Flers, *Fisherman's Cottage, Dieppe* (also known as *Normandy Shore*), c. 1848. Musée Thomas Henry, Cherbourg. Photo © Bridgeman Images.

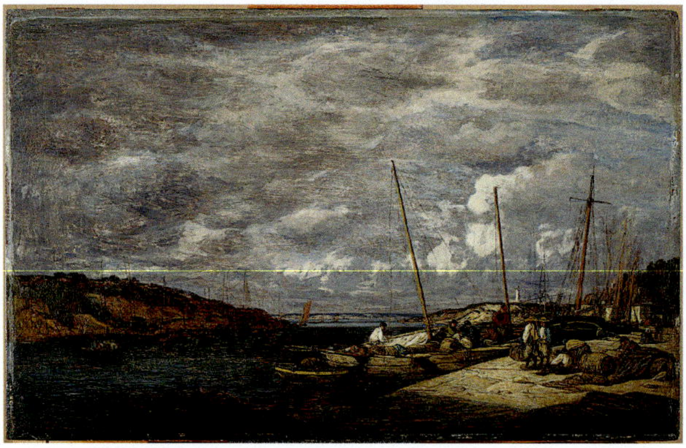

Plate 7 Eugene Louis Boudin, *Fishing Boats at the Dock, Douarnenez,* 1855. Oil on cradled panel. Private Collection. Photo © Christie's Images/Bridgeman Images.

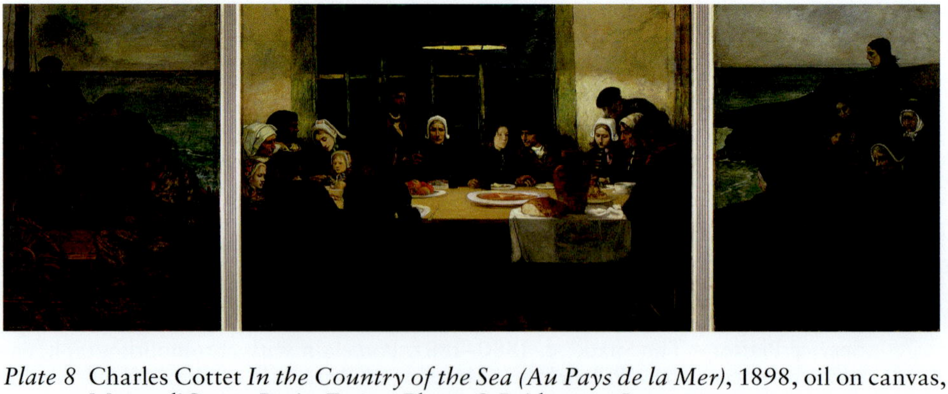

Plate 8 Charles Cottet *In the Country of the Sea (Au Pays de la Mer)*, 1898, oil on canvas, Musee d'Orsay, Paris, France Photo © Bridgeman Images.

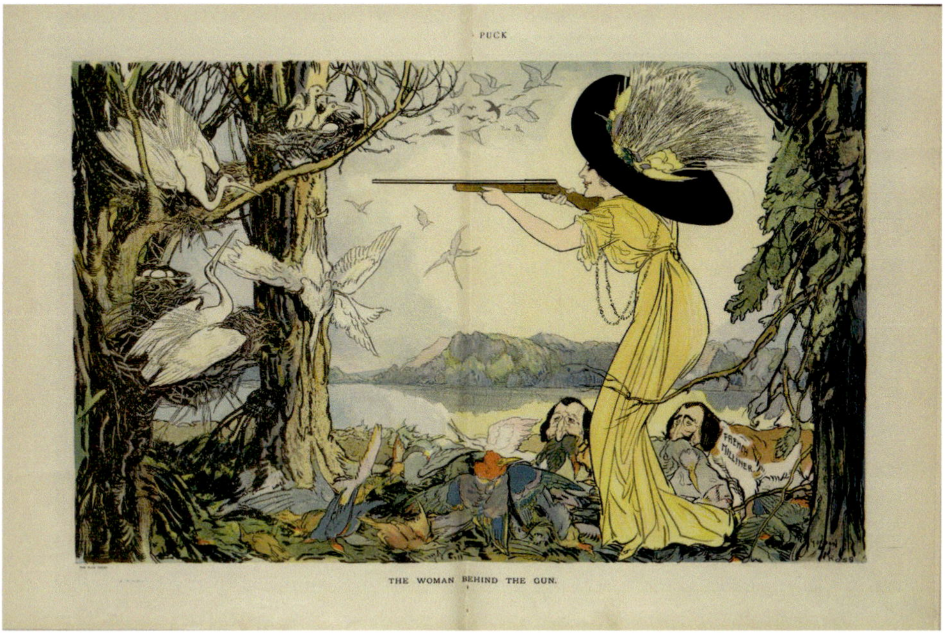

Plate 9 Gordon Ross, “The Woman Behind the Gun.” *Puck Magazine*, Vol. 69, No. 1786 (May 24, 1911). Image courtesy Library of Congress Prints and Photographs Division.

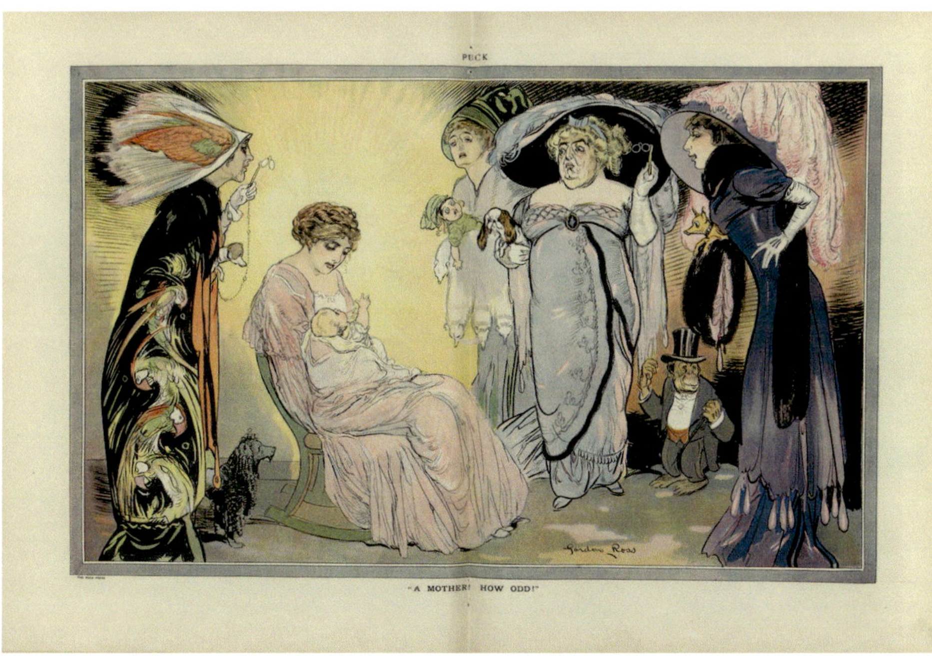

Plate 10 Gordon Ross, “A Mother! How Odd.” *Puck Magazine*, v. 68, no. 1764, December 21 1910. Image courtesy of the Library of Congress Prints and Photographs Division.

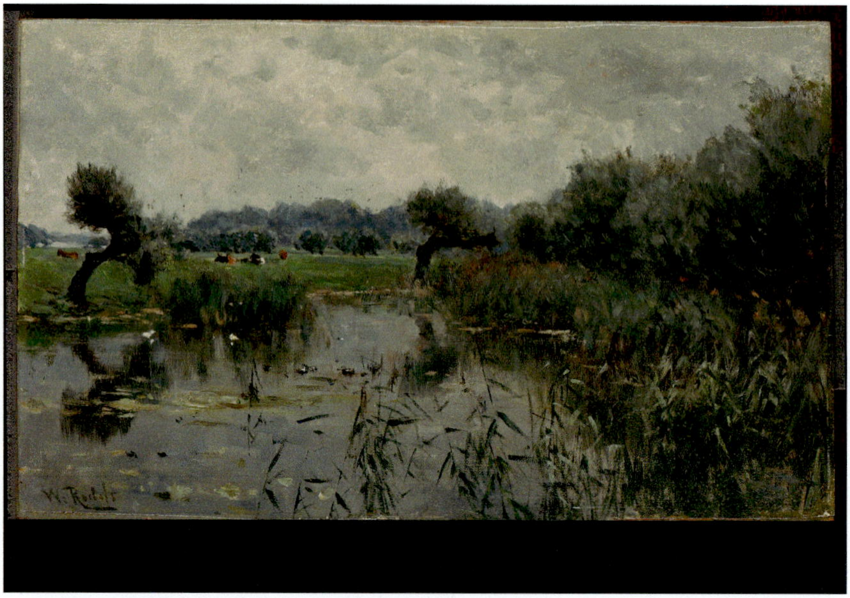

Plate 11 Willem Roelofs, *In the Floodplains of the River Ijssel*, 1870–1897, oil on canvas, Rijksmuseum, Amsterdam.

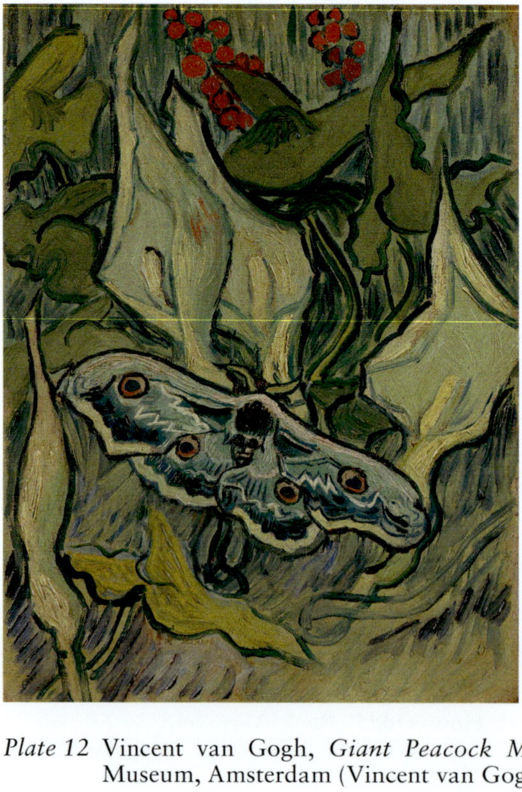

Plate 12 Vincent van Gogh, *Giant Peacock Moth*, 1889, oil on canvas, Van Gogh Museum, Amsterdam (Vincent van Gogh Foundation).

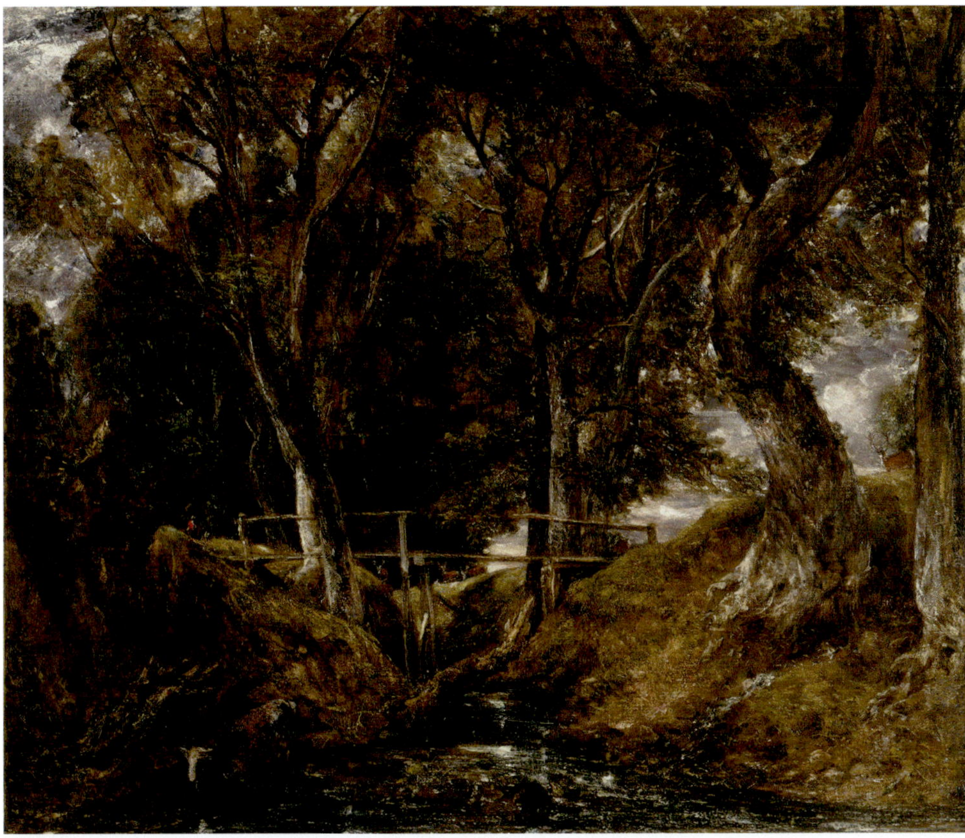

Plate 13 John Constable, *The Dell at Helmingham Park*, 1830, oil on canvas, The Nelson-Atkins Museum of Art, Kansas City, Missouri. Purchase: William Rockhill Nelson Trust. Photo courtesy of Nelson-Atkins Media Services/E.G. Schempf.

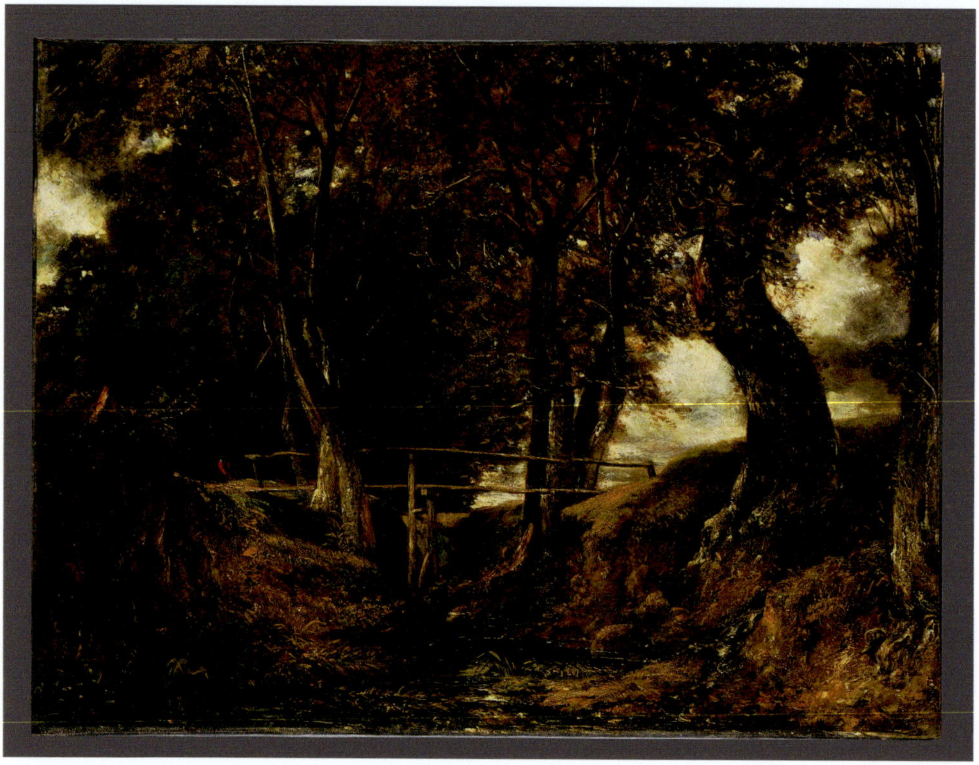

Plate 14 John Constable, *Dell at Helmingham Park*, 1825 or 1826 (retouched 1833), oil on canvas, Philadelphia Museum of Art, John G. Johnson Collection.

11 Shifting Baselines, or Reading Art through Fish

Maura Coughlin

Shorelines and seacoasts are provident ecological spaces of intense change, oppositions, peril and pleasure. Paintings, prints and other images depicting the French Atlantic coast and its waters have long been viewed in the social history of art through the lens of a modernist aesthetic that takes its cues from the social and economic concerns of tourism. Unfortunately, this anthropocentric approach to coastal visual culture in Anglophone art history has led to a repetition of the same canonical artists such as Claude Monet in iconic sites such as Etretat and Belle-Isle to represent a later nineteenth-century modern encounter with the edges of the Atlantic (e.g. Groom 2003, 35–54). However, an environmental turn in the art history of the French Atlantic shore could make room for considering lesser-known artists who were more attuned to material nuances of place, site and ecological systems and who worked at sites both outside the beach-front holiday and the established artist colony. A further ecocritical engagement with visual and material culture of the coast (and the ecology of coastal tourism) might also offer a return to the political agency that social historians of art in the 1970s found in the visual arts of nineteenth-century France. If we view the vital entanglements (rather than division) of human and nonhuman on the coast and sea by looking anew at Naturalist artists whose works were widely popular in their day, we may better make an ecological study of "human maritime communities" as they interact with "marine biological communities" (Bolster 2014, 6). As I will argue in this chapter, engagement with these past visual and material relationships provides a better understanding of our own overfished seas, plastic oceans and lost shores. Partaking in the ecocritical possibilities afforded by these images, we may retrace encounters of artists with people and environments in the hope that the stories they tell may offer imaginative new models or ethics of promoting ecological resilience.

Apart from canonical Impressionist imagery of coastline tourism, there is a great deal of lesser-known, later nineteenth-century French visual culture that visualizes intensely local perceptions of tide, geology, beach morphology and marine botany (Coughlin 2015). Foregrounding material ecology, which is not the same thing as naively claiming transparency or indexicality in naturalist imagery, broadens the politically engaged study of nineteenth-century art to include works previously relegated to local interest. This expanded field affords a rethinking of the ecological and thereby the social agency of nineteenth-century visual culture. Ecofeminist "trans-corporeality," Stacy Alaimo's useful iteration of "material interchanges across human bodies, animal bodies, and the wider material world," further opens possibilities of reading the shoreline as an ideal space for the enactment of networked relationships between tide, animals, people, plants and stone (Alaimo 2012, 476).

Thinking ecologically about the matter of the shoreline and the sea does not simply mean revealing images of damaged ecosystems or documenting decline. As political scientist Jane Bennett explains, "the image of dead or thoroughly instrumentalized matter feeds human hubris and our earth-destroying fantasies of conquest and consumption. It does so by preventing us from detecting ... a fuller range of non-human powers ... which can aid or destroy, enrich or disable, ennoble or degrade us, [but] in any case call for our attentiveness, or even 'respect'" (Bennett 2010, ix). Finding agency in material relationships, such as the flows of fish and animals, seaweed, people, sand, stones, boats and other actors that moved across and through the strand, permits us, in the words of Elspeth Probyn, "to more widely imagine what sustainability could be" (Probyn 2016, 34). The global economic history of fishing the oceans is a long-haul story about human populations using up wild resources. I am interested in looking anew at images that help us to understand these relationships. After describing the ecological and material conditions of nineteenth-century fishing practices and communities on shore and in global waters and examining a few painters who worked on the French Atlantic, this chapter will turn to summarizing the ecological acuity of a series of recent exhibitions that examined the culture and consequences of cod fishing.

Tourism and Dwelling

Over the course of the nineteenth and early twentieth centuries, train travel brought urban and international travelers to the English Channel, Normandy and, later, Brittany. Towns such as Dieppe, Etretat, Trouville, Berck-sur-mer, Le Croisic and Houlgate, all along the Atlantic shoreline, became desirable summer destinations. The shoreline of former fishing ports and villages was gradually cleared of fishing culture and the ephemeral structures of those who worked the shore. As articulated by anthropologist Tim Ingold, a "dwelling perspective" on landscape can shed light on the way that "the landscape is constituted as an enduring record of—and testimony to—the lives and works of past generations who have dwelt within it, and in so doing, have left there something of themselves" (Ingold 2000, 189).

Older forms of vernacular architecture and shoreline village life that would vanish a few decades later in the modernization of Dieppe's harbor are pictured in Camille Flers's painting *Fisherman's Cottage, Dieppe* (c. 1848) where barnyard animals live beside fishermen, nets, boats and thatched structures (Plate 6). The browns of thatch, shore, cows and muck interweave with the blues and white of houses, sky and reflections to create a fully enmeshed site of living with and on the land and shore. These details are echoed in Eugène Isabey's many sketches, prints and paintings of fishing villages and ports clinging to the side of the coast in all weather conditions, especially in the area of Dieppe. As demonstrated in the range of mid-nineteenth-century works collected in the encyclopedic volume *Les Peintres à Dieppe*, so much human life preceded both industrial ports and the vacationer's ideally empty shore. Collectively, these images attest to the way that today's historic port towns of the coast have been stripped of their fishermen's villages (Delarue et al. 2009). As many recent critiques of Nature Writing have asserted, the exaltation of a prelapsarian "pristine" state of nature in any given landscape is not very useful – for can one really point to one way-back, uninhabited moment as correct (Morton 2007, 2)? As a corollary question, how can we use the tools of ecocriticism to visualize a "dwelling perspective" of human and nonhuman maritime communities without resorting to the patronizing, nostalgic and unproductive language of primitivism?

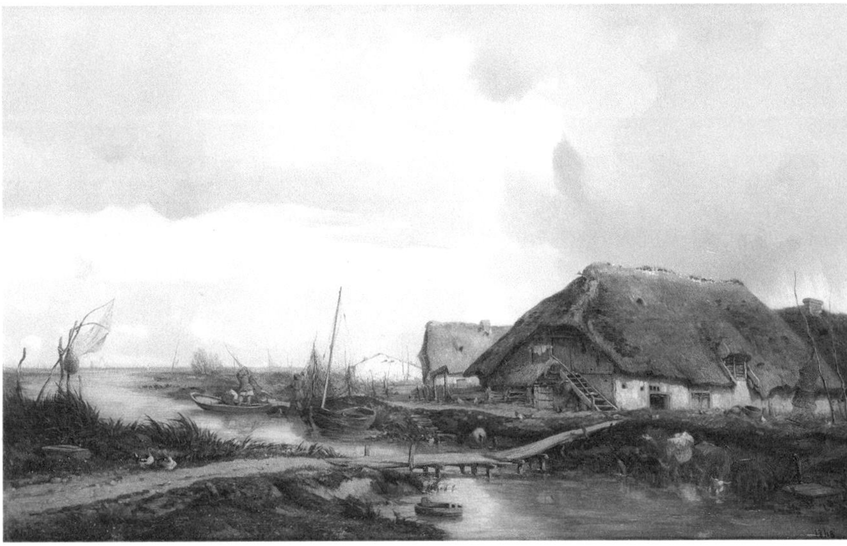

Plate 6 Camille Flers, *Fisherman's Cottage, Dieppe* (also known as *Normandy Shore*), c. 1848. Musée Thomas Henry, Cherbourg. Photo © Bridgeman Images.

As this body of imagery articulates, nineteenth-century human maritime settlements intensively worked and exploited the resources of Atlantic French beaches that would later be colonized for summer sea bathing. Peasants picked shellfish and caught shrimp at and below the low tide; in small boats, they worked lobster and crab traps; they harvested kelp, burned it on the beach and sold its ash for its iodine and soda content. They dug, sold and transported calcium-rich coralline algae (*maërl*) beach sand (in former fishing boats) to amend farmers' fields, and they grazed their cows, sheep and goats on beach plants and seaweed (Coughlin 2015). This year-round life and activity did not cease with the growth of beach tourism, but it became a secondary use in the high season when the beach and town spaces, demanded by monetized vacations, were prioritized by the local business owners who stood to profit from them. As the volume of summer tourism grew steadily over the course of the nineteenth century to people the beaches of the Channel and Normandy coasts (and later Brittany), residents found new ways to make a living in service industries: as bathing instructors, chamber maids, laundresses, artist's models, etc. (Hopkin 2007, 35). Peasant fishing and gleaning from the shore, newly viewed as anachronistic, became favorite themes of coastal painters who sold their images to visitors, sent their images to market elsewhere and, in turn, helped to promote coastal spots. Picture postcards and popular illustration from the 1890s onward show how tourists came to emulate the less strenuous forms of this work, such as shrimping, as vacation pastimes.

Dieppe and other towns in Upper Normandy were developed as seaside resorts as early as the 1820s. By the 1870s, former fishing villages had become summer destinations in Normandy and on the Channel – from Trouville (positioned south of the estuary of the Seine) north to Boulogne-sur-Mer. The Brittany coast, however, was comparatively inaccessible to Parisian visitors until the 1880s; after this, artists and other visitors continued to view it as relatively well-preserved in a "primitive" past.[1] But Brittany was not isolated or insular: modernity did not

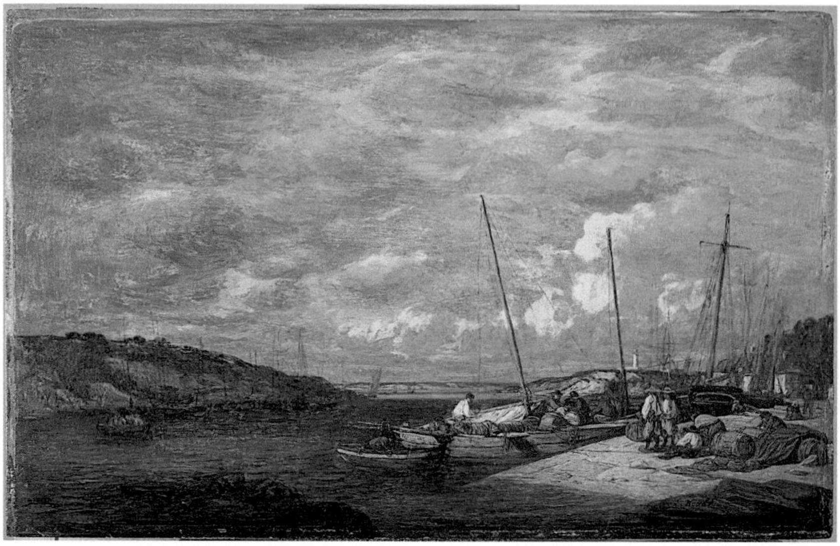

Plate 7 Eugene Louis Boudin, *Fishing Boats at the Dock, Douarnenez,* 1855. Oil on cradled panel. Private Collection. Photo © Christie's Images/Bridgeman Images.

suddenly arrive when French rail lines expanded, for its sea ports had always been in a network with other ocean ports. Throughout the nineteenth century, steam-powered packet boats followed the coastline, calling in from harbor to harbor, transporting people, food, fish and other goods. Travel along the coast also happened more informally (especially to the off-shore islands) on the boats of fishermen. Boudin, a native of Honfleur and the son of a fisherman and steam-boat captain, first traveled by packet boat along the Brittany coast, producing images of the ports (such as Douarnenez) and peasant fishing culture of Brittany as early as 1855 (Plate 7), long before the region had attracted other artists and the summer tourists who followed (Delouche 2000, 11).

Visitors to well-known destinations on the French Atlantic coast, who sought the "authentic" life of the peasantry, were often disappointed and complained of the touristic colonization of the beach or the industrialization of the fishing industry. In Brittany, they lamented the loss of local costume and religious customs that had been depicted in romantic-era travelogues and illustration (Young 2012, 87). This response closely parallels a phenomenon in fisheries management science, when an observer of a marine environment makes the naïve assumption that his or her first memories or perceptions of a place represent an "untouched" or "natural" state that precedes the depleted, reduced, corrupted or well-trod present; this is referred to as a "shifting baseline syndrome" (Pauly 1995, 430). "Shifting baselines" is an apt term to use when thinking about the ecological complexities of artistic encounters on the coast. Mourning the loss of "traditional fisheries," as historian Jeffrey Bolster describes, is a "shorthand for preindustrial activity [that] remains a mythic trope obscuring historic changes in marine ecosystems. It plays to the indefensible but commonplace assumption that the ocean has existed outside of history" (Bolster 2014, 6). Peasant residents did not share in nostalgia for an impoverished time past – at least not immediately, for, as historian David Hopkin remarks, theirs was a difficult existence requiring ingenuity and flexibility for survival (Hopkin 2007, 33). Seasonal work was variable and coastal working people profited (as they always had) by seizing opportunity and recovering quickly from disaster, even if disaster came in the form of seasonal tourism. Coastal people took the opportunities that came their way, performing heritage spectacles for tourists or even landing fish on the beach or docks (as depicted many times in Concarneau by Alfred Guillou) for the benefit of a watching crowd (Young 2012, 87). The other side of seasonal business was, of course, the long winter of shuttered homes and empty restaurants that had replaced the earlier fishing villages on the shore that now only served the summer crowds.

Fishing Communities and Ecologies

André Cariou notes that fishing was a subject mostly absent from the paintings that were made by summer visitors to Brittany until the later nineteenth century; in many accounts, tourists generally found French Atlantic fishing towns populated by only women, children and old men (Cariou 2011, 214). The relative invisibility of this human relationship with the sea in coastal artwork is explained by the seasonal patterns of life on the coast. For centuries, seasonal sardine, herring and tuna fishing had taken men and boys off-shore on short trips to pelagic fishing grounds. Furthermore as Hopkin notes, nineteenth-century viewers were uninterested in the hardship and alienated labor of fishing: the monotony, deprivation, dirt, loneliness, abuse, cold and hunger that individuals suffered at sea (Hopkin 2007, 32).

The modern scarcity of fish, that has only accelerated to the present, is a constant fact of human fishing of the oceans, known as "the tragedy of the commons." It is not the result of a rupture with a more sustainable past (or steady baseline) that was initiated by industrial modernity (Bolster 2014, 87). Before it became a year-round, defining *métier*, rather than a seasonal activity in the mid-nineteenth century, fishing had been just one aspect of peasant seasonal labors that added up to survival on the coast (Hopkin 2007, 32). The industrialization of fishing that added freezing to early conservation techniques such as drying, salting or canning (this would reach its pinnacle in twentieth-century factory trawlers) affected ocean ecosystems and the fishing populations that depended on the sea. But this was hardly a new result of modern patterns of consumption: in Atlantic France, cod stocks close to the shore were already overfished by the sixteenth century, which sent fishermen to sea for longer and farther voyages. As popular historian Mark Kurlansky has shown, codfish has a centuries-long environmental history of driving transatlantic commerce as a staple of both the European diet and the Atlantic slave trade (Kurlansky 1997, 82). A fish with little oil in its flesh, it was first air-dried, then salt-cured. By the eighteenth century, in the European North Atlantic, fisheries on the Dogger Banks had already boomed and then collapsed. With the depletion of coastal fish stocks, cod boats from ports on the French Atlantic such as Granville, St. Brieuc and St. Malo made longer journeys across the Atlantic – for half the year – to fish the cold waters off Iceland and Newfoundland. Men and boys who departed on trips of six months or more were treated as disposable labor, performing dangerous and alienating work for this global market (Cushing 1988, 60; Thoulet and Jamieson, 2005).

Although neither had been raised near the sea, both Francis Tattegrain and Charles Cottet, while living year-round among in-shore fishing villages, came to paint the lives

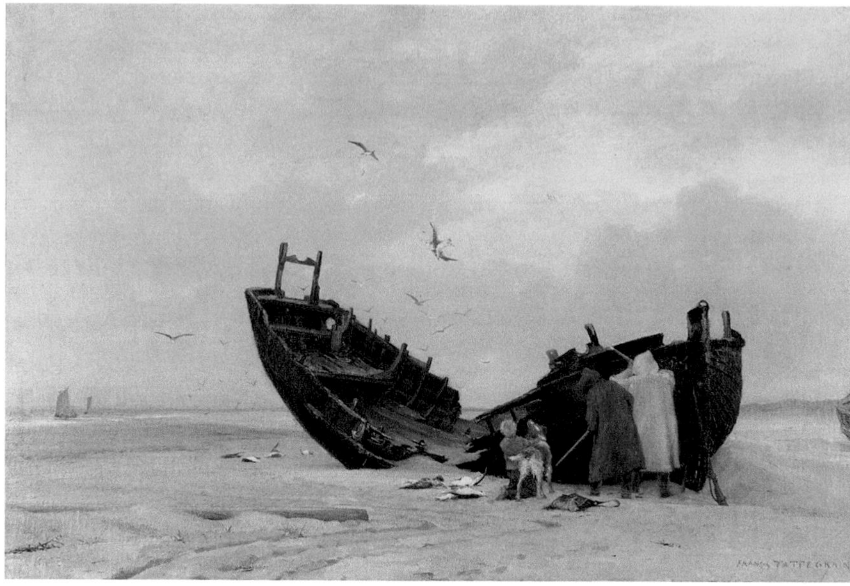

Figure 11.1 Francis Tattegrain, *Hunters on the Beach*, c. 1885, oil on canvas, Musée d'Opale-Sud.

that they intimately witnessed.[2] Tattegrain was a decade Cottet's senior; both were successful naturalist painters who regularly exhibited to acclaim in Paris in the later nineteenth and early twentieth centuries. In the 1870s, Tattegrain was one of the first artists to install himself in Berck-sur-Mer (Pas-de-Calais) where he painted for decades and eventually lamented the loss of its local community. As he observed (during his decades in residence in the later nineteenth century), seaside villas, thalassotherapy stations and tourist hotels came to replace the former fishing village (Montaigne 2007, 49).

In addition to depicting the harrowing voyages of fishermen who sold their catch on a national or even global market, Tattegrain's perceptive paintings also closely focus on the details of subsistence living (hunting, gathering and salvage) on the coast, done throughout the year (Figure 11.1). The techniques and ingenuity of coastal peasant culture, and its methods of recuperating waste (such as provident sea-borne detritus) and extracting value from the sea and shore are demonstrated in his images of hunters in the sand dunes at Berck, women digging for bait worms, and women on the strand unloading boats, wreck-picking girls, as well as wrecked, salvaged and re-used boats. The dire consequences of living by and on the sea are underscored in images of drowned bodies of the lost at sea, washed up on the beach like the flotsam of their former ships (Montaigne 2007, 109). This was intensive, hard-scrabble pre-industrial existence that could only sustain a minimal standard of living for residents of the coast. To invoke Bolster's apt phrases again, Tattegrain's coastal paintings picture an entire ecology of human maritime and marine biological communities (Bolster 2014, 6).

Cottet settled in the port of Camaret-sur-Mer (Finistère, Brittany) in 1886 and, until he fell fatally ill in 1913, he frequently traveled to the outer islands of Ouessant and Sein on fishing boats (Cariou 1988). His often dark, evocative images of the culture of mourning on the Brittany coast won him praise in the 1890s (Coughlin 2014). In Camaret-sur-mer, on the Crozon peninsula in Finistère (*finis terra*), it felt like the end of the earth. Cottet set up a long-term artistic practice and dwelling there, in sight of the sea and became a familiar part of the locale. During the Third Republic, at the same time that the artist colony at Pont Aven drew an international crowd, Douarnenez and Camaret also attracted throngs of summer artists (Douarnenez was dubbed the "Breton Barbizon"). But whereas Pont Aven (a river port) had a strong agricultural base (and the labors of the surrounding fields attracted artists), the economies of Douarnenez and Camaret were based on sardine and tuna fishing. Because the fishing fleets and the burgeoning canning industry were heavily invested and focused on these particular species, and factory workers, mostly women drawn from the impoverished countryside, were increasingly concentrated in their crowded centers, they were very vulnerable to the cholera outbreaks of 1890s and the Sardine Famine of 1903. These were the dark subjects that Cottet painted in the 1890s and early 1900s, when images of death and mourning filled his work.

Cottet, like Tattegrain, knew how fisher-people lived from and on the coast, and his perception of the land was, like those he lived among, informed by traveling on the sea. The coastline's undulating presence in many of his paintings conveys the sensuous way that land-forms seem to shift as one moves on the water; his paintings permit viewers to contemplate the everyday lives of coastal people. Maritime mobility can collapse the perception of distance: Camaret, Douarnenez and Brest, for instance, are closely linked by water, but are quite far apart for those who must use roads or rails. Traveling the coast of Finistère to better understand his subjects, I have been tempted to find or to name the features of the Brittany coast that he describes, but Cottet's landscapes are not maps. Geographer Edward S. Casey articulates the distinctive nature of his geographical sense nicely: "maps aim (at least officially) at representing the exact contours of land and sea masses and the precise distances between them—a cartographic concern—whereas landscape paintings attempt to convey the sensuous aspects of the environing place-world" (Casey 2002, xiv). A sense of environing, or comprehension of a coastal person's embodied dwelling experience, permeates Cottet's best-known work, the triptych *In the Country of the Sea* (1898, Paris, Musée d'Orsay; Plate 8). We see the day fishermen, on the left, who depart on a small boat before dawn, on the right are the women, children and older people on shore who were their connections to the land: they watch, wait and mourn. When members of this tribe gather together – as in the center panel for a meal prior to another departure – it is a sad and uncomfortable affair.

In the Country of the Sea of 1898 pictures the flow of people and materials from the coast to the sea. On the north coast of Brittany, boys shipped out very young to the fishing banks of Iceland or Newfoundland, apprenticed to a life on the sea; those from fishing towns worked as cabin boys (*mousses*), those who came from inland villages were condemned to work as *gravières,* endlessly flipping drying cod on the stony shore. They were treated as expendable maritime resources grown on land – like the hemp and linen and wood that made ropes, sails and boats. Fishing for lobster, tuna and sardines in Cottet's region of Brittany, Finistère, was done on much shorter voyages of about two weeks. Although a lobster or crab trap (woven, like a basket) is the foreground of the left panel, indicating an in-shore fishing trip not a long sea journey, regardless of the duration, working at sea was hard on the body and for each trip out, the risk of not returning was very high. As I have explored elsewhere, Cottet's work was received by critics in the context of a Symbolist-inflected obsession with mourning

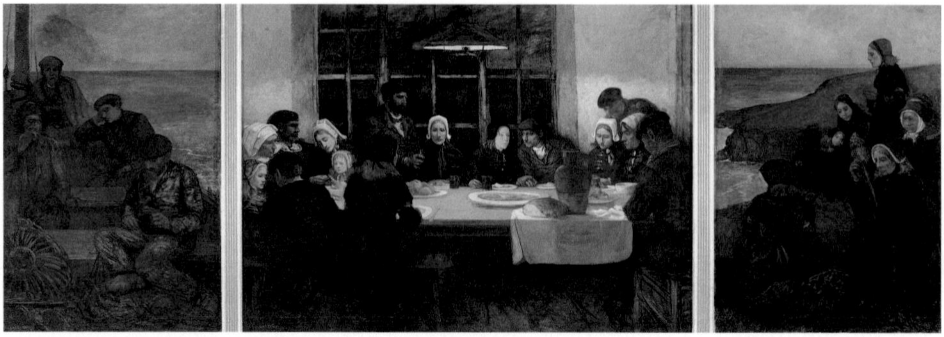

Plate 8 Charles Cottet *In the Country of the Sea (Au Pays de la Mer)*, 1898, oil on canvas, Musee d'Orsay, Paris, France Photo © Bridgeman Images.

and death (Coughlin 2014). Like Tattegrain, Cottet is one of the few artists of his generation who hints at the hardships, rather than the romance of fishing – the monotony, deprivation, dirt, loneliness, cold and hunger. This was a form of peasant realism that had not been addressed by earlier generations of artists, and critics of his day noticed this fact, likening Cottet to Millet or Courbet. Because of their long-term immersion in their chosen towns and the bonds of friendship they forged there, both artists pictured the formerly unseen lives and experiences of fisher-people. At this time, the ecological and material evidence of fishing communities were gradually being cleared from the shoreline because of the "pristine" aesthetic of leisure that tourism demands be extracted from a vacation landscape (McKenzie 2010).

Material Actors of Global Fishing

Writing long before the dramatic and dangerous life of the fisherman was romanticized and became a favorite staple of *fin de siècle* painting and popular culture, fishing was conceived of as rationalized, routine labor. This is given exemplary depiction in Henri-Louis Duhamel du Monceau and Jean-Louis de la Marre's Enlightenment-era, encyclopedia of fishing, *Traité général des pesches* (Duhamel du Monceau and Marre 1764). In this text, the technologically intensive and demanding nature of global fishing practices is represented as a network of material and ecological actors: lines were made of horse hair, flax or hemp – floats were cork, bladders kept nets afloat and lead was used for sinkers. One illustration (Figure 11.2) depicts French fishermen working the Grand Banks, east of Cape Breton and Newfoundland. The line fishers stand, protected from the weather and sea, in identical wooden barrels that have been lashed to the ship's deck. They efficiently remove the entrails for bait and send the fillets to be salted to the next group of workers. Dried and salted cod fish was processed on boats and in fishing bases in Newfoundland, especially on the remaining French territory in Atlantic Canada: the "French shore" (until 1904) and the archipelago of Saint-Pierre-and-Miquelon, the sole colonial territory that France has retained in Atlantic Canada.

In Duhamel's text, fishing boats (and the sites where fish were dried and salted on the shore) are consistently described as sites of resource extraction, populated by interchangeable workers. Up to the present, Duhamel's images are frequently marshaled

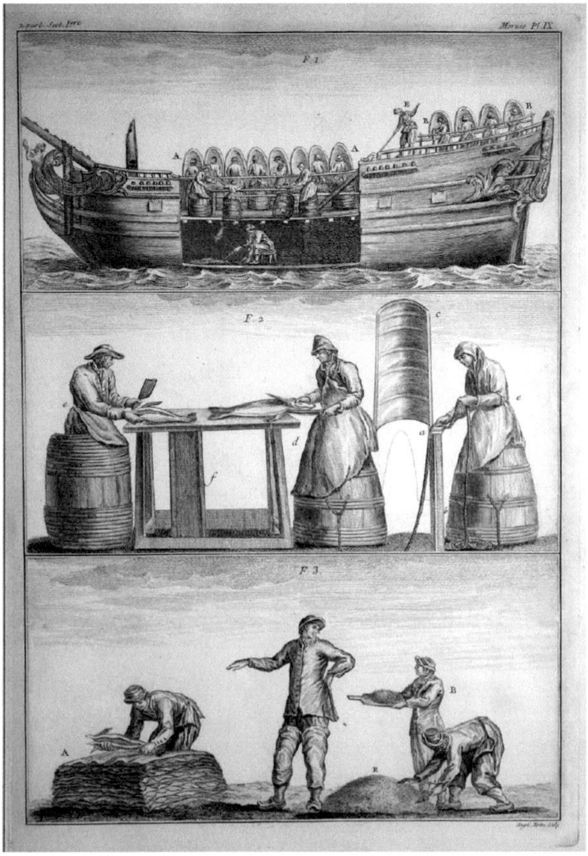

Figure 11.2 *Cod Fishing*, plate from Duhamel du Monceau, *Traité Général des Pesches*. Part 2, Section 1, Plate 9. 1769–1782.

to illustrate past fishing practices in historical texts (Cushing 1988) and museum displays. Several recent French exhibitions have notably used these images to initiate new dialogues about co-existence, resilience and biodiversity. Almost every large town in Atlantic France has a local museum or ecomuseum displaying the material things of sea culture: model or actual boats, nets, rigging, clogs, costumes, family photographs and baskets have been saved to speak of past sea-faring relationships. Many coastal museums encourage their visitors to consider how fishing is a materially intensive form of food hunting. In the summer of 2012, the Port Museum in Douarnenez presented an exhibition, *Fibres Marines* (Marine Fibers), as part of a larger regional theme *Chanvre et Lin* (Hemp and Linen) that was shared that season across multiple sites and that recounted the history and importance of natural fibers in Brittany's past and in global maritime history. For centuries, these crops were grown, processed and finished in Brittany and were the materials from which nets, lines, ropes and sails were made. Viewers were encouraged to look differently at these materials as actors that enabled relationships of the human and nonhuman worlds. Viewers had to think about the way that these plants made sea travel and shipping possible. Seen through Jane Bennett's emphasis on "a vibrant materiality that runs alongside and inside humans to see how analyses of political events might change if we gave the force of things more due" (Bennett 2010, viii) when these materials were put at the center of the exhibition, and representations (photographs, objects, films) served as supplementary, contextual evidence, the entire terms of the museum-visit shifted. The humble matter itself (from hemp plants under grow lights to bales of fragrant raw fibers) assumed an agency (that arguably it had all along) in the telling of local history intersecting with the global. Demonstrating their use for ropes and ship rigging up into the World Wars of the twentieth century, the show concluded pointedly: there is too much plastic in the ocean – and much of this comes from degraded or discarded fishing gear and lightweight boats. Linen and hemp, finally, were offered as much more sustainable alternatives to plastics. In concert with other fibers and composites, they can potentially replace petrochemical resins in promising ways, to make such things as boots or biodegradable boats.

In a similar focus on materiality, visual and material artifacts of French maritime culture were brought together in an ambitious, cooperative exhibition series *Terre Neuve/Terre Neuvas* that was on view from 2013 to 2014 at four museums in Brittany and Normandy (first Rennes and St. Brieuc, then Saint-Malo and Granville) (Chanas et al. 2013). These were curatorially innovative assemblages of visual and material culture that told the centuries-long story of the French fleet that traveled across the Atlantic to fish for cod. Although organized and exhibited at the Champs Libres/Musée de Bretagne cultural center in Rennes, the three other sites were chosen because they were significant ports for centuries in the transatlantic cod trade. This series of exhibitions were significant collaborations between art and history museums (although the catalog does not feature or reproduce many of the art works exhibited). The fifth and final presentation of this show, retitled *Dans les Mailles du Fillet* (*Caught in the Net*), was on view at the Musée de la Marine in Paris from 2015 to 2016. The description that follows comes from my visits to Rennes and St. Brieuc.

As one entered the show at the cavernous, postmodern Champs Libres building in Rennes, hundreds of illustrations from Duhamel's text had been transformed into a remarkable short cutout-animation film by the animation company, Virement Lundi from Rennes. This digital video was accompanied by a very large and present taxidermied codfish and piles of dried cod, amateur paintings from nineteenth-century Newfoundland, cod jigs, fishing gear and maps. The Rennes show primarily displayed objects (including dories, clothing, advertisements and packages of dried cod), photography and video to trace a chronology of cod fishing in the economic life of Atlantic France up through the twentieth century, and then into current times of overfishing. Concurrent to this, at the much smaller St. Brieuc Art and History museum, a broad range of visual and material popular culture included Salon paintings, silent film, letters sent between fishermen and their families, diaries, maps, portraits, clothing, fishing gear, ex-voto paintings, cemetery plaques and votive boats. As in Rennes, historical media (including theatrical silent films) ran on many walls, surrounded by material witnesses. In St. Brieuc, two early anonymous ex-voto paintings spoke to the visual culture of fisherfolk's religious practices of prayers, pardons and thanks for salvation from shipwrecks.

Votive boats were paraded through the streets of a port town during sailors' and fishermen's pardons and blessings of the fleet and are often represented in paintings and popular imagery of Breton festive religious culture. Although some of these ritual objects are housed in regional museums or public historical collections, most are still found in coastal chapels devoted to Mary, Virgin of the sea, and Saint Anne, who were both thought to be protectors of fishermen. Sailors typically made pilgrimages to these sites for religious pardons before long voyages, or marched in penance and thanks after a safe return because of "a vow made during a storm at sea, or other occasions of danger, to make a pilgrimage to one of her shrines if they are saved. Very often the vow (*voeu*) included going barefoot and bare-headed, dressed only in shirt and trousers, and fasting on bread and water" (Anson 1965, 69). Pilgrimages such as this could be taken on singly, but maritime pardons were large-scale events that increasingly came to be seen as spectacles of heritage in the early twentieth century, rather than expressions of piety and thanksgiving for safe returns. Ex-voto paintings that came from this tradition of faith and salvation were on display: one from Notre Dame de Grace in Honfleur (Calvados, Normandy), and another from the Chapel of Notre Dame de la Cour, Lantic (North Coast, Brittany), a Gothic chapel that was the site of pilgrimage for sailors and fishermen of the region. An enormous painting by Rouen painter Albert Démarest, *The Vow* (1894, Nantes), depicting a fishermen's religious pardon also accompanied the wooden votive boats that were carried in these processions.

Many of the academic naturalist painters included in these shows, such as Emma Herland, Henri Royer, Henri Dabadie, Démarest and Henri Rudaux depicted the separation, loss and mourning of fishing families. Dabadie's massive *Departure of the Iceland Fishermen*, 1900 (Nantes Musée des Beaux Arts), a Salon crowd-pleasing concoction of Impressionism, academic painting and *japonisme*, was framed by objects such as the sturdy leather and wood boots worn by Breton fishermen on these long journeys, sea chests and wool long underwear that had been repeatedly mended, one would have to guess, during the many boring hours spent in transit. A review in *Le Télégramme* of 26 October 2013 singled out the latter in St. Brieuc: "un caleçon de laine rouge reprisé comme une vieille chaussette..." (a worn-out set of red long underwear, mended like an ancient sock) as the most moving object of the show.

Emma Herland's painting, *Gaud Mevel* (1887 Musée de Laval), depicting the fictional character of Gaud, from Pierre Loti's wildly successful novel *Iceland Fishermen* (1886), was accompanied (in the St-Brieuc exhibition) by cemetery plaques that were similar to those in the painting. In Loti's tale, Gaud goes daily to the village cemetery and the nearby chapel, expecting bad news of her husband's death at sea, obsessively reading the family names on similar plaques commemorating the lost at sea that repeat for several generations. In her daily, repetitive visits to this site of memory and mourning, the premonition seizes her that a new plaque will soon be added for her missing husband. As Aurélie Maguet, Assistant Director of the St. Brieuc museum, explained in a public lecture during the run of the show, melodramatic images such as this or Dabadie's *Departure of the Iceland Fishermen* may have been dramatically epic, but they were not necessarily untruthful. From many ports, boats departed and women looked sorrowfully to sea; and yet these images were significantly formed and burned into the popular imagination by popular imagery in the illustrated news and novels such as Loti's. Tourists' spectacles in coastal towns took advantage of this fictionalized celebrity: in Paimpol, for instance, a winter pardon dedicated to cod fishermen departing for the Iceland sea (that had begun in 1855) became a secular summer heritage festival in 1904 in response to the popularity of Loti's novel. To counter the persistent force of this mythic representation, Maguet explained, the curators installed a section containing letters, diaries, affective objects and images of religious devotional culture that served to give a voice to the fishermen and their families who are so often represented but silent.

Conclusion

I have argued that an ecological reading of coastal paintings and maritime material culture can articulate relationships between the human and more-than-human worlds of the nineteenth-century shore and sea. Objects that enabled sea travel, the catching of fish, the separation of families and the expressions of grief and faith have contemporary stories to tell. These affective images and things, when presented in dialogue, exceed the status of mere contextual filler to an exhibition. They give voices to the peasant class of the sea and shore who rarely represented themselves to the outside world, enable visitors to picture the human and environmental cost of eating from the ocean, and may provide models of future engagement with ocean materiality and ecological sustainability. When nineteenth-century visual culture is presented to the public in an ecological context that enables the viewer to understand the vast project of taking fish from the global oceans, rather than focusing on artistic personalities, art criticism, art markets or modernist canons, important new networks of understanding and empathy can be cultivated. The shifting baselines of nostalgia for times gone by are inadequate in our age of biological scarcity, pollution of our oceans and the commodification and gentrification of shorelines. Visualizing past ecological relationships enables us to better negotiate the present and reimagine the future of living by and with the sea.

Notes

1 In Brittany, train lines were extended to the central city of Rennes in 1857, then to Guingamp and Brest in 1865. The Paris to Nantes line was extended to Quimper in 1863. Train travel brought tourists *en masse* to Brittany, but it also enabled the shipment of fish, produce and meat to Paris and other markets. Train lines also enabled and accelerated the rural exodus of Breton peasants (Coughlin 2013, 67–69).

2 Conversely, the *intimiste* painter Henri Le Sidaner, son of fisher-people from Isle of Bréhat and St. Malo who lived and worked along the Atlantic coast rarely painted references to fishing apart from the representation of docked boats.

Works Cited

Alaimo, Stacy. 2012. "States of Suspension: Trans-corporeality at Sea." *ISLE Interdisciplinary Studies in Literature and Environment*. 19, 3: 476–493.

Anson, Peter. 1965. *Fisher Folk-lore; Old Customs, Taboos and Superstitions among Fisher Folk, Especially in Brittany and Normandy and on the East Coast of Scotland*. London: Faith Press.

Bennett, Jane. 2010. *Vibrant Matter: A Political Ecology of Things*. Durham, NC: Duke University Press.

Bolster, W. Jeffrey. 2014. *The Mortal Sea: Fishing the Atlantic in the Age of Sail*. Cambridge, MA: Harvard University Press.

Cariou, André. 1988. *Charles Cottet et la Bretagne*. Baillé: Ursa.

———. 2011. *De Turner à Monet: la découverte de la Bretagne par les peintres paysagistes*. Quimper: Musée des beaux-arts de Quimper.

Casey, Edward S. 2002. *Representing Place: Landscape Painting and Maps*. Minneapolis: University of Minnesota Press.

Chanas, Céline, Barré, Éric, and Jean-Philippe Roze. 2013. *Terre-Neuve, Terre-Neuvas*. Musée de Bretagne, Musée d'histoire (Saint-Brieuc, France), and Musée du Vieux Granville. Tourgéville: Illustria.

Coughlin, Maura. 2013. "Place Myths of the Breton Landscape." In Simon Kelly and April M. Watson, eds., *Impressionist France Visions of Nation from Le Gray to Monet*. New Haven, CT: Yale University Press: 67–79.

———. 2014. "Death at Sea: Symbolism and Charles's Cottet's Subjective Realism." In Marja Härmänmaa and Christopher Nissen, eds., *Decadence, Degeneration and the End: Studies in the European Fin de Siècle*, New York: Palgrave: 203–223.

———. 2015. "Biotopes and Ecotones: Slippery Images on the Edge of the French Atlantic." *Landscapes: The Journal of the International Centre for Landscape and Language*. 7, 1, https://ro.ecu.edu.au/landscapes/.

Cushing, David H. 1988. *The Provident Sea*. Cambridge: Cambridge University Press.

Delarue, Bruno et al. 2009. *Les Peintres À Dieppe Et Ses Environs: Varengeville, Pourville Et Arques-La-Bataille*. Yport: Delarue.

Delouche, Denise, and Eugène Boudin. 2000. *Eugène Boudin et la Bretagne: une aventure picturale à travers le thème breton*. Quimper: Éditions Palantines.

Duhamel du Monceau, Henri-Louis and Dè la Marre. 1764. *Traité général des Peches & histoire des Poissons qu'elles fournissent, tand pour la subsistence des hommes que pour plusieurs autres usages qui ont rapport aux arts & aux commerce*. Paris: Sallant e Nyon.

Groom, Gloria. 2003. "The Sea as Metaphor in Nineteenth-Century France." In Juliet Wilson-Bareau, David C. Degener, and Lloyd DeWitt, eds., *Manet and the Sea*, Philadelphia, PA: Philadelphia Museum of Art: 35–53.

Hopkin, David. 2007. "Fishermen, Tourists and Artists in the Nineteenth Century: The View from the Beach." In John House, ed., *Impressionists by the Sea*, London: Royal Academy of Arts: 31–37.

Ingold, Timothy. 2000. *The Perception of the Environment: Essays on Livelihood, Dwelling and Skill*. London: Routledge.

Kurlansky, Mark. 1997. *Cod: A Biography of the Fish That Changed the World*. New York: Penguin.

McKenzie, Matthew G. 2010. *Clearing the Coastline: The Nineteenth-century Ecological & Cultural Transformation of Cape Cod*. Hanover: University Press of New England.

Montaigne, Claire. 2007. *Francis Tattegrain, 1852–1915*. Berck-sur-Mer: Musée de Berck sur Mer.

Morton, Timothy. 2007. *Ecology without Nature*. Cambridge, MA: Harvard University Press.

Pauly, Daniel. 1995. "Anecdotes and the Shifting Baseline Syndrome of Fisheries." *Trends in Ecology & Evolution*. 10, 10: 430.

Probyn, Elspeth. 2016. *Eating the Ocean*. Durham, NC: Duke University Press.

Thoulet, Julien, and Scott Jamieson. 2005. *A Voyage to Newfoundland*. Montréal: McGill-Queen's University Press.

Young, Patrick, 2012. *Enacting Brittany: Tourism and Culture in Provincial France, 1871–1939*. Aldershot: Ashgate Publishing Group.

12 "A Better Acquaintanceship with Our Fellows of the Wild"

George Shiras and the Limits of Trap Camera Photography

Jessica Landau

In the Foreword to the second edition of George Shiras's formative 1896 guidebook *Hunting Wild Life with Camera and Flashlight*, Edward W. Nelson, the former chief of the U.S. Biological Survey, writes: "The great auk, passenger pigeon, and Carolina paroquet, as well as other birds, have gone forever, while the bison, prong-horned antelope, wapiti, and grizzly bear in the United States have been reduced from their former vast numbers to the danger point. These signs point to the urgent need of making field and photographic studies of our wild life before it becomes too late" (Shiras 1936, xi). A senator, amateur naturalist and good friend of Teddy Roosevelt, Shiras believed in the power of the photograph, if not to prevent extinction, then to conserve threatened animal populations and slow the process of decimation. As cultural historian Finis Dunaway points out, for early wildlife photographers, like Shiras, the camera had the ability to stop time, functioning, at the very least, as a technology of memory to record and preserve threatened landscapes and animals (Dunaway 2000). These wildlife and landscape images functioned much like Carleton Watkins's landscapes or even Edward Curtis's photographs of American Indians did, as an attempt to preserve what white settler audiences understood as innately vanishing places, people and animals (Egan 2006; Hutchinson 2009). In addition to photography, Shiras also took decisive action to preserve wild lands and animals, including promoting early land preservation legislation and wildlife management.

After witnessing extreme declines in white tailed deer populations in the Northeast and Midwest of the United States, beginning in the mid-nineteenth century,[1] Shiras, along with other nature photographers, saw photography, specifically camera hunting, as a conservation-minded replacement to gun hunting, and, as the quotation in the title suggests, a means toward a "better acquaintanceship" with nonhuman animals. After several years of experimenting with photographic techniques drawn from hunting practices, Shiras developed an automatic photography technology.[2] The predecessor of today's trap cameras which use sensors and infrared technology, Shiras's method utilized trip wires to set off the camera shutter when touched by an animal. In his text, Shiras refers to this technique both as automatic photography and as camera or flashlight trapping, but to further emphasize its linkage, both discursively and physically to hunting, I primarily use the terms trap camera or camera traps. His first trap image, from 1891, was published in his book with the caption, "This was the first wild animal to take its own picture" (Shiras 1936, 133) (Figure 12.1). The image, produced when the doe tripped a wire attached to a camera mounted on a stake or a tree, played a part in Shiras's attempts to reveal specifics about deer life by seemingly removing the human agency in the production of the photograph. As this essay

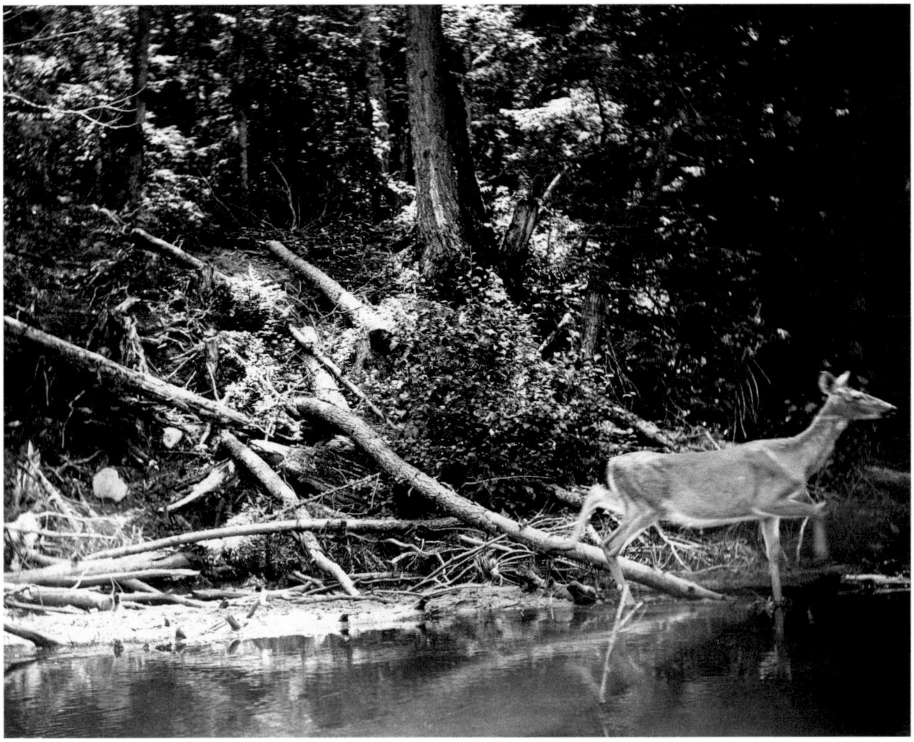

Figure 12.1 "This was the first wild animal to take its own picture." c. 1890, from *Hunting Wild Life with Camera and Flashlight,* 1936, George Shiras, NG Image Collection.

argues, despite attempts to understand the interconnections between humans and deer, efforts by conservationists like Shiras ironically served to destroy the ecosystems of which deer were a part, through equally damaging acts such as the mass-slaughter of wolves. This dialectic of conservation and devastation reveals the limits of Shiras's emerging ecology – one which emphasizes the human linkage to some species while denying such affinity to others.

In this essay, "the first wild animal to take its own picture" is compared to a Shiras photograph of an animal in a more traditional trap: an image of a large gray wolf ensnared, defeated and near-collapsed, with one leg caught in a hunting trap (Figure 12.2). In a text otherwise devoted to advocating wildlife photography, which presumes the liveliness of the animals captured on film, this image is the only one depicting a live wolf in his book. The other representations of wolves are either of hanging carcasses or of a taxidermied animal, set up and photographed out of doors, yet tellingly stiff. Together, the doe and the wolf – understood as participants in a natural ecosystem as well as in an ecology of photography – break down the naivety of Shiras's ostensible conservation aesthetic and demonstrate the limits of trap camera photography.

Trap camera photography masquerades as an accurate depiction of wildlife, recording how animals behave when humans are not around, but this is partially a fantasy,

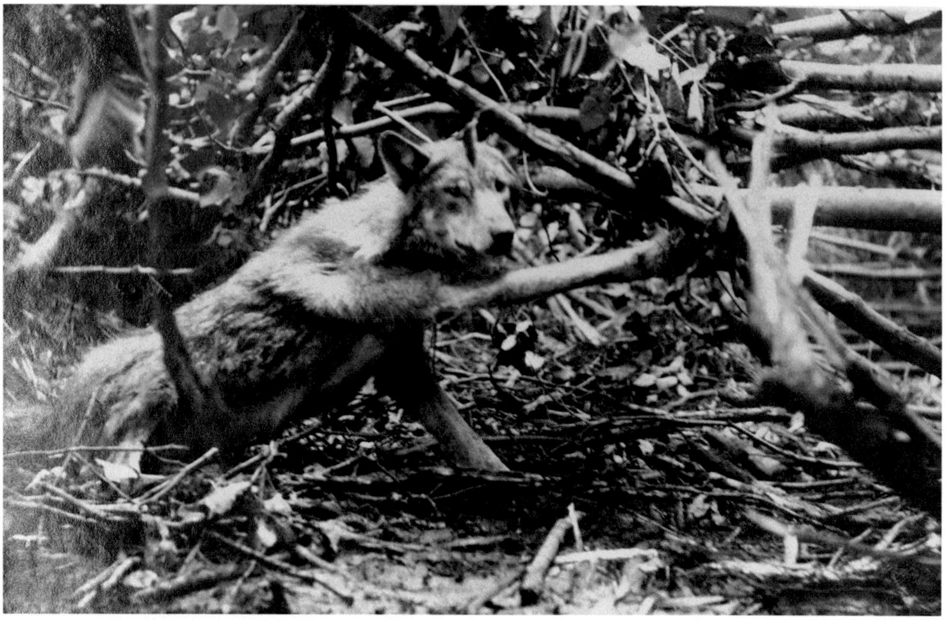

Figure 12.2 "An 85 pound timber wolf was caught on a deer runway," 1907, from *Hunting Wild Life with Camera and Flashlight,* 1936, George Shiras, NG Image Collection.

of course. As in many photographic images, what is *not* pictured here is as important as what *is*. This is what photo-historian Shawn Michelle Smith interprets as 'the optical unconscious,' borrowing the term from Walter Benjamin (Smith 2013, 6). Put simply, the optical unconscious is what cannot be seen with the human eye, but what is revealed by the camera. This divulges a relationship to the nonhuman world which is multi-valent and crosses species lines, begging questions of how we view animals, not only photographically, but in the wild as well. Understanding Shiras's trap camera images as an attempt at "better acquaintanceship," or even something Donna Haraway would later call ethical relating, his photograph of the wolf troubles an otherwise ecologic way of seeing (Haraway 2003, 50). In an attempt to reckon with this challenge, I read Shiras's photographs not only through the lens suggested by Smith, but also with the help of photo-theorist Ariella Azoulay, extending her development of photography's 'civil contract' to include other species in order to create an ecology of photography aimed at changing the way we look at nonhuman others (Azoulay 2008).

Like many elite men of the second half of the nineteenth century, Shiras fell in love with the outdoors as a child during his family's Northern Michigan vacations from their Pennsylvania home. Eventually serving as a lawyer and U.S. representative, Shiras was an active participant in Roosevelt's masculinist vision of the Strenuous Life – the idea that the United States will build a stronger nation if the country can raise a generation of men educated in patriotic duty through rugged experiences in sport and outdoor recreation (Roosevelt 1899). A one-time avid hunter, Shiras ultimately saw photography as the superior hunting technology, preserving animal life and the thrill of the chase simultaneously. Camera hunting became a popular pastime by the nineteenth century's end, allowing sportsmen to fulfill all aspects of a hunt, but instead of killing an animal, they were in pursuit of the perfect "snapshot" of it, a phrase borrowed directly from hunting terminology (Coe and Gates 1977, 6).

Shiras's early experiments with wildlife photography involved constructing blinds, in which he could hide at a distance from a passing animal and pull a wire when one approached. While he captured images of unassuming deer this way, they were distant and small, and often out of focus. Shiras then mimicked the hunting practice of jacklighting, which had been outlawed in most states due to its cruelty. Jacklighting was the practice of paddling along a lake shore at night, silently searching for animal eyes along the shore. When spotted, an assistant to the hunter would shine a spotlight or bright flash at the animal – which would momentarily stun it and allow the hunter to shoot it at point-blank range.[3] Shiras copied this practice photographically in the Northern Michigan lake country. While he certainly captured stunning images of animals, illuminated in high contrast against the blackness of a lake, he also ran into many technical troubles, including setting his boat on fire.

This drove him to experiment further – searching for what he understood as a purer way to picture animal life. Working from the assumption that nonhuman animals in the wilderness were separate from humanity and civilization, Shiras imagined a way to photograph animals that would seemingly remove human intervention from the creative act. Possibly inspired by Eadweard Muybridge's contemporary work with animal locomotion, Shiras invented a mechanism connected to a trip wire that would set off a flash and then camera shutter when tripped. Thus, he began setting up camera traps alongside riverbeds, lakeshores and other game trails throughout the forests, cleverly anticipating deer movement. While he often captioned these images as if the deer had been imaged entirely free from human mediation, he just as frequently baited the traps with salted vegetables to draw the deer in front of his lens. The resulting images are largely concerned with preserving the fantasy of untouched wildlife; the animals, for the most part, seem unaware of the camera, with the viewer positioned almost more as voyeur than hunter.[4]

In this way, "the first wild animal…" acknowledges the photographer's skills to track and picture an animal, while also seemingly removing him from the equation linking animal to image. The picture is blurry, almost shaky and lacks the careful composition of a studied photographer's hand. The doe has tripped the wire just as she begins to exit the frame and the fallen logs behind her are in much better focus than she is. Without the caption, the viewer might assume this to be a poorly timed, clumsy amateur photograph. But since the image was published in his text, *Hunting Wild Life…*, it is important to consider the caption as well, and in it, Shiras all but attributes the image to the doe herself. In response to the threat of decreasing deer populations, this doe is not just saved from the hunter's rifle, but she marks her physical presence, as an agent, through making a visual record. Thus, the photograph begins to illustrate Shiras's emerging ecological view, attributing a preservative attempt that originates from the deer themselves.

In her study of photography, *At the Edge of Sight: Photography and the Unseen*, Shawn Michelle Smith argues that photography has a special ability to reveal the unseen, operating at the limits of our cultural vision. Smith warns, however, that while photography has the ability to expand our sight, it is also has the power to limit it. She writes that "as it extended the realm of the visible, photography also suggested that some things would remain forever out of sight" (Smith 2013, 8). While Smith's analysis is centered around images that she interprets as representations of race in America, including the violent 2003 images of the tortures at Abu Ghraib, her arguments extend well to early trap camera photography, a method which works hard to photograph the unseen, and in doing so, reveals its own limits.

These limits are also unveiled through an ethical viewing of the images. In *The Civil Contract of Photography*, photo-theorist Ariella Azoulay identifies the indexical qualities of the medium that create a connection between the photographed subject, the photographer and the viewer. She writes: "Every photograph of others bears the traces of the meeting between the photographed persons and the photographer, neither of whom can, on their own, determine how this meeting will be inscribed in the resulting image" (Azoulay 2008, 11). This idea of the photograph as meeting means that images can be read as events with multiple participants. For Azoulay, photographs – and particularly photographs of suffering – are contracts that require the responsibility of the viewer, upon whom the viewed can make demands, despite distances of space or time. One might, by analogy, include distances between species.

In order to bridge this distance between species, Azoulay's ethics of photography can intersect with Donna Haraway's ethics of relating to the nonhuman world (Haraway 1984–1985, 2003). By understanding animals as co-constituting *naturecultures* alongside humans, Haraway understands nonhumans as beings to think *with* rather than *through*. In terms of animal photography, I contend that this co-constitution needs to be placed in dialogue with photographic events in which the viewer, photographer and subject are all simultaneous actors. In a sense, what I am developing is not so much a 'civil contract' of photography, but rather an ecology. This ecology both reveals and implicates the viewer in the violent history of hunting as well as in the imagery that was both corrective and corollary to its practices.

"The first wild animal..." photo is, at first glance, almost certainly not an image of violence. Unlike early camera hunting images in which trophies are displayed as dead animals alongside the hunter, or cornered animals about to be shot, the knowledge that this is a tripwire-triggered image removes any threat of gun violence to the animal body. Yet the violence and bloodiness of the hunt remain lurking in what Smith calls the optical unconscious. Because of the camera placement, the viewer is positioned as a hunter, with the wide flank of the deer revealed to us, almost as if through a rifle sight. One could imagine this as the moment just before a fatal shot is fired. Yet the discourses regarding camera hunting often sought, unsuccessfully, to erase the violence inherent in the production of the image.[5]

Despite the confusion in terms linking 'trap' photography to its history as a means of animal slaughter, the method can be read as an attempt to mask this history of animal violence, particularly by its avowed removal of the human presence. By claiming that the deer took her own picture, Shiras, in his proto-ecological vision, can be seen as suppressing the anxiety about the human impact on the environment, specifically here, concerning the dwindling deer population. The poorly framed image is not just evidence of the lack of human consciousness governing the composition; since the deer's nose just touches the left edge of the frame, it is proof that she will leave unharmed. The doe in "the first wild animal..." quells human fears simply by getting away. But not all animals were so lucky.

In the heavily illustrated second edition of *Hunting Wild Life with Camera and Flashlight*, only four photographs depict wolves and only one shows a live wolf, caught in a trap. The caption for this image reads: "This cruel marauder was captured on July 29, 1907, near the author's camp in Northern Michigan. Immediately after being photographed the creature collapsed from fear and overexertion and lay in a dying condition as a bullet ended its life" (Shiras 1936, 264). In response to the decreasing deer populations, early twentieth-century conservation policy agreed that one of the best solutions to save deer was to eradicate their greatest natural predator: the wolf. With support from ranching and farming communities, and increasing bounties on their pelts, wolf slaughter became good business, continuing well into the late twentieth century, and some would argue until today. While wolves and their advocates are still battling for protection, intermittently added to and removed from the endangered species list, deer populations have, as we know, rebounded tremendously. Without understanding the importance of an apex predator to an ecosystem's health that such demographic dynamics ably demonstrate, Shiras readily advocated the killing of wolves.

For this unfortunate wolf, being photographed was a step in the extermination process. Shiras provides details about the capture of this animal, who, after being trapped overnight, dragged herself more than 20 yards as she struggled to free herself. Upon finding the wolf, Shiras writes that "at the first snap of the camera, the animal collapsed and refused to move" (Shiras 1936, 262). Here, the camera is analogous to the gun, delivering the fatal shot as quickly as his rifle. But, Shiras continues: "I certainly had never expected to feel sympathy for the plight of such a marauder, but this animal's bloodshot eyes, protruding tongue, and entire lack of resistance would have appealed to its most relentless enemy, and I hastened its end with a shot" (Shiras 1936, 262). Unlike the doe, the hunter is evident in this image, not only in Shiras's narrative, but also in the direct gaze of the wolf, who, through the twisted underbrush and her taut, trapped forelegs, makes direct eye contact with the viewer. We are, in Shiras's narrative, in the place of the hunter who delivers the fatal shot. By directly acknowledging us, the wolf, despite her exhaustion, will not let viewers lose sight of the presence of her killer.

This photograph is, in some respects, the optical unconscious of "the first wild animal to take its own picture." The image of the wolf, about to be killed, is the repressed aspect, the inverse, of the humanless forest ecosystem Shiras attempted to picture. It enacts a specific violence that Shiras otherwise resisted picturing. Men like Shiras believed in the necessity of wolf culling to maintain game populations precisely to preserve their own way of life. In fact, wolf eradication was an integral part of the process of nation building and extermination of indigenous human populations in the United States.[6] Beginning in the seventeenth century, wolves, which were often compared to Native Americans, were seen as a threat to civilization and prosperity.[7] As writer Barry Lopez has noted, troublesome equivalences characterized attitudes toward the nineteenth-century Indian Wars and wolf eradication programs.[8] The detailed history of this war on people and wolves alike is hard to stomach, especially for twenty-first-century ears, unused to such direct canine violence. Wolf slaughter was undiscriminating, killing any and all wolves, throughout most of the nineteenth and twentieth centuries. Wolfers' favored weapon of choice was strychnine poisoning, usually left inside a large carcass of wolves' typical prey. These poisoned carcasses did not attract only wolves, and thousands of domestic dogs, foxes, coyotes and other scavengers died.[9]

Shiras's chosen method of a steel wolf trap was perhaps more responsible, if no more humane. Most traps by the turn of the century were Newhouse #14s, newly remodeled with sharp teeth, more difficult for the wolf to tear loose (Lopez 1978, 190). The trap was usually attached to a six-foot steel chain with a drag hook – making this particular wolf's 20-yard journey once trapped even more of a herculean effort. While men like Shiras were interested in camera hunting because of the way it maintained the challenges of sport hunting, trapping wolves was not considered a sporting business. Traps were buried in a shallow hole, and concealed with earth, and then often baited with rotting meat. Wolves, like the one pictured, would often suffer in traps for hours, if not days, to be shot at close range or die of exposure. The death of the animal in Shiras's photograph was unsporting and easy, yet she stared the hunter in the face, forcing him to confront her, in the perpetuity of photographic afterlife. In short, this is a violent and disturbing image. But, as Shiras himself noted, the process of photographing the wolf creates an otherwise unattainable sympathy and closeness with the creature.

When a photograph is understood as an event operating within Haraway's ethical way of relating with nonhuman others, an image can function in terms of an ecology, an interconnected web of photographic occurrences, subjects, photographers and viewers. Instead of a contract between human citizens of the image, the photograph expands to embrace multiple actors, across time and space. Here, both of Shiras's images are set in relation to one other, as well as to the photographer, the viewer and the animals pictured. Shiras's 'trapped' deer cannot take her own picture without the concurrent violent implementation of wolf eradication programs. As in an ecology, all actors are interconnected, and while some may possess more agency than others, each one collaborates in the production of meaning of the image.

As part of this ecological view, the wolf, in the photographic event just preceding her death, connects directly with the photographer and the future viewers of the image. Of this participation, Azoulay writes: "the photographed subjects of numerous photographs participate actively in the photographic act and view both this act and the photographer facing them as a framework that offers an alternative – weak though it may be – to the institutional structures that have abandoned and injured them..." (Azoulay 2008, 18). Though the photographer tries to suppress it, Shiras's image cannot deny the interconnectedness and therefore ecological impact shaping the existence of both humans and wolves. And this is perhaps why Shiras had so few images of wolves, particularly live ones, in his book – because he feared direct confrontation and the responsibility it placed on him for their wanton slaughter. Perhaps an emergent sense of an ecological thinking indicated his awareness of this inescapable connection between humans, wildlife policy and their physical implications on wolves and deer.

If these photographs are understood as functioning within an ecocritical paradigm, they cannot be seen in isolation but instead are perennially imbricated with one another. Additionally, if the photograph is understood as an ongoing event, then the viewer is also always implicated in the image and its interpretation. While Shiras often attempted to distance his viewer from his subjects, particularly in the case of the wolf, looking at his images of animals together as part of a web of meaning unveils their complicated optical unconscious: a murky story of human violence against the nonhuman world, committed even in the name of preservation. Acknowledging the influence of viewers on how we look at nonhuman animals may shift how we see wolves, and alter the policy we write about their treatment, ostensibly moving beyond the limits of the image.

Notes

1 While it seems unimaginable today, whitetail deer were threatened in the second half of the nineteenth century, primarily because of mass hunting. Like their endangered contemporaries, the passenger pigeon and bison, whitetails faced extinction as urbanization destroyed their habitats and hunters systematically turned deer carcasses into trophies, evidence, that is, of patriarchal domination of the natural world. The species not only rebounded thanks to the concerted efforts of conservationists like Shiras, but were also allowed to overpopulate, due in part to conservation efforts like the eradication of natural predators, specifically wolves. (See Stolzenburg 2008, 103–104.)

2 Shiras's camera techniques began as emulations of jacklighting, discussed later, and setting hunting blinds. For this reason, in his survey of American hunting photography, Mathew Brower considers Shiras among other photographers using a hunting blind to hind behind while waiting for an animal to cross their camera lens, much as they would a rifle sight. While Shiras was certainly within this sphere of hunting photography, I argue that his use of automatic camera traps creates an additional level of mediation and specific relationship to the animal that cannot be considered only within a limited category of the photographic blind (Brower 2011).

3 For further discussion of jacklighting and other eventually outlawed inhumane hunting practices and their representations in material culture, including Shiras, see Annie Ronan's (2017) study of Winslow Homer in *American Art*.

4 This spectatorial model contrasts with the interpretation of contemporary material culture presented in Donna Haraway's 1984–1985 essay "Teddy Bear Patriarchy." Haraway understands the taxidermy displays at the American Museum of Natural History as putting the viewer in the role of a surrogate game hunter, in direct encounter with the wild animal, in order to create a virtual experience of the Strenuous Life.

5 Allen G. and Mary Wallihan, the husband and wife team who published one of the first animal photobooks in 1894, for instance, often photographed cornered animals just before they shot them, shifting away from trophy pictures of a carcass. This was common practice for another husband and wife animal photographer team, Martin and Osa Johnson, who also frequently shot animals immediately after taking their pictures on African safaris.

6 Several scholars have discussed a similar connection between policies concerning the treatment of animals or resources and indigenous populations, as well as their linkage to nation building in the United States including William Cronon, Alan Braddock, Nancy Anderson, Alex Nemerov and Angela Miller

7 Historian Michael Wise writes about the ways in which wolves and Blackfoot Indians were often equated as predators, at odds with the productive nature of capitalism, like the beef and whiskey trades in the Rockies (Wise 2016).

8 Lopez quotes the parallels in the expansion westward, laying out poisoned carcasses with distributing blankets poisoned with smallpox, and digging out dens to kill pups with kidnapping children and forcing them into boarding schools (Lopez 1978, 170).

9 Dying wolves also salivated heavily in their last moments, frequently poisoning the grass in the areas they died, later killing any ungulates, like horses, ponies and bison, who grazed on the poisoned grass.

Works Cited

Azoulay, Ariella. 2008. *The Civil Contract of Photography*. New York: Zone Books.

Brower, Matthew. 2011. *Developing Animals: Wildlife and Early American Photography*. Minneapolis: University of Minnesota Press.

Coe, Brian, and Paul Gates. 1977. *The Snapshot Photograph: The Rise of Popular Photography 1888–1939*. London: Ash and Grant.

Dunaway, Finis. 2000. "Hunting with the Camera: Nature Photography, Manliness, and Modern Memory, 1890–1930." *Journal of American Studies*. 34, 2: 207–230.

Egan, Shannon. 2006. "'Yet in a Primitive Condition': Edward S. Curtis' The North American Indian." *American Art*. 20, 3: 58–83.

Haraway, Donna. 1984–1985. "Teddy Bear Patriarchy: Taxidermy in the Garden of Eden, New York, 1908–1936." *Social Text*. 11: 20–64.

———. 2003. *The Companion Species Manifesto: Dogs, People, and Significant Otherness*. Chicago, IL: Prickly Paradigm Press.

Hutchinson, Elizabeth. 2009. "They Might Be Giants: Galen Clark, Carleton Watkins, and the Big Tree." In Alan Braddock and Christoph Irmscher, eds., *A Keener Perception*. Tuscaloosa: University of Alabama Press: 110–128.

Lopez, Barry. 1978. *Of Wolves and Men*. New York: Charles Scribner's Sons.

Ronan, Annie. 2017. "Capturing Cruelty: Camera Hunting, Water Killing, and Winslow Homer's Adirondack Deer." *American Art*. 31, 3: 52–79.

Roosevelt, Theodore. 1899. "The Strenuous Life." Speech before the Hamilton Club, Chicago. April 10.

Shiras, George. 1906. "Photographing Wild Game with Flashlight and Camera," *The National Geographic Magazine*. 17, 7: 367–426.

———. 1936. *Hunting Wild Life with Camera and Flashlight*. Washington, DC: National Geographic Society.

Smith, Shawn Michelle. 2013. *At the Edge of Sight: Photography and the Unseen*. Durham, NC: Duke University Press.

Stolzenburg, William. 2008. *Where the Wild Things Were*. New York: Bloomsbury.

Wise, Michael. 2016. *Producing Predators: Wolves, Work, and Conquest in the Northern Rockies*. Lincoln: University of Nebraska Press.

Part 4

Natural Histories/Animal Agencies

In 1977, when art historian John Berger posed the critical question "Why Look at Animals?" in his book *About Looking* he underscored that our relationships with animals have long been profoundly interwoven and sparked deep ontological questions regarding their artistic representation as subjects and symbols. In the nineteenth century, he notes, the stakes of animal and human existence changed, becoming more "dysfunctional," as capitalist modernization intensified. Many animals ceased to be seen as creaturely "co-constituents" and became instead exploited objects, relegated to categories of domesticated servants, rarified commodities or wild resources subject to human whim and demand. Animal bodies were both subjects and agents of display, enacting a mutually transformative process with their beholders by which, as Donna Haraway notes, "beings constitute each other and themselves" (Haraway 2003, 6). In this section, the essays delve into critical animal studies and post-humanist methodologies to explore the ethical stakes of interconnectedness across the biosphere. Drawing motivation from scholars of the other-than-human such as Haraway, Stacy Alaimo, Vinciane Despret and Cary Wolfe, the writers see animal intercessors as crucial agents whose co-relations with humans reveal the "polymorphous, heterogeneous world" all entities on earth participate in making (Wolfe 2003, xvii).

The case studies of these five scholars posit the formative roles animals played in the creation of nineteenth-century visual culture – from serving as inspiration to becoming collaborators – and from motivating conservation movements to destabilizing hierarchical models of knowledge. Re-imagined within a dispersed array in which human and nonhuman animals together demonstrate cosmo-political symbiosis, the diverse products of visual culture emerge not as discrete things but as assemblages, or in Bennett's terms "ad hoc groupings," that coordinate to produce experiences of meaningfulness (Bennett 2009, 23).

Across all four essays, traditional forms of art and expanded products of visual culture do not affirm ontologies as much as demonstrate their propensity to shift, depending upon a beholder's vantage point. Annie Ronan considers how American sculptor Albert Laessle's 1914 bronze goat *Billy*, a public sculpture that occupied Philadelphia's Rittenhouse Square, has been brought to a paradoxical state of completion by deteriorating. Laessle's *Billy* nurtured communitarian participation by welcoming contact and demonstrating the affective power of nonhuman entities. Caressed, kissed and hugged by over a century's worth of visitors, and only recently removed from its site, *Billy* solicited human contact in its form and presentation, and thus, it serves as an example of visual arts' unique capacity to make inter-species relatings physical, palpable and embodied. Unlike the allegorical beasts made during the turn-of-the-century "City Beautiful" movement, Laessle's animal sculptures also refuse to endorse hegemonic models of geo-political progress. Instead, Ronan shows how these sculptures produce a destabilized and non-hierarchical vision of natural and national orders that bespeak the physical and emotional intimacy we share with our other-than-human animal companions.

Kelly Bushnell shows how the display of animals entwined the discourses of natural history and economic interdependence. She calls upon the emerging literature of a "blue ecomaterialism," bringing the ethical concerns of the present to bear upon studies of Victorian visual culture in the treatment of a captive cetacean. By reconstructing the life of the first "white whale" publicly shown at the Royal Aquarium in Westminster, England in 1877, Bushnell shows how the Cartesian duality severing nature from culture was complicated when nature *became* a spectacularized form of culture. Amid the discourses of capitalism and science that held the Westminster whale captive, an exemplar of what Rosi Braidotti has termed the "zoo-proletariat," beholders contended with the troublesome destabilized hierarchies produced when the captive whale looked back (Braidotti 2013, 70). By also considering the problems of containment unique to ocean-dwelling creatures, Bushnell correlates her object of historical study with the ongoing struggle for animal liberation at aquariums and marine parks today.

Iovino and Opperman have stressed the need to "redraw the maps of ecological interactions, restructuring ethics and politics in the complex, nonlinear, co-evolutionary interplay of human and nonhuman agency" (Iovino and Oppermann 2012, 451). In this spirit, Emily Gephart and Michael Rossi corroborate animal-centered ethics and politics as they explore the aesthetic logic that underwrote both fashion and nascent conservation movements in the United States. At the close of the nineteenth century, even as proponents of bird protection condemned plume-adorned millinery, demand for fashionable bird hats was unabated. These hybrid objects are paradoxical entities in which avian life and death intertwine, and thus call for "transcorporeal" (Alaimo 2010) reevaluation of their vitality and visuality. Yet, far from just being about saving birds, the authors argue that the denunciation of animal cruelty for fashion brought a conspicuous material garment into dialogue with coalescing Darwinian narratives about evolution and behavior, and with parallel notions about population survival, fitness and the 'proper' orders of society. Ultimately, they argue that millinery fashion's ostensible violations set in motion renegotiation of the modern valences of aesthetic presentation, predation, protection and other affinities shared between the human and other-than-human worlds.

Shifting from a focus on animals as aesthetic agents to their formative role in the visualization of natural histories, the miniature but expansive realm of insects provides the vibrant substructure of Joan Greer's focus on two Dutch artists who brought together the visual arts and entomological science: Willem Roelofs and Vincent van Gogh. She not only invites us to think about the landscape of insect ecologies, but also challenges traditional analyses of artistic responses to the natural world by evaluating these Dutch artists together, in contrast to customary juxtapositions of their very different styles. Greer illuminates how both worked within the dynamic intersection of art production and scientific study, and as such were particularly attentive to inter-species relational systems. Troubling the anthropocentrism that circumscribes nonhuman animal worlds as separate from human societies, Greer borrows the Indigenous concept of a "grammar of animacy" from First Nations scholar Robin Wall Kimmerer (2013) to stress the expressive and communitarian value of non-linguistic and non-verbal forms of communication. Framing her analysis as a crucial engagement between art and the extinction crises of the Anthropocene, Greer articulates an 'ethology' of inter-related life forms in these artists' work, and investigates their common "ecological envisioning": a category of image-making and close looking, whose complexities call for reconsideration.

Works Cited

Alaimo, Stacy. 2010. *Bodily Natures: Science, Environment, and the Material Self.* Bloomington: Indiana University Press.

Bennett, Jane. 2009. *Vibrant Matter.* Durham, NC: Duke University Press.

Berger, John. 1977. *About Looking.* London: Penguin.

Braidotti, Rosi. 2013. *The Posthuman.* Cambridge: Polity Press.

Haraway, Donna. 2003. *The Companion Species Manifesto: Dogs, People, and Significant Otherness.* Chicago, IL: Prickly Paradigm Press.

Iovino, Serenella, and Serpil Oppermann. 2012. "Theorizing Material Ecocriticism: A Diptych." *Ecozon.* 3, 1: 75–91.

Kimmerer, Robin Wall. 2013. *Braiding Sweetgrass: Indigenous Wisdom, Scientific Knowledge, and the Teachings of Plants.* Minneapolis: Milkweed Editions: 48–59.

Wolfe, Cary, ed. 2003. *Zoontologies: The Question of the Animal.* Minneapolis: University of Minnesota Press.

13 Petting *Billy*

Albert Laessle's Significant Other(ness)

Annie Ronan

Too Much Love

In September 2018, city officials quietly removed *Billy* (Figure 13.1), a life-sized bronze goat, from Philadelphia's Rittenhouse Square Park. Despite being a relatively small sculpture, no larger than the impeccably groomed neighborhood dogs which strut through the park each morning, *Billy* had become a much-beloved public menace. The original cast's horns, blunted by over a century of outdoor display, had only grown shorter and sharper with each passing year. Other regions of the body were so dangerously thinned out that they threatened to open into raw wounds, razor-edged apertures into the cast's dark empty center. Much of the original work had been lost already. The finer details and furry textures on the goat's trunk had been scrubbed down to a silky, freshly burnished finish. By 2018, the bronze's original green patina only endured in the deepest of the crevices and crannies which, like rapidly healing

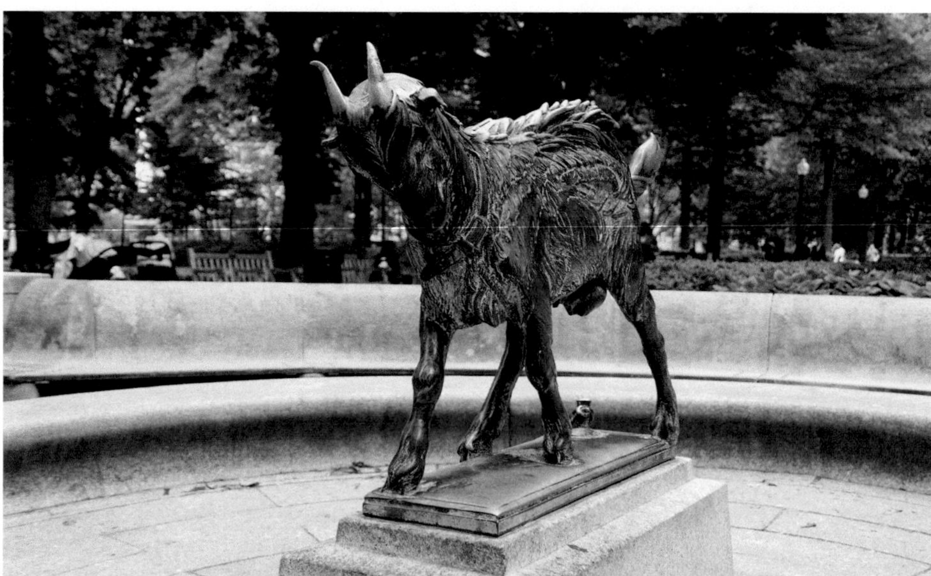

Figure 13.1 Albert Laessle, *Billy* (1914). Photo by Alec Rogers © 2015. Courtesy of Association for Public Art (APA).

scars, still stubbornly resisted being smoothed into oblivion. Even the signature of *Billy's* creator, the Philadelphia-based sculptor Albert Laessle, had been burnished away by time. Most glaringly of all, the taut bronze rope that once connected the goat's collar to the base had evidently been snapped off long ago. For decades, Laessle's little goat had been straining to free himself from an invisible tether.

Although Laessle is little discussed today, largely relegated to the margins of America's art history, the deterioration of his Rittenhouse *Billy* was not the result of neglect. Rather, the problem with this sculpture had always been an excess of attention. From the beginning, Laessle's goat served as a community pet as much as a piece of public art. When first installed in Rittenhouse Square in 1915, the piece, a newspaper reported, was almost instantly mobbed, "surrounded by children who were fondling its rough coat and pretty head, climbing upon its back and quite crazy with delight at their new playfellow" (*Philadelphia Inquirer* 1915). Ten years later, another local Philadelphia paper noted that eager children were forced to wait in line for a chance to ride Laessle's sculpture, "and sometimes," the article noted, "the demand is so brisk that [*Billy*] has to take three at a time along his spiny ridgepole" ("Billy the Goat" 1925). Over the years, Laessle's sculpture has been buffeted by generations of tiny hands, each new rider, each hug, pat and caress wearing him away bit by bit. A 2016 *Philadelphia Inquirer* article discussing whether the Rittenhouse *Billy* should be removed and "put in the barn" begins, "can a goat be loved too much?" (Lubrano 2016).

Destructive though it may be, this sculpture's capacity to attract attention, affection and touch is hardly accidental. From the beginning, Albert Laessle was content if not downright pleased to watch his most recent artistic achievement endure young Philadelphia's punishing adoration. As early as 1925, a local reporter observed that, although it was "probably more climbed-over and sat-upon than any other work of the sort in America," Laessle "accepts the burnishing process of clambering children as a token of *Billy's* popularity, and is much pleased thereby" ("Billy, the Beloved Goat of Rittenhouse Square" 1925). The artist's personal scrapbooks attest to this fact, encrusted as they are with decades of newspaper clippings that show children playing with *Billy*; posing with him; and often offering him food, affection and care. Unauthorized though such interactions may seem, Laessle not only "accept[ed]" this re-processing of his cast but, through the piece's child-friendly scale and lively approachability, positively invited it. Albert Laessle sculpted *Billy*, but was apparently content with if not intent upon the children of Philadelphia finishing it.

Billy, in its creation as well as in its slow disappearance, marks the culmination of a sustained and remarkably consistent artistic project, a career-long striving to create "pet-able" sculptures capable of figuring the complex co-constitutive entanglements that define our relations with other companion species. Informed by the artist's own sustained and sensitive observation of individual animal models, Laessle's animal sculptures explicitly draw viewers into intimate, proximate and sympathetic relationships with nonhuman collaborators. However, as *Billy* so aptly demonstrates, companionship has consequences. The sculpture's compelling liveliness also ensured its mortality; its pet-ability guaranteed its dissolution.

Thus, beyond simply memorializing a specific nonhuman historical subject – a goat who Laessle viewed as both companion and collaborator – in its slow collective finishing, *Billy* performs the mutually transformative process by which, as Donna Haraway has described it, "through their reaching into each other, through their 'prehensions' or graspings, beings constitute each other and themselves" (Haraway 2003, 6).

Billy offers an example of the visual arts' unique capacity to make these complex relatings physical, palpable and thoroughly relatable. Like so many of the other stone and bronze critters that began to populate the American "City Beautiful" at the turn of the century, *Billy* was on a civilizing mission to improve the lives and tastes of urban masses. Yet, unlike the guardian lions and big game proliferating in urban spaces at that time, the small-scale and insistently un-allegorical nature of Laessle's sculpture hails viewers into an ethic of companionability, not a logic of domination. Demanding prehension, frustrating comprehension, *Billy* continues to offer a radical vision of public art's capacity to civilize, one in which significant otherness is a thing to be continuously embraced, not cleansed away.

Touching Turtles

On the one hand, Albert Laessle fits comfortably among the ranks of the many sculptors who, at the turn of the century, on either side of the Atlantic, specialized in the forms of creaturely life. During the final decades of the nineteenth century, the market in contemporary sculpture was completely flooded with zoological subjects. "So much has animal sculpture come into vogue," one early twentieth-century critic observed, "that one can scarcely go to one of the large zoological gardens...without seeing a man or woman with clay and modeling tools in front of a cage" (Brush 1913, 1028). Like these other *animaliers* of his period, American sculptors such as Edward Kemeys, Alexander Phimister Proctor, Edward Clark Potter, Anna Hyatt Huntington and Arthur Putnam, Laessle devoted himself almost exclusively to the careful study and accurate rendering of animal bodies. For these artists, the animal body itself, often standing alone, often presented without any sense of geographical, historical or narrative context, was the primary focus of their artistic practice. Whether installed on the street or sitting on the shelf, their bronzes and Laessle's seem to coexist rather peacefully with one another.

However, while exhibitions, expositions and public works projects increasingly put turn-of-the-century sculptors to work making buffalo, lions and other beastly exemplifications of progress, power and martial virtue, Laessle's gaze was fixed downward, captivated by the smaller, muckier creatures – the frogs and lizards, bugs and crabs – that still wallowed in the primordial ooze. As a student at the Pennsylvania Academy of Fine Arts, he modeled decorative turtles to adorn the base of the Buffalo Pan-American Exposition's "Fountain of Man," an installation designed by his teacher, Charles Grafly, to express, according to sculpture plan director Karl Bitter, "the progress of man, his institutions and his development from the savage state up to the Age of Enlightenment" (Gray 1901, 61).[1] Rejecting this vision of biological and historical "progress," Laessle would continue to focus on humbler animal subjects, the abject critters that the glorious ascent of man would necessarily leave behind. After the Pan-American project, Laessle first gained individual recognition for *Turtle and Crab* (1901) and *Turtle and Lizards* (1902), two sculptures of an especially captivating ten-pound snapper which had, when Laessle first encountered it, been destined for Mrs. Grafly's stewpot (Miller 1924, 24). He continued to exhibit a preference for cheaper, humbler species. When he sculpted a predator on the prowl, as in his 1908 tabletop bronze *The Hunter*, his model isn't a jungle cat or roaring beast but a quiet, common toad. His 1919 *La Source* actively works against the francophone pretensions of its vague title, depicting nothing more than a duck hatchling, blindly screeching as it reaches upward and forward from the base, straining away from the cracked remains of its egg and into the space of the viewer.

The roots of *Billy's* problem – its degenerative condition brought on by too much love, too much attention, too much touching – can be traced to the research and development that Laessle's early sculptures represent. For example, his *Turning Turtle* (Figure 13.2) forcefully compels the human viewer to engage with the animal subject. In this piece, a life-sized gopher tortoise has been tipped on his side, and its life, quite literally, hangs in the balance. Its head and feet dig into the ground, its thin appendages struggling to flip the rest of the massive, dome-like body over. The trail of claw marks leading away from the front bottom foot and the limbs at the top of its body flailing out in search of something to grip suggest the tortoise's tremendous efforts and their possible futility. The animal, with its open mouth, even seems to shout and scream with frustration. This turtle seems to positively beg for human interaction. The front leg straining upward even seems to end in a curiously hand-like claw. It reaches out, grasping at the air for contact. With the body's off-center placement on the base, the sculpture's imbalanced quality makes the need to touch the tipping animal, to help it right itself, both a humane issue and an aesthetic one. The only way to right the turtle and the sculpture is to sympathetically intervene. Furthermore, if one

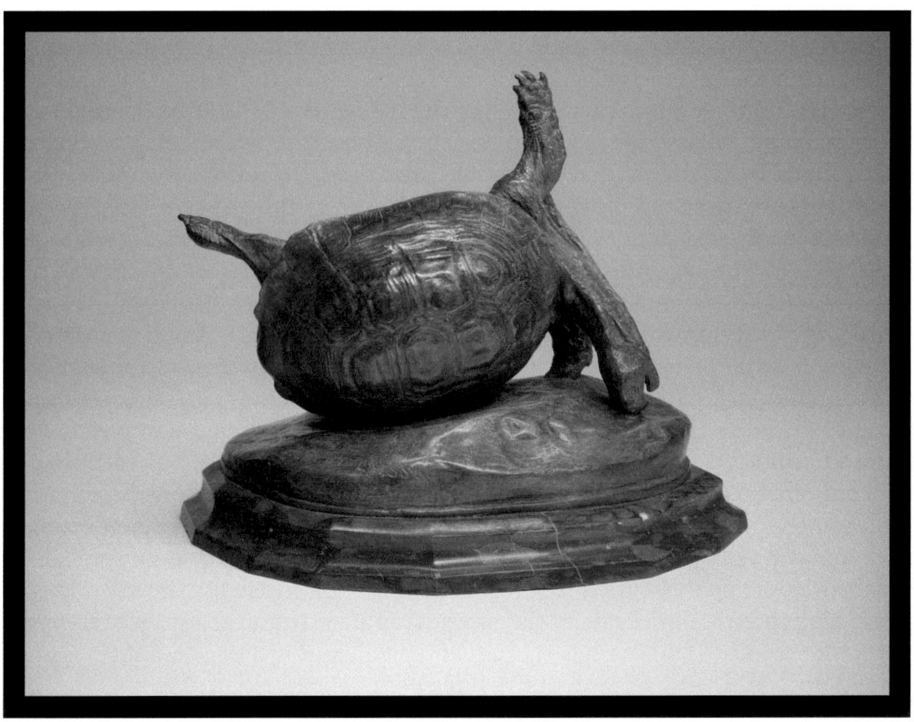

Figure 13.2 Albert Laessle, *Turning Turtle*, 1905; cast 1917. Rogers Fund, 1917. The Metropolitan Museum of Art, New York, NY, U.S.A. Image copyright © The Metropolitan Museum of Art.

Image Source: Art Resource, NY.

were feeling more sadistically inclined, wanting to, say, tip the turtle over into a truly helpless position, it's body would fall off the side leaving nothing but an empty base behind. Laessle's composition will not allow it. Cast in bronze, the turtle is eternally locked in this precarious pose, a moment in which the viewer's touch is not only desired but necessary.

In a similar fashion, *Billy* not only invites human contact but shapes the nature of that interaction. On the one hand, his back appears to strain upward for our touch, while his spiky fur, slightly flattened and pointing toward his tail, suggests the preferred trajectory of any future caresses. Furthermore, thanks to its small scale and low pedestal, the sculpture offers very little to the viewer who refuses to get down to the ground; it absolutely demands close proximity. According to Laessle's sketches, the artist deliberately chose to present *Billy* on a shorter pedestal than his other outdoor sculptures. For children, as a reporter observed, the goat's low height ensures that "it is very easy to climb aboard him, even if you are shod in roller skates" ("Billy the Goat" 1925). For taller and presumably older viewers, the result is that, in order to see the sculpture properly, one must be willing to maneuver themselves outside of their comfort zone and engage with *Billy* at goat-level. The face, turned downward, is particularly difficult for the average-sized adult to visually access, insisting that they squat, stoop or bend just to get a good look. The goat's face, then, serves as a sort of lure, a tease, coyly hiding in order to bring the viewer closer and incite them to change their position and perspective. The visual pleasure of *Billy* is only available to those willing to adjust their bodies to accommodate the goat's needs.

In this sense of mutual give and take, in which human viewer and animal sculpture meet for mutually co-shaping transformation, *Billy,* more so than Laessle's tabletop bronzes, figures what Haraway has described as the "dance of relating" that binds humans and other species together in a shared natureculture, "a bestiary of agencies, kinds of relatings, and scores of time [that] trump the imaginings of even the most baroque cosmologists" (Haraway 2003, 25, 2008, 6). In her *Companion Species Manifesto* and later 2008 book *When Species Meet*, Haraway has emphasized how, in our relationships with companion species, "we are training each other in acts of communication we barely understand...mak[ing] each other up, in the flesh" (2003, 2). Within Haraway's vision, in which "the world is a knot in motion," the boundaries between seemingly distinct individuals and species interpenetrate; the notion of a singular, isolated, "pure" or *a priori* form of being is inconceivable and unwanted (2003, 6). *Billy*, in inviting proximity and prehension that will physically alter viewer and sculpture alike, stimulates precisely the sort of conscious reaching out to "significant otherness" that Haraway views as necessary to an ethics of "get[ting] on together," namely "vulnerable, on-the-ground work that cobbles together non-harmonious agencies and ways of living that are accountable both to their disparate inherited histories and to their barely possible but absolutely necessary joint futures" (2003, 7). In contrast to public sculpture that celebrated "the progress of man from the savage state up to the Age of Enlightenment," *Billy* suggests that enlightenment is only accessible through the messy and uncertain work of intra-active interspecies engagement.

Working with Animals

These works provoke the very same sort of human-animal relations that informed their production. Among other *animaliers*, Laessle stands out for his commitment to not only studying the living animal, but to studying the same individual specimen over an extended period of time. Invasive anatomical studies are nowhere to be found within what survives of his personal papers. Instead, one finds countless references to the nonhuman subjects who, whether through the protocols of the archive or the humanist habits of art historical work, are typically cleansed from our view of the past. As if anxious to assert the presence of his nonhuman collaborators, in interview after interview, Laessle detailed the relationships he developed with his models, many of whom were incorporated into his household as family pets. In many of his anecdotes, Laessle relates how he worked with these animals, often changing his artistic vision to suit the models' behavior and personality. For instance, in discussing the creation of his *Bronze Turkey,* Laessle studied a good deal more than the physical makeup of the bird; he viewed it as his mission to figure out what a turkey desires and, in turn, do whatever it took to ensure that "the turkey might be pleased [enough]...to strut in lordly fashion," as reported in a 1924 profile of the artist. He had a particularly long history with his model for *Billy,* a disobedient little goat who reportedly "ravage[d]" the sculptor's home and neighborhood (Miller 1924, 27). The manner in which the finished piece is forever straining against his tether, hooves digging into the seemingly soft surface of the base, ensures that *Billy* is locked in an appropriately defiant attitude forever. Laessle's relationship to his models was explicitly more intimate than artist and model, resembling something more like the emotionally companionate and physically proximate character of a person's relationship with their pet. It was a process of negotiation, not domination. In turn, the sculptures capture the attitude of the individual creature, immortalizing 'Mr. Turkey' rather than an inert and idealized image of a turkey.

The relations Laessle immortalized in bronze are quite different than those suggested by the imposing, dominant and emotionless guardian lions that, at the turn of the century, were rapidly proliferating within American public spaces. "Just looking at a creature from outside iron bars won't give you much information...you have to get on intimate terms with your animal," Laessle explained in an interview, joking "maybe that's why I haven't made a lion!" (Grafly). Tellingly, when Laessle would eventually try his hand at sculpting lions, an animal that could not so easily be incorporated into his studio, home and life, he seemed incapable of making the king of the jungle exude anything but genial amiability. Bearing garlands of flowers at the portal of the Panama-Pacific Exposition's Court of Abundance, guidebooks described Laessle's lions as "friendly-looking" and "charmed into friendliness," cuddly creatures whose gentle amiability seemed a kind of overcompensation, an awkward attempt to conjure the "intimate terms" that Laessle himself was unable to experience (Perry 1915, 51; Todd 1921, 289).

The City Naturecultural

As opposed to the civic sculpture advocated by the City Beautiful movement, sculpture that celebrated the "progress of man" through allegorical form, Laessle's animal bronzes engage with all viewers regardless of their level of erudition or background. In her book *Public Sculpture and the Civic Ideal in New York City,* Michele H. Bogart has examined the politics of turn-of-the-century public art, tracing a history in which proponents of the "City Beautiful" used political and economic power to lay claim to public land, and sponsor works of public art that advanced a racially and culturally specific vision of American identity (1989, 62–66). As Bogart points out, the complex thematic programs favored within the preferred "didactic approach" were only decipherable to those already familiar with the conventions and visual codes of the cultural elite. In their impenetrability, these works reified existing inequities of power, "put[ting] those with less culture or education in a socially and politically dependent position in relation to the artist and the educated individuals who had delegated to themselves the task of articulating values" (1989, 229, 230). For sculptors working in this mode, embracing and perpetuating these forms of cultural hegemony gave them the opportunity to capitalize upon their European training and elevate the status of their profession. Viewed in relation to the dominant mode of public sculpture at his time, it's clear that Laessle's work was going for something very different.

Instead of civilizing through the imposition of traditional, authoritative meaning, Laessle's animals replicated the dynamics of hands-on observation and experience. These sculptures don't simply allow themselves to be touched, they often, as in the case of *Turning Turtle*, explicitly reach out to us, even refusing to obey the traditional boundaries between viewer and artwork. Laessle's goat civilized through means other than the imposition of traditional symbols, or the preservation of a noble past. Rather, it engaged with viewers directly in a lively and continuing present, so much so that the Rittenhouse *Billy* was utterly transformed and practically destroyed by a century of being handled and loved like a pet instead of an artwork. While City Beautiful sculptors used animals to affirm the certainty of the natural order and, by extension, the received hierarchies of the social realm, Laessle's rebellious animals celebrate the virtues of disorder, of co-shaping collaboration between heterogenous actors.

In describing the work of contemporary sculptor Andy Goldsworthy, Haraway notes that "his sculptures endure for seconds, sometimes for decades; but mortality and change are never out of consciousness" since "process and dissolution – and agencies both human and non-human, as well as animate and inanimate – are his partners and materials, not just his themes" (Haraway 2003, 23). Albert Laessle's *Billy,* in the deterioration that's a consequence of its prehensile and sympathetic relatings, similarly performs the processes of dissolution that characterize the "knot" of natureculture. In so doing, it reveals the mortality and mutability of all bodies, even those cast in bronze. Despite other public sculptures' claims to monumental, metallic permanence, they too, *Billy* tells us, are enmeshed in nature, ever-changing.

In its new home, the Philadelphia City Institute Library's Children's Reading Room, the ongoing transformation of *Billy* has been somewhat arrested. However, it is not "restored." Rather, *Billy* continues to bear the traces of this "dance of relating" between artist, animal and community. The parts of the statue which endured the most human contact, the graspable tail and horns, the pet-able back, the kissable lips, all now glitter with raw, exposed bronze. Burnished by affection, the original *Billy* has been made golden, made glorious. The freshly cast replica which currently stands in Rittenhouse Square Park, recently unveiled to the public with a speech from the mayor and a celebratory petting zoo, will undoubtedly come to meet a similar fate.

Note

1 For more on Laessle's involved with the project, see *The Philadelphia Inquirer*, 1900. Albert Laessle Papers, Smithsonian Institution.

Works Cited

Archives

The Albert Laessle Papers (1897–1971). Smithsonian Institution, Washington, DC.
Untitled Article. 1900. *The Philadelphia Inquirer.* November 25.
Newspaper clipping 1903. Dated Jan 24. [microfilm reel 74].
Untitled scrapbook clipping. 1915. The *Philadelphia Inquirer* June 27.
"Billy the Goat." 1925. *Evening Public Ledger* (Philadelphia). October 17 [microfilm Reel 74].
Eustice Fiske. 1925. "To Albert Laessle." May 14.
Dorothy Grafly. Undated. "Albert Laessle, Sculptor, Has a Persuasive Way With Animals." Newsclipping from unidentified source [Microform Reel 74].
T. A. Daly. Undated. "'Billy of Rittenhouse Square."

Published Sources

"Billy, the Beloved Goat of Rittenhouse Square." 1925. *The Christian Science Monitor.* August 17.
Bogart, Michele H. 1989. *Public Sculpture and the Civic Ideal in New York City, 1890–1930.* Chicago, IL: University of Chicago Press.
Brush, Edward Hale. 1913. "Animal Sculpture among the Ancients. *Art and* Progress. 4, 9: 1028.
Gray, David, ed. 1901. *Art Hand-Book, Sculpture, Architecture, Painting: Catalog to the Pan-American Exposition.* Buffalo, NY: D. Gray.
Haraway, Donna. 2003. *The Companion Species Manifesto: Dogs, People, and Significant Otherness.* Chicago, IL: Prickly Paradigm Press.
———. 2008. *When Species Meet.* Minneapolis: University of Minnesota Press.
Lubrano, Alfred. 2016. "Original Rittenhouse Square Billy Goat to be Put in the Barn?" *The Philadelphia Inquirer,* October 4. Accessed January 27, 2018. www.philly.com/philly/news/20161005_Rittenhouse_Square_billy_goat_to_be_put _out_to_pasture_.html
Miller, R. Roy. 1924. "A Sculptor of Animal Life." *International Studio.* 27.
Perry, Stella G. S. 1915. *The Sculpture and Murals of the Panama-Pacific International Exhibition: The Official Handbook Giving the Symbolism, Meaning and Location of All the Works.* San Francisco, CA: The Wahlgreen Company: 51.
Todd, Frank Morton. 1921. *The Story of the Exposition: Being the Official History of the International Celebration Held at San Francisco in 1915 to Commemorate the Discovery of the Pacific Ocean and the Construction of the Panama Canal.* Vol. 2. New York: The Knickerbocker Press: 289.

14 Looking at Leviathan

The First Live Cetaceans in Britain

Kelly P. Bushnell

"On making an incision into the lung, out came the truth—cause of death palpable—plastic grey pneumonia," wrote an 1877 observer of the autopsy of the first whale to be exhibited live in Britain (*The Fishing Gazette* 1877, 9). One hundred forty years after the death of this whale, American aquarium chain SeaWorld announced an end to their captive orca breeding program, but not before their youngest whale suffered the same fatal illness: "Park officials suspect the three-month-old calf died from pneumonia, the same infection that killed the infamous captive whale Tillikum" (Gibbens 2017).

In the intervening decades, the display of live cetaceans reached its zenith, but gradually the tide has begun to turn (hastened by the 2013 documentary *Blackfish*) toward recognizing the cruelty of keeping such intelligent creatures captive. Though the emotional and intellectual sensitivities of whales are no longer in question, we still wrestle with some of the fundamental issues of treating a sentient creature as visual culture. What does it mean to look at a captive whale and to see her look back at us? How do we conceive of nature and culture when nature *becomes* culture? (Or, perhaps, thanks to Donna Haraway (2003), *natureculture*?) And how is the visual culture of captivity unique for creatures of the sea?

This chapter seeks to expand the ecocritical study of Victorian visual and material culture by considering the live, intact, material bodies of whales as visual culture. While whales' constituent parts were used for decorative and practical items such as scrimshaw and corset stays, Britons of 1877 onward also experienced the whale *as* whale in several highly publicized live whale exhibitions. Regarding the whale as imbricated in visual culture also speaks to the fundamental anxiety of the island nation: Britannia rules the waves, but whales (and what else?) rule the depths.

Claude Levi-Strauss, redirecting the notion that animals are "good to eat," asserts in *Totemism* that animals are "good to think [with]" (Levi-Strauss 1964). Ann Colley, Harriet Ritvo (1987), Deborah Denenholz Morse and Martin Danahay (2007) and Donna Haraway (2003) have shown in the past 20 years, as Sarah Amato puts it, "how human relationships to and understanding of animals are historically and culturally contingent" (Amato 2015, 12). Haraway's concept of *natureculture* has particularly emphasized the semiotic inextricability of nature and culture, and John Berger's *Ways of Seeing* adds the specter of capitalism which has uneasily accompanied science since the earliest commodification of animals, reminding us that "the nineteenth century, in western Europe and North America, saw the beginning of a process, today being completed by twentieth century corporate capitalism, by which every tradition which was previously mediated between man and nature was broken" (Berger 1977, 2). Ecocriticism often attends to the tension between the pastoral and the wilderness, the latter being a point of overlap for ecocriticism and animal studies because an inescapable component of wilderness is the potential beasts therein. Greg Garrard reminds us that "wilderness" comes from the Anglo-Saxon *wilddeoren*, where "*deoren*" (beasts) "existed beyond the boundaries of cultivation" (Garrard 2011, 66–67). Humanistic studies of the sea and its creatures are a prime field for interrogating beasts beyond the cultivation, control, understanding and comfort of man after – as Byron famously writes – "our control / Stops at the shore" (Byron 2008, 179).

In 2015, Jesse Oak Taylor remarked that the "striking thing about Victorian ecocriticism is that there is so little of it" (Taylor 2015, 877). There is especially not yet a Victorian *oceanic* ecocriticism dedicated to the unique questions and problems posed by the sea and its creatures in this period. Daniel Brayton and Steve Mentz have pointed out the "terrestrial bias" in the environmental humanities, calling for a "blue" turn in green studies (Mentz 2009). Victorian studies has long embraced material approaches to this intensely visual, "stuff"-centered period and has recently enjoyed its "animal turn" as well. A "blue" or oceanic ecomaterialism in Victorian studies informed by the ways in which Serenella Iovino and Serpil Opperman have framed ecological thought may also help to decentralize the human and instead privilege the agency of what Iovino and Oppermann call "storied matter." They conceive of the "world's material phenomena [as] knots in a vast network of agencies, which can be 'read' and interpreted as forming narrative, stories" (Iovino and Oppermann 2014, 1). Thus, "*material* ecocriticism examines matter both *in* texts and *as* a text, trying to shed light on the way bodily nature and discursive forces *express* their interaction whether in representations or in their concrete reality." Building upon Haraway's *naturecultures*, Iovino and Oppermann position bodies in particular as "living texts that recount *naturalcultural* stories" (Iovino and Oppermann 2011, 1, 2 & 6). For Tim Morton, *nature* is generally a "transcendental term in a material mask," and the "question of animals... radically disrupts any idea of a single, independent, solid environment" (Morton 2007, 14 & 99).[1] The sea is a prime location for this encounter between the theoretical and the material in the nineteenth century because of the myriad ways in which public and private actors used bodies and *things* to try to domesticate the unfathomable sea, exemplified by traveling whale carcasses, marine fossils, sea shore collecting and the Victorian aquarium craze.

The stakes of "blue" ecocriticism are especially high in the Victorian period, as the *eco of* ecology is, of course, rooted in the Greek *oikos*: home. The tiny Atlantic island taking over the world desperately needed the sea environment to be part of its imperial, cultural and scientific ecosystem – its *home-system*. However, as Morton writes in *Ecology Without Nature*, "the idea of 'our' environment becomes especially tricky when it starts to slither, swim, and lurch toward us." In 1877, the first whale was transported to London, and she began to swim toward us in ways that forever altered human-cetacean relations.

Though so much of daily life in Britain depended on whales – they both literally and figuratively illuminated and lubricated the Industrial Revolution – it was rare for someone other than a whaleman, fisherman or sailor to see a live whale. This changed in September 1877 with the arrival of the first "white whale" at the Royal Aquarium in Westminster, just blocks from the Houses of Parliament. The proprietor of the Royal Aquarium was William Leonard Hunt, an American showman and promoter better known as "The Great Farini" for his early career as a daredevil which included a trip over Niagara Falls in a barrel. Marine life was just one facet of the Royal Aquarium; Farini (as he was known in London) envisioned it as a complex for various arts and entertainments. The aquarium theatre put on productions of "School for Scandal" and W.S. Gilbert's adaptation of *Great Expectations* in addition to sideshow-style acts like Zazel the Human Cannonball and a young girl called Krao covered in hair whom Farini promoted as the "Missing Link." In 1877, he announced the arrival of his most anticipated exhibit yet: a live "white whale" from Labrador.

The "white whale" obviously alluded to *Moby-Dick*, though Melville's outsize villain was a sperm whale, and Farini's specimen was a nine-foot-long beluga whale – a species known for its charming song and playful nature. The life of the first Westminster whale was short and cruel, but *seeing* her, in the middle of London, catalyzed a discussion about the nature of animal captivity. What does she still have to tell us about the visual cultures of nature in Victorian Britain, and how can she help us think about similar issues of cetacean captivity today? When Philip Armstrong considers the literal and literary "rendering" of whales in the nineteenth century, he focuses on *Moby-Dick*, yet his central question (which he positions as the central question of Melville's novel) – "What do whales mean?" – rings true. He answers, "Critical replies to this have mostly concentrated upon reading cetaceans as a screen for the projection of human meanings, but attended only incidentally to what else they might mean, or how they might mean otherwise—that is, the ways in which whales trouble or escape human representation" (Armstrong 2005, 101). The captive whale can no longer "escape human representation" but she certainly "trouble[s]" it. And her arrival at the Royal Aquarium on September 27 troubles this representation as poignantly today as in 1877.

The *Fishing Gazette* reported on September 28, 1877: "THE LIVE WHALE has arrived at Westminster. She is apparently in good health, and feeds well upon live eels. A full account of her Whaleship will appear in our next Number, with a page of Illustrations." Her Whaleship had been captured in a seine net off the coast of Labrador by Zach Coup, who had procured P.T. Barnum's belugas, one of whom "became so tame it would allow itself to be harnessed to a car in which it pulls a young lady round the tank" (Lee 1878, 8–9). (Farini, no doubt, had similar hopes for his.) From Labrador, the whale was transported by steamer to Montreal then by train to New York, where she was kept in a reservoir at Coney Island before being loaded onto the German Lloyd's steamer *Oder* for passage across the Atlantic. Off the coast, she was transferred to a tender which brought her to the Southampton Docks where she was loaded into the South-Western Railway bound for London. The zoologist Henry Lee writes of her arrival at the aquarium: "About noon, a large wooden box, twelve feet in length, was lifted out of a van and placed at the side of the tank by a score or so of laborers. Within it lay the whale, half embedded in sea-weed—which smelt of anything but ozone—and breathing at intervals of about twenty-three seconds" (Lee 1978, 2). The narrative and representation of her transportation is a crucial part of her imagery. Amato reminds us that "Victorians used new consumer amenities to complete transactions involving animals, including new railway systems, which revolutionized possibilities of domestic transport, travel, and retail within Britain, allowing for everyone and everything to be put in motion" (Amato 2015, 11). The whale traveled by carriage, ship (sail *and* steam), on the shoulder of men in a box, and locomotive.[2] Her successful capture and swift carriage could not have been possible even ten years earlier. She is also, then, a testament to new technology, an organic attestation to industry, an avatar and an advertisement.

Though the 45,000-gallon tank, set into the concrete floor of the aquarium, was the largest ever constructed in England, it was still painfully small for her at just 40 feet in length by 20 feet wide and 6 feet deep. Lee marvels at the engineering: "The weight of the iron plates alone is thirty tons. Thirty thousand holes had to be drilled in them to receive the 15,000 rivets, and yet this immense receptacle was commenced and finished within eleven days." The basement was fitted as a sort of arena: "Tiers of raised seats were erected at each end of the tank, from which visitors could look down upon the whale" (Lee 1878, 1 & 2). Between her various modes of transportation and the feat-of-Victorian-engineering tank which would hold her, the visual of the whale is inseparable from industrial technology. Once removed from the sea, she will always be contained in something made by humans. She has become *natureculture*: a semiotic entanglement of industry and environment. Any viewer's first sight of the whale will not be the whale but her constraints: the net, the box, the tank. The visual culture of the whale in Victorian Britain is impossible to disentangle from the visual culture of her captivity – created by humans.

Once subsumed into the Victorian visual culture of display, she is no longer an agent unto herself; she is inextricable from her manufactured context. Yet, unexpectedly, she does retain an ecological agency. According to David Abram, the material beings which comprise *naturecultures* "have the ability to communicate something of themselves to other beings" (Abram 2010). When she was finally lowered into the tank,

> all eyes were agog to see what kind of creature, what animal in the animal series—"a fish-like mammal" could be; and the first impression was at once favorable. The outline of this whale... is undoubtedly graceful in its curves, while the color is something between that of the purer kinds of vulcanized india rubber such as is used for children's dolls, and a white horse—a creamy white, with a dash of grey in it. Its motions were as graceful as its body, as it glided round the tank, apparently with an enquiring eye as to the extent and depth of its new domain (*Fishing Gazette*, 5 October 1877).

Though the initial perspective is human – "all eyes were agog" – and the correspondent describes her as a sort of hybrid of wilderness and domesticity (an *assemblage*, perhaps, in Jane Bennett's terminology in *Vibrant Matter*), the ultimate perspective is that of the whale: her "enquiring eye" inspects the tank even as she is being inspected. When the human stares "agog" at the whale, the whale stares back, mammal to mammal, forever altering the visual culture of the watery abyss when, to borrow from Nietzsche, "the abyss stares back at you" (Nietzsche 2003) with a sentience as yet unavailable in Britain. Though the London Zoological Society and various London menageries had granted access to land mammals, the particular oceanic and imaginative "abyss" in which the whale resided added an extra layer of obscurity. In Berger's particular "way": "the animal scrutinizes [the viewer] across a narrow abyss of noncomprehension" (Berger 1977, 5). The print media surrounding the whale narrowed this abyss of noncomprehension even further with notes on her similitude. On October 5, 1877, the *Lichfield Mercury* remarked that the whale is "said to be very fond of shrimps, and in this respect resemble[s] the Londoners themselves" and the *Liverpool Mail* notes that she "flounders about, and 'spouts' with as much vigor as the most loquacious politician" (29 September 1877).

And though she appeared healthy upon her arrival on Wednesday, Westminster's first whale died on Saturday morning after just four days on display at the Aquarium. Her residence in Britain was so short-lived that in some rural newspapers, the notices of her arrival and her death were printed in the same issue. The aquarium issued an official announcement of her death in the London *Daily News* on October 1: "It is with regret the announcement is made that the Whale died THIS MORNING." She was, of course, exhibited anyway. According to Lee: "During the day some sixteen hundred visitors, who had come to see the live whale, inspected it as it lay dead, and they saw a 'show' which I, as a naturalist, may perhaps be thought to over-value, but was one to see which I would have travelled a long distance, if necessary" (Lee 1878, 5). Despite Lee's protest that the main objective of the whale's presence is scientific, even he reduces her to a "show." The *Fishing Gazette* quantifies it thus: "On Saturday, some hundreds paid 6d each in addition to the 1s admission to the Aquarium, to inspect the dead body of this whale" (5 October 1877).

The eminent "Fish Culturist to the Queen" Frank Buckland was immediately recalled from Scotland while his secretary made a plaster cast of the animal to be painted by the renowned "fish artist" H.L. Rolfe. The following morning, Lee assisted with the necropsy, noting "we had been requested to avoid injuring the skin or skeleton in any way which would prevent the one being stuffed and the other articulated for future exhibition" (Lee 1878, 5). The cause of death was found to be pneumonia, likely contracted during the transatlantic passage in which she was regularly sluiced with sea water in freezing temperatures on deck. Her 63-ounce brain went to the Hunterian at the Royal College of Surgeons, her cast to Buckland and her skeleton and skin to Farini.[3] Though the visual display of live animals is finite (especially in a period in which veterinary care was so rudimentary), the animal's post-mortem career was permanent and three-fold: the skeleton was articulated, the skin preserved for stuffing and the cast painted to look like the live, healthy specimen. The act of looking at the live whale for the Victorian viewer, then, is also informed by the knowledge that one's children will potentially see the same whale after its death, stuffed and/or articulated. These second "lives" provide a sort of insurance and return on investment in a business in which the star employee will likely expire, but they also endow the creature with an agency that will outlive even its captors.[4]

The print culture that sensationalized the first Westminster whale's demise reinforced her position in relation to humans. The illustration from the October 5 issue of the *Fishing Gazette* detailing the necropsy places the whale in the context of whaling, with the inclusion of decorative harpoons (Figure 14.1). The five-panel illustration shows (1) a beached beluga whale in its Arctic habitations; (2) a right whale and a sperm whale, respectively, also in the Arctic; (3) the beluga in her tank at the Royal Aquarium surrounded by patrons; (4) the beluga being introduced to her tank; (5) and the skeleton of a right whale. The main panel, showing the captive beluga at the aquarium, is flanked by two bracketing vignettes of the implements of the whale hunt: harpoons, lances and rope. To place the captive live beluga in the context of the industry around cetacean killing creates a sort of invisible tank as well, presenting even the wild whale as contained within reach of men.

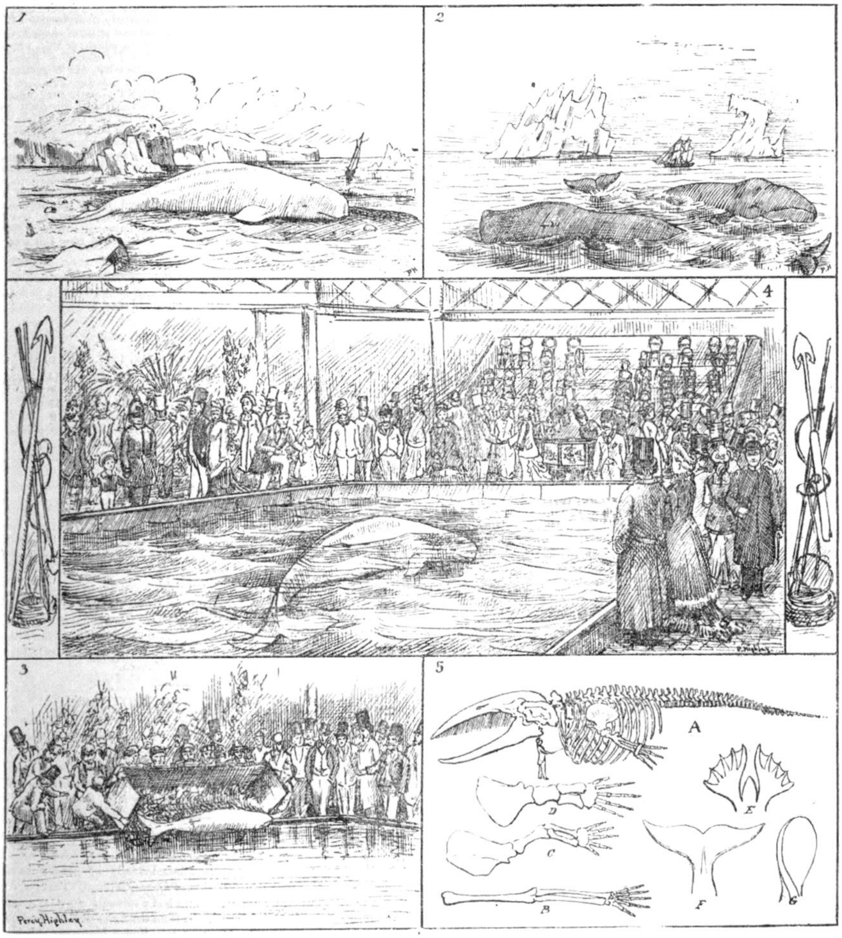

Figure 14.1 The Fishing Gazette, 5 October 1877.

The visual culture of any captive whale so long as the whale hunt continued was (and remains) a visual culture of death, and the first Westminster whale's death was in many ways even more of a spectacle than its arrival. George Reeve Smith, proprietor of the Brighton Aquarium, "was so sorry when he heard that the Westminster whale was dead that he shed floods of tears (of course, salt ones)" and ordered the flag at the Brighton Aquarium to be flown at half-mast (*Illustrated London News*, 6 October 1877). The *Illustrated London News* commentator "G.A.S." summarized her life as such:

> For some days the "wooden walls of old England" (I mean the ubiquitous Mr. Willing's hoardings) have been covered with proclamations—"The Live Whale is Coming!" He came—a poor little white fellow, not much more important in size than a large porpoise—to the Westminster Aquarium. They put him into a tank, and gave him eels to eat; but there was something the matter with the whale, or the tank, or the water; and the poor lilliputian leviathan died. And then the eels tastened [sic] on the fins of the deceased and began to eat *him!* Which is the way of the world.

The invocation of the "wooden walls of old England" as hoardings pasted with advertisements both reinforces and rewrites the centrality of the sea in British life, setting the context for Farini's beluga. ("Mr. Willing" refers to Willing and Co., a London advertising firm whose images often covered the wooden hoardings and walls throughout London). The "wooden walls of old England" were first conceived as Royal Navy warships in Henry Green's 1773 naval ode:

> Thine Oaks descending to the main,
> With floating forts shall stem the tides,
> Asserting Britain's wat'ry reign
> Where'er her thundering Navy rides:
> Nor less to peaceful arts inclin'd,
> Where Commerce opens all her stores,
> In social bands will league mankind,
> And join the sea-divided shores:
> Spread then thy sails where Naval Glory calls:
> Britain's best bulwarks are her WOODEN WALLS (Green 1773).

The transformation of Britain's "wooden walls" from warship to billboard cuts to the heart of the ecological and imperial relationship with the sea in the latter part of the nineteenth century. The Royal Navy fleet, built of British oak, creates the visual metaphor that it is Britain's *nature* which will protect her *culture* from invasion (and vice versa). The epithet remained in use through the nineteenth century. There are a number of paintings entitled "The Wooden Walls of Old England" – the most famous by George Clarkson Stanfield depicts a British Man of War next to a fishing boat and rowboats, and a fisherman on shore gazing out at sea (thus reinforcing the importance of all of Britain's maritime industries). The *naval* wooden walls of old England protect the island nation from invasion. In 1877, however, well into the *Pax Britannica*, the term has been humorously co-opted to depict the ubiquity of London construction ("Where Commerce opens all her stores") as physical substrate for visual culture advertising the whale – a domestication of the wild sea. These new commercial "wooden walls" also invoke the whale's containment within the aquarium walls. Each version of the "wooden walls of old England" (warship, construction hoarding advertising the captive whale and aquarium walls, respectively) thus asserts the same thing: *Britannia rules the waves.*

The whale, however, did not ultimately submit to this domestication, and her death occasioned commentary from across London. Animal advocates accused Farini of ill treatment, but Henry Lee deemed these "inaccurate and unwise suggestions," assuring readers "As a looker-on of some experience, I am satisfied that in the treatment of this Westminster Whale everything was done that foresight could dictate in the *existing state of knowledge and skill* in the carriage of living animals" (Lee 1878, 9, emphasis mine). The *Times* confirms: "When once the animal was safely deposited in the tank its surroundings were fully as favourable as those of most other creatures when deprived of their natural liberty. The supposed marks of ill-usage on the dead body were the consequences of the eels in the tank having after its death nibbled the edges of its fins" (The *Times*, 3 October 1877) (Figure 14.2).

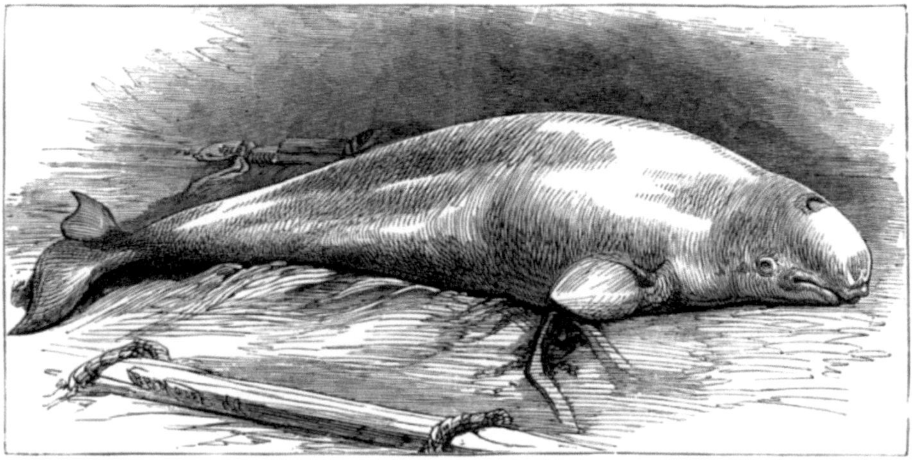

THE DEAD WHALE AT THE ROYAL AQUARIUM.

Figure 14.2 Illustrated London News, 6 October 1877. British Library via Mary Evans Images.

Others called into question whether such a creature should be held captive in the first place (either until more could be learned about how to care for a captive whale, or at all). Francis Francis, of the Brighton Aquarium, wrote: "When we can get a good open pond, in a nice breezy spot, and from 150 to 200 feet long, 50 or 60 feet wide, and a dozen feet deep, with a constant stream of salt water flowing in and out, then I will confidently undertake the introduction of whales with a full expectation of keeping them alive" (*Illustrated London News*, 1877). This debate over tank size that began in the Victorian era continues to be of paramount concern today in the conversation on captive whales: before SeaWorld announced the phasing out of live cetacean shows, park administration attempted to appease critics by making plans for expansions of its cetacean tanks, and animal welfare groups have suggested that all cetaceans currently in captivity who cannot be released should live out the rest of their lives in expansive coastal sanctuaries.

For other critics, however, it was not a question of *when* or *how* a whale should be kept, but *if*. The well-known Bishop of St. Albans, Piers Claughton, wrote to the *Times* to express his concern, citing that the whale, "the creature of which the Psalmist speaks," is "placed in its element [the sea] by the Great Creator." He calls out the hubris of men who would remove such a creature to a subterranean "suffocating tank" not even deep enough for it to dive. He reframed the debate, for the first time as a "question of morality." On October 6, 1877, The *Hampshire Telegraph and Sussex Chronicle* took Claughton's side, lamenting, "The unlucky little whale is dead at last. It could not possibly have lived more than a few months, and it is a very great question whether the attempts to exhibit the whale in an aquarium at all does not amount to something very like cruelty" (*Hampshire Telegraph and Sussex* Chronicle, 1877). The article goes on to call the whale's tank "nothing more than a cistern" and her captivity "a species of Barnumism… little allied to true science."

ILN's G.A.S. agreed, invoking both Keats and Spenser to drive home the cruelty of the whale's captivity:

> For my part, I am of good Bishop Piers Claughton's opinion... and would let Leviathan alone. John Keats used to revel in that magnificent Spenserian epithet, "the sea-shouldering whale." What would the author of "Endymion" have thought of a "tank-shouldering whale?"

If what delighted the aging Romantic Keats was the scale associated with the whale's element, then he would not likely have been impressed by the tank. As Wordsworth cautioned, "Sweet is the lore which Nature brings; / Our meddling intellect / Mis-shapes the beauteous forms of things:— / We murder to dissect." Though Romanticism theoretically resisted the containment of nature by culture, Morton reminds us that environmentalism is, at its heart, consumerism and that Romantic consumerism "produced subjective states that eventually became technically reproducible commodities." After all, "*Ecology* derived from the Enlightenment view of the *economy of nature*" (Morton 2007, 94 & 80).

And though Farini and his aquarium staff did not physically "murder to dissect" that first beluga, they dissected it nonetheless, and the uneasy bond of Victorian science and capitalism forged in the tank indeed brought harm to individual creatures in the name of science and often under the guise of public education. When Lee eulogizes the Westminster whale, he mourns both the "plucky" plans made by its industrious proprietors ("I cordially sympathized with its owners and the authorities of the Aquarium in their loss"), but for the educational potential:

> The public, too, were deprived of a great sight, from an educational point of view. Thousands of persons who had opportunities of seeing the porpoises in the Brighton Aquarium arena then for the first time to appreciate the fact that the cetacea are no fishes. They read with their own eye from Nature's own book, far better than any printed page could teach then, that the whales breathe by lungs and not by gills; that they propel themselves by vertical movements of the tail and not by their pectoral fins; that they never spout water from their spiracle; and many other details of their movements, habits, and mode of life. And as they looked and learned, their interest increased with their understanding of the subject, they were taught that they have before them a warm-blooded animal that suckles its young, and one of high intelligence, almost, if not quite, equal in brain power to the seals. And if this whale had lived, thousands more who never went to the Brighton, and whose means or duties would not, perhaps, permit them to do so, would have learned similar lessons. Popular fallacies would have been dissipated, and popular knowledge increased (Lee 1878, 4).

Lee's argument is still common from proponents of captive cetaceans: when ordinary people can learn about the animals by seeing or interacting with them (to "read with their own eye from Nature's own book"), they become stakeholders in saving them.

This model of environmental education, in which keeper and spectator willingly implicate themselves in the captivity of the animal (as an individual) they are supposedly engaged in liberating (as a species), thus has its roots in these early aquaria in which unabashed capitalism (Farini) was underwritten by well-meaning naturalism

(Victorian scientists like Lee). Sarah Amato considers this transformation of the animal to commodity: "As living beings, animals were animate possessions and unique commodities. Unlike other consumer goods, they were not produced by human craftsmanship, but they were subject to various manipulations. Like other goods, animals could take on and express social and cultural meaning through acquisition, use, and other consumer processes. In the Victorian period they became commodities and possessions at a particularly significant moment in the development of consumer society" (Amato 2015, 9). She tells us that this relationship was especially fraught during this period: "For Victorians, the materiality of living animals was a matter of common sense and sometimes problematic. Their relationships to pets and zoo animals were tangible, smelly, messy, disconcerting, comforting, and sometimes tasty; representations could be equally troubling" (Amato 2015, 15). This relationship is *especially* "disconcerting" for sea animals, whose wild lives are even further removed from human view, and whose captive lives must be mediated twice: through the surface of the enclosure *and* through the surface of the water.

The commodification of the Westminster Whale's death confirmed for Farini the profitability of live cetaceans, and "no sooner was the breath out of [the whale's] body than [Farini] exclaimed, 'If I live till next year I will bring a dozen White Whales here if I want them" (Lee 1978, 9). On May 18, 1878, four whales left America on the Allan Line Steamer *Circassian* and arrived at Liverpool on May 27. One suffocated en route when it turned over in its crate in rough weather, and of the three who survived, one was sent to Blackpool, one to Pomona Gardens in Manchester and the third to the empty tank at the Royal Aquarium in Westminster (Figure 14.3).[5]

The Whitsuntide holiday of 1878 was particularly busy for the second whale. Remarked the *Times*:

> The Royal Aquarium was one of the few places of indoor amusement that did not suffer by the fine weather, and this success must be largely attributed to the attractions of the whale. It was long since discovered that nothing is so fascinating to an English crowd as a sea-monster... The whale was visited by 36,000 people last week, and by great numbers yesterday (The *Times* 11 June 1878, 8).

Cashing in the fact that "nothing is so fascinating as a sea monster" (especially a sea monster like the whale, with whom visitors know they share a certain mammalian kinship), Farini and his underwriters more than recouped their investment with their second whale. But just ten days after the second whale's banner holiday weekend, it died too. Fearing more bad publicity, Farini and his fellow proprietors denied that the whale had died and quickly sent for a replacement from Blackpool which they unsuccessfully attempted to pass off as the same whale (*Edinburgh Evening News*, 20 June 1878). The third Westminster Whale did not live long either. The Royal Aquarium's days were numbered too, and by 1890, the exhibits had largely shifted to seedier attractions such as tattooed American women illumined after dark by the light of the tanks which remained full of water but devoid of fish (Sands 2011).[6] The Royal Aquarium was demolished in 1903 and is now the site of the London Methodist Central Hall.

The Westminster whales bring the ethical concerns of the present to bear upon studies of Victorian visual culture and vice versa. These creatures constituted an early anxious alliance between those who would profit purely monetarily from exhibiting

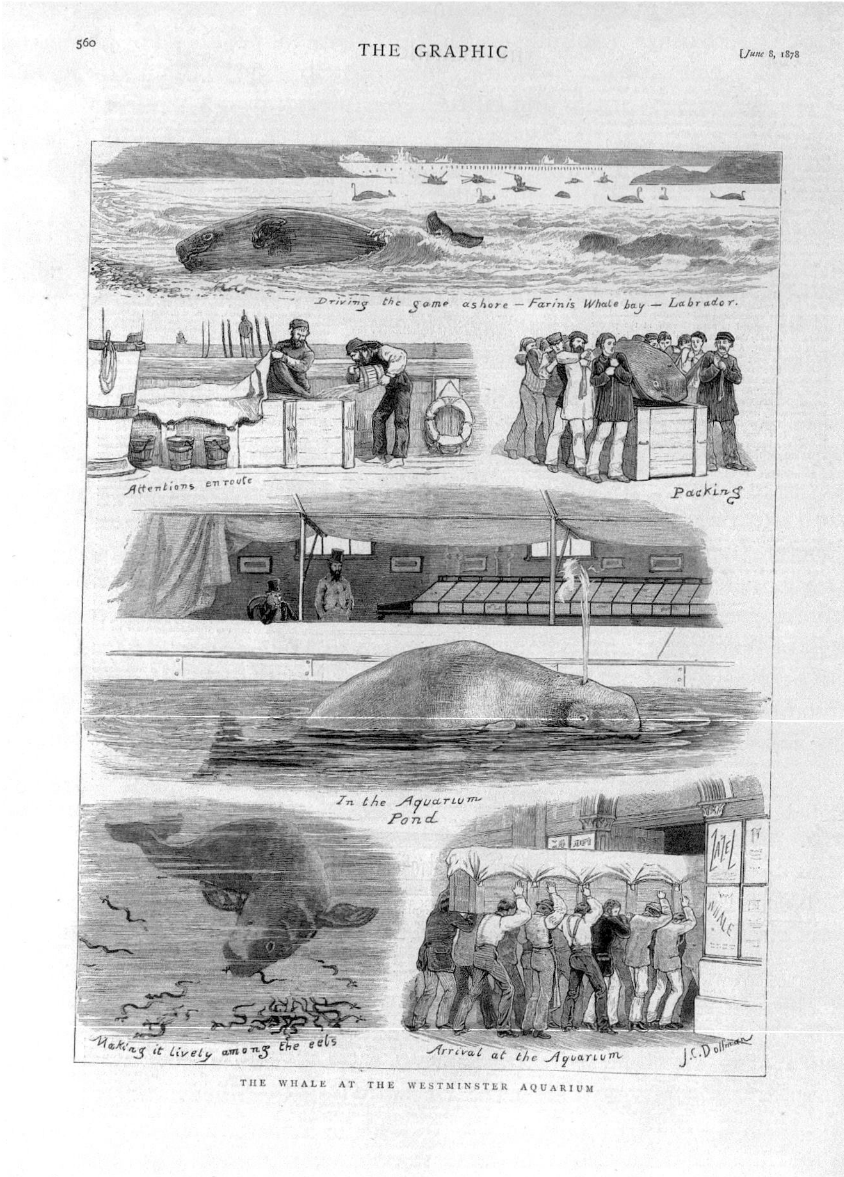

Figure 14.3 The Graphic, June 8, 1878. British Library via Mary Evans Images.

live cetaceans, and those whose research would profit and who would educate the public by way of live whale. "As animals were subject to different managements, manipulations, and interpretations" in the Victorian era, writes Amato, "they took on different social roles, and this irrevocably changed the lives of both animals and humans" (Amato 2015, 6). In terms of the whale's agentic lived experience, where once her worst fate might have been at the end of a harpoon, in the new Victorian economy of live cetacean display, she could be harpooned but not killed, and as that wound healed, she might be subject to a slow death at the hands of inexperienced caretakers who would confine her first to a coffin-sized box in transit across the very ocean from which she was removed and then to a pool no deeper than she is long. Iovino and Opperman posit that "what lies behind the nodes of the ecological crisis—pollutions, mass extinction, poverty, enslavement of humans and animals, and many other forms of oppression—are tangles of natures and cultures that can be unraveled only by interpreting them as narratives about the way humans and their agentic partners intersect in the making of the world" (Iovino and Opperman 2011, 6). The ecological and humanitarian crisis of containing large intelligent mammals like cetaceans has its roots in the nineteenth-century aquarium (a very different cultural construction in Britain than in America), and can be "untangled" only through an interdisciplinary understanding which also involves the "narratives" Iovino and Opperman propose.

These narratives indeed "[make] the world," and I have attempted to bring the narrative of the first Westminster whale to bear upon our contemporary relationship to its figurative descendants in captivity today. Though there were once as many as 36 dolphinaria and live cetacean shows in the United Kingdom, the last closed in 1993 after new legislation which did not ban cetacean captivity but required prohibitively expensive expansions to existing tanks.[7] There are currently, in 2018, over 3,000 whales and dolphins in captivity world-wide. In March 2016, under mounting pressure from animal welfare groups, SeaWorld agreed to phase out its orca shows (San Diego in 2017, Orlando and San Antonio in 2019). The Georgia Aquarium has agreed not to remove any more whales or dolphins from the wild (after an unsuccessful legal challenge to import 18 belugas from Russia), though it has no plans to end captive breeding. The Vancouver Aquarium has announced that it will no longer display captive cetaceans, though the recent death of its two beluga whales has left the aquarium with just one white-sided dolphin called Helen. SeaWorld's last captive orca birth – a calf called Kyara – was born at the San Antonio park in April 2017 but died of pneumonia in July, just like the first beluga at Westminster (and the infamous Tillukum, who killed trainer Dawn Brancheau in 2010). And though SeaWorld has ceased its captive breeding orcas, the parks will continue breeding belugas and dolphins; a baby beluga (the same species as the Westminster whales) born July 2017 at SeaWorld Orland died just moments after birth. We have perhaps learned less than we thought in the 140 years since The Great Farini engaged his first "liliputian leviathan."

Notes

1 "Sometimes I wonder whether [the question of animals] is *the* question," Morton writes, alluding to Cary Wolfe's *Zoontologies: The Question of the Animal* (Wolfe 2003).

2 Ann Colley reminds us in *Wild Animal Skins in Nineteenth Century Britain* that: "If the specimens survived the journey across a foreign land and reached the ship returning to England, there was the probability that they would not survive the voyage. Shipwrecks and fires on board destroyed collections... High winds pitched crates into the sea; unruly animals on board were tossed into the ocean; and live specimens died as a result of close confinement on board, improper diets, trauma, injury, illness, and cannibalism" (Colley 2014, 18).

3 The brain is still at the Hunterian: Ref. no. RCSHM/D521, Brain of *Delphinapterus leucas,* 1877 (1,791 grammes). This was a way for Farini to recoup some of his investment, as although she was also the first whale to be insured, her policy only lasted until she arrived at the aquarium. As the *Fishing Gazette* put it: "The Insurance for £500 was effected from the time it was lifted on board

the *Oder*, to the moment of its being deposited in the tank at Westminster. Had it died one minute sooner, the loss would have fallen on the Paris Marine Insurance Company, one minute later—or as now, upon the Messrs Morris, Farini, and Robertson." The *London Illustrated News* reports that the live whale was worth upward of £1,000 (as opposed to the £20–30 she would fetch for her blubber in Labrador).

4 A well-known British example is Guy the Western Lowland gorilla, who lived at the London Zoo from 1947 until his death in 1978. After his death, his skin was donated to the Natural History Museum, where he was kept in cold storage until being stuffed and briefly displayed in 1982. His display struck too morbid a chord with many (The Viscount Anthony Chaplin, Honorary Secretary of the Zoological Society of London asked, "Are all future Hon. Secs, Presidents, etc. of the society to be stuffed and exhibited in a museum?"), and Guy was removed to Scientific Collections until his 2012 return to permanent display in the museum's new Cadogan Treasures Gallery. When I visited Guy just after his installation, a man looking at the gorilla with his young son remarked to me that as a child, his father had taken him to see Guy at the zoo.

5 In the event of the death of the second round of whales, Lee writes in 1878: "I am informed that Mr. Farini's agents have six more of these whales alive in a sea-pond on the coast of Labrador. They can be sent to him at short notice" (Lee 1878, 10). "If the whale should die before reaching England, Barnum declares he will ship half a dozen at one time next summer, hoping that one or two may survive" (*Illustrating Sporting and Dramatic News* 15 September 1877).

6 Writes Erroll Sherson in *London's Lost Theatres of the Nineteenth Century:* "The Royal Aquarium, in short, was intended to be a sort of Crystal Palace in London within easy reach of Charing Cross, a covered-in promenade for the wet weather, with the glass cases of live fish thrown in. In truth, the attractions of the place soon began to be very 'fishy' indeed" (Sherson 1925).

7 See *Dolphinaria: Report of the Steering* Committee, prepared for the Department of the Environment (now DEFRA), Crown copyright (1988), which is based on Klinowiska and Brown's "A Review of Dolphinaria" (1985). The Department of Environment, Food, and Rural Affairs's stance was confirmed in a 2007 House of Commons debate by the DEFRA Secretary: "It is not illegal to keep cetaceans in this country, the Zoo Licensing Act 1981 (as amended) aims to ensure that, should cetaceans be kept at an establishment for exhibition to the public, the establishment is licensed and the animals kept in accordance with strict standards relating to their health and welfare requirements" ("Cetaceans." Unpublished debates, House of Commons, 8 May 2007, Column 31W-21W).

Works Cited

Abram, David. 2010. *Becoming Animal: An Earthly Cosmology*. New York: Pantheon.

Amato, Sarah. 2015. *Beastly Possessions: Animals in Victorian Consumer Culture*. Toronto: University of Toronto Press.

Armstrong, Philip. 2005. "What Animals Mean, in Moby-Dick, for Example." *Textual Practice*. 19, 1: 93–111.

Bennett, Jane. 2010. *Vibrant Matter: A Political Ecology of Things*. Durham, NC: Duke University Press.

Berger, John. 1977. *About Looking*. London: Penguin.

"Cetaceans." Unpublished debates, House of Commons, 8 May 2007, Column 31W-21W.

Colley, Ann C. 2014. *Wild Animal Skins in Victorian Britain*. Burlington, VT and Farnham, Surrey: Ashgate.

Daily News (London). 1 October 1877. *British Newspaper Archive*, British Library, October 1. Accessed 2016.

"Dolphinaria: Report of the Steering Committee," prepared for the Department of the Environment (Crown copyright, 1988).

Edinburgh Evening News. 1878. *British Newspaper Archive*, British Library, June 20. Accessed 2016.

Fishing Gazette (London). 1877. *British Newspaper Archive*, British Library, October 5. Accessed 2016.
Garrard, Greg. 2011. *Ecocriticism*. New York and London: Routledge.
Gibbens, Sarah. 2017. "SeaWorld's Last Captive-Born Baby Orca Dies." *National Geographic*, July 25.
Green, Henry. "The Wooden Walls of Old England." *Morning Chronicle and London Advertiser* (25 June 1773).
Hampshire Telegraph and Sussex Chronicle. 1877. *British Newspaper Archive*, British Library, October 6. Accessed 2016.
Haraway, Donna J. 2003. *The Companion Species Manifesto: Dogs, People, and Significant Otherness*. Chicago, IL: Prickly Paradigm Press.
Illustrated London News. 1877. *British Newspaper Archive*, British Library, October 6. Accessed 2016.
Iovino, Serenella, and Serpil Opperman. 2011. *Material Ecocriticism*. Indianapolis: Indiana University Press.
Klinowiska, M. and S. Brown. "A Review of Dolphinaria," prepared for the Department of the Environment (Crown copyright, 1985).
Lee, Henry. 1878. *The White Whale*. London: K. Burt and Co.
Levi-Strauss, Claude. 1964. *Totemism*, translated by Rodney Needham. London: Merlin Press.
Lichfield Mercury. 1877. *British Newspaper Archive*, British Library, October 5. Accessed 2016.
Liverpool Mail. 1877. *British Newspaper Archive*, British Library, September 29. Accessed 2016.
Lord Byron. 2008. "Childe Harold's Pilgrimage." (IV.179). In Jerome McGann, ed., *The Major Works*. Oxford: Oxford University Press: 199.
Mentz, Steve. 2009. *At the Bottom of Shakespeare's Ocean*. London: Continuum.
Morse, Deborah Denenholz, and Martin Danahay. 2007. *Victorian Animal Dreams: Representations of Animals in Victorian Literature and Culture*. Burlington, VT and Farnham, Surrey: Ashgate.
Morton, Tim. 2007. *Ecology without Nature*. Cambridge, MA: Harvard University Press.
Nietzsche, Friedrich. 2003. *Beyond Good and Evil*, edited by Michael Tanner, translated by R. J. Hollingdale. New York and London: Penguin.
Ritvo, Harriet. 1987. *The Animal Estate: The English and Other Creatures in the Victorian Age*. Cambridge, MA: Harvard University Press.
Sands, John. 2011. "Sullivan and the Royal Aquarium" (*The Gilbert and Sullivan Archive*, Boise State University, Accessed 2013. https://diamond.boisestate.edu/gas/articles/sull_aquarium/index.html.
Sherson, Erroll. 1925. *London's Lost Theatres of the Nineteenth Century*. London: John Lane.
Taylor, Jesse Oak. 2015. 'Where Is Victorian Ecocriticism?' *Victorian Literature and Culture*. 43, 4: 877–894.
The Times (London). 1877. *British Newspaper Archive*, British Library, October 3. Accessed 2016.
The Times (London). 1878. *British Newspaper Archive*, British Library, June 11. Accessed 2016.
Wolfe, Cary. 2003. *Zoontoloties: The Question of the Animal*. Minneapolis: University of Minnesota Press.

15 How to Wear the Feather

Bird Hats and Ecocritical Aesthetics

Emily Gephart and Michael Rossi

Introduction

Fashion may be cruel, but it is not without its own logic. This was the point of the illustration (Figure 15.1) accompanying an editorial in the November 1883 issue of *Frank Leslie's Illustrated Newspaper.* The target of the text and picture was the burgeoning trend in women's hats adorned with the feathers; body parts; and sometimes even the full, taxidermied bodies of birds. Enumerating the players held responsible for "the Cruelties of Fashion," the drawing depicts a Via Dolorosa of frivolous avian death: from a hunter shooting down a lark at upper left to a taxidermist eviscerating a bird in his studio at upper right to a barking dog chasing a feathery victim at lower right. But it is the central figure in the image who is also the most unlikely agent of carnage: a slender, modish and pretty young woman who flirts with the viewer from under the brim of an artfully bird-covered bonnet. It was she who demanded blood sacrifice, the accompanying editorial confirmed, "in order that her hat, her coat, her cuffs may be adorned with the gloriously-colored plumage."

On the one hand, the picture is an indictment of womankind's fatuity: blind to the suffering caused in pursuit of superficial egoism, she is undistracted by the chaos that her obsession with fashionable display wreaks all around her. Such an interpretation undergirds the fascinating and well-told history of how bird hats served as a forceful prompt to turn-of-the-century conservation movements (Doughty 1975; Price 1999; Moore-Colyer 2000; Weidensaul 2008). The passionate defense of birds – led, more often than not by middle-class women and pursued through groups like the Audubon Society – eventually led to restrictions on the global trade in plumes and avian ornament, and wholesale changes in fashionable tastes. In the literature tracking conservation's rise, hats have proven an appealing and materially grounded demonstration of how progressive coalitions of right-minded citizens, convinced of the self-evident merits of conservation, came to a consensus about the appropriate relationship between nature and society: one in which responsible stewardship bolstered both the natural and the cultural. Historians have conscientiously noted how class and gender inflected these movements. Defensively paternalistic and self-righteous, they constructed "nature" as an ideal and prelapsarian state in need of protection from the unbridled greed of the ignorant masses easily swayed by fashion (Mason 2002; Merchant 2010).

On the other hand, even as the caption below the illustration decries the "cruelties of fashion," the composition of the figures tacitly naturalized an essentially gendered impulse to extravagant visual display. If it was woman who centrally bore the blame for a need to adorn herself in accordance with the adage "fine feathers make fine birds" – as

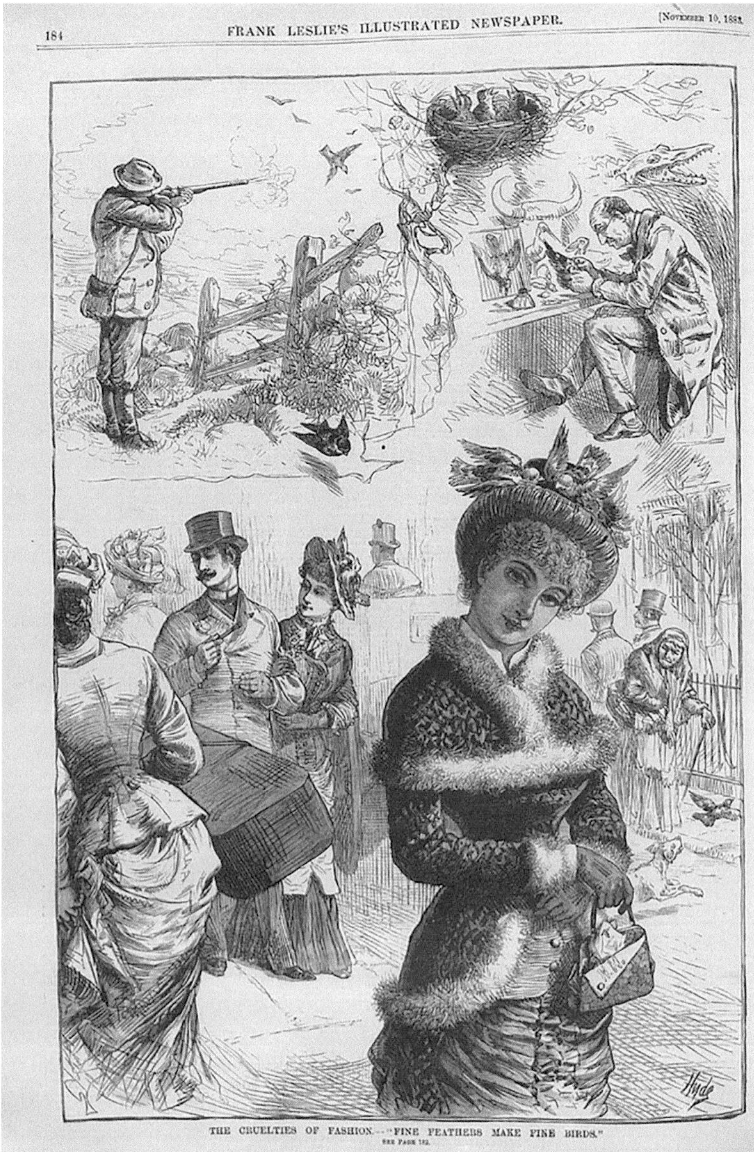

Figure 15.1 John N. Hyde. "The Cruelties of Fashion – Fine Feathers Make Fine Birds." *Frank Leslie's Illustrated Newspaper*, November 10, 1883. Image courtesy of Library of Congress Prints and Photographs Division.

the caption suggested – she did so according to a sort of natural law. Included among the hunter, the dog and the taxidermist who orbit the central woman, we find a dapper gentleman who gazes admiringly at other hat-wearing ladies, affirming the primary function of all fashionable attire as commonly held by *fin de siècle* conservationists and fashionable society alike – winning attention and thereby securing a mate. In this way, the illustration reckons not only with a history of conservationism based on utilitarian moralism, but also with a parallel narrative – one in which wearing hat-feathers recapitulated, in distinctly modern terms, habits of aesthetic display within the (nonhuman) animal world. Indeed, if death in the name of fashion was cruel, it was not entirely purposeless, nor even unnatural, since both display and predation were behaviors common to all animals. Within this framework, hat-wearing women served not only as predator but also as a kind of prey. Her ostensibly natural desires are both the primary cause of – and the justification for – the need to protect birds (and, by extension, all of nature) from the depredations of human kind. Amid such aesthetic contradictions, modern conservationism took flight.

Bird hats call for ecocritical reevaluation of their peculiar animal-and-human hybridity, their vitality and their visuality. Far from simply being about saving birds, condemnation of fashion's cruelty also mobilized new arguments about the natural and the cultural, the social and the scientific, the monstrous and the alluring. We argue that these dichotomies brought a prominent garment into dialogue with notions of population survival, fitness, the 'proper' order of society and a peculiarly progressive form of aesthetic vitalism, as we track the inter-related pragmatic, economic and moral justifications for bird conservation to which illustrators provided visual form. Ultimately, we argue that millinery fashion's ostensible violations set in motion wholesale renegotiation of the modern valences of aesthetic presentation, predation, protection and other behaviors shared 'transcorporeally' between the human and other-than-human worlds (Alaimo 2010).

Economy

Bird hats are paradoxical entities. Although composed of the lifeless bodies of birds – and responsible for the near-extinction of species such as egrets – the dead birds and disembodied plumes on hats enhanced the wearer's allure by reference to the living animals' vitality. The celebrated grace, elegance and splendor of birds lent themselves to the women who wore them atop their heads. "We do not wonder," remarked one commentator, "that women should wish to adorn themselves with a masterpiece from the hand of nature which is more delicate than lace and as graceful as the lines of frost upon the window pane..." ("Women and Egret Plumes" 1895, 4). The more exotic the animal, the greater appeal of its plumes, and the more compelling the analogy to human traits: a 1914 issue of the illustrated humor magazine *Puck* compared its feather-wearing cover girl to a tropical Bird of Paradise, which were by that time hunted to the verge of extinction. She preens like a rare specimen flashing her natural finery, her face framed by sweeping plumes instead of hair (Figure 15.2).

Feathers and bird parts could be seen on all kinds of hats, from close-fitting bonnets and sporty boaters, to outlandishly wide-brimmed 'picture hats.' In some years, ornate toques sported the entire bodies of grebes or parakeets, posed as if preparing to nest, in a perverse imitation of suspended vitality. At other times, the heads of hundreds of jewel-toned hummingbirds embellished the bandeaux of 'picture hats,' flanked by sprays of fluttering quills descending from their upturned brims. Along the eastern seaboard of the United States, entire nesting colonies of common terns were wiped out, their wings applied whole to 'Mercury' hats, named for their resemblance to the fleet-footed Roman god's helmet (Chapman 1899, 205). A craze for 'Chanticleer' hats, referencing playwright Edmond de Rostand's stunningly popular

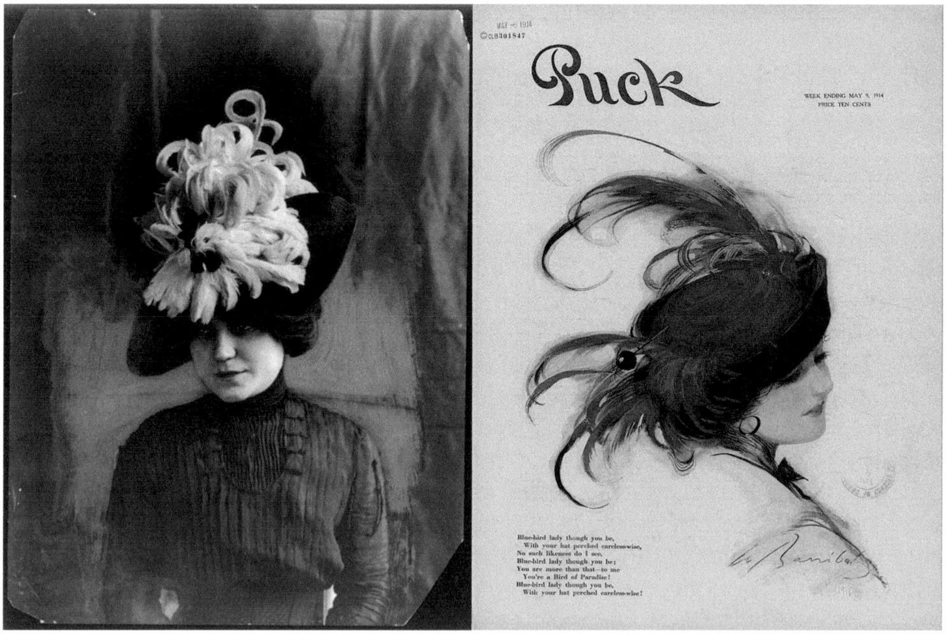

Figure 15.2 Left: Bain News Service. Model wearing 'Chanticleer' hat of bird feathers, ca. 1912. George Grantham Bain Collection. Right: William Barribal, "Blue-bird lady though you be." *Puck Magazine* cover, Vol. 75, No. 1940, May 2, 1914. Both images courtesy of Library of Congress Prints and Photographs Division.

1910 avian fable, flaunted extravagant cascades of plumes (Figure 15.2). America's foremost trade journals tracked such fluctuating trends, and carefully listed the provenance of bird parts used. *Illustrated Milliner*, for example, observed that the leading autumn styles of 1902 featured "elaborate fancy breast effects ... of three or more different kinds of plumage," including "the herron [sic], nashwah, the pheasant in lovely copper tints, and ... paradise tails" ("A Few of the Novelties Shown in New York," 1902).

Hats were mandatory attire for upper-middle-class women who wanted to demonstrate fashionable acumen: in 1904, as one journalist observed, "that every woman who would be considered well dressed must have separate and distinct hats for various occasions is as fixed as the laws of the Medes and the Persians" ("Midsummer Fashions" 1904, 688). After trends in ornament were set in Paris, American fashion journals such as *Vogue* and *Harper's Bazar* relayed the popular millinery modes that were sure to secure the notice of their middle-class female subscribers. If a season's leading styles called for feathers – as most did, even amid rising criticism of the plume trade – women displayed their sophisticated aesthetic tastes by wearing bird parts. In her textbook on millinery history and manufacture, Charlotte Rankin Aikin noted that, "feathers are suitable for all seasons as they are always attractive" (Aikin 1918, 72). Unlike seasonally specific fur or flowers, multihued feathers coordinated with any attire.

But this genus of fashion extracted considerable cost: ornithologist T. Gilbert Pearson reflected on the alarming rise of the American plume trade in the decade after 1880. "People seemed to go mad on the subject of wearing feathers... Never in this country have birds been worn in such numbers," he observed, and worried whether "the call for feather finery rings so loudly in the hearts of women that it must never cease to be heard"(Pearson 1916, 254 & 258). In a vivid and widely repeated anecdote, fellow ornithologist Frank Chapman spent two afternoons in 1881 walking through Manhattan, spotting no fewer than 40 species of birds atop the hats of shopping ladies. He published a list of "the species destroyed" in *Forest and Stream* – a conservation-minded magazine (Chapman 1886, 84). In 1905, meanwhile, *Harper's Weekly* reported the plumes of 192,960 dead aigrettes sold out of one London auction house alone (Clemens 1905, 658).

The force of outrage at the wanton waste of bird life for fashion's whims poured forth most vocally from what one observer termed "a small coterie of American women ... the acknowledged social leaders in our principal cities" ("Reintroduction of Feathered Millinery" 1888, 208). After reading an 1896 essay concerning the massacre of Florida egrets and the plight of their abandoned offspring, Boston philanthropist Harriet Hemenway dedicated herself to their protection. She rallied support from her elite friends calling upon their sentiments as well as their sense of social entitlement, and – along with ornithologist William Brewster – revived the Audubon Society, a largely defunct 1886 conservation club (Graham 1990; Mitchell 1996).[1] Thanks to the efforts of many similar societies devoted to bird preservation that arose in the 1890s, regulations slowed the trade in poached wild birds. The Lacey Act of 1900 – called the "bird" bill for its specific protections of "game birds and other wild birds" – offered the first federal attempt to control the trade in wildlife in the United States (Cart 1973, 4). Even so, federal laws could do little about the global market for exotic plumes, and illicit hunting remained a concern.

At the root of this narration of bird conservation – from moral outrage to legislative action – was a tension between different ideas about the practical and moral uses of nature. On the one hand, hunting for food and clothing was an acknowledged aspect of the human and natural order. Nature was a part of human economy, and birds were a resource that supported human beings. In 1913, Missouri Senator James Reed argued in favor of hunting egrets for their plumes, wondering "[w]hy there should be any sympathy or sentiment about a long-legged, long-beaked, long-necked bird that lives in swamps and eats tadpoles?" Cessation of the plume trade, after all, threatened to put thousands out of work: as many as 83,000 people were involved with the collection of feathered parts for hats, according to some annual employment estimates ("A Bad Showing –for Missouri" 1913, 173). What sort of civilization valued tadpole-eating birds over human beings?

In riposte, bird advocates offered a sort of affective ecology of birds and humans, in which the realization of mutual benefit was a cornerstone of notional civilization. As a writer in *Audubon Magazine* observed in 1888, "birds contribute so greatly to man's well being by their services in preying on insects and small rodents, and to his enjoyment by their beauty, vivacity, and song, that their annihilation would be a crime against humanity"("Reintroduction" 1888, 208). According to this interpretation, the lives of birds were so beneficial to humans that their loss – regardless of the economic cost – redounded to a crime: not merely an ecological issue, but a juridical one. A "sister of Dr. C. Hart Merriam" (a prominent ornithologist) put the matter more bluntly when she wrote to her fellow conservationists about the duty that they owed the natural world:

> Foremost in the circle of lives among which we move is that of the bird-world. [...] This beautiful bird-world is always about us, or rather, we are always in its midst, and, absorbed in our own thoughts and feelings, often pass along and think nothing of it. But we do more than this. We wrong the bird-world more actively, and though thoughtlessly, it is no less a wrong. [...] Every time that we buy a wing or a head, and part of once held a happy bird's life, in order to add to our own attractiveness, we are not only committing a crime against the bird-world, not only violating our own best natures, but we are retarding the progress of civilization by an act of barbarity ("Birds and Bonnets" 1885, 385).

For this writer, it was not merely the shocking destruction of the natural world that should motivate her fellow bird lovers, but its threat of subverting a presumed hierarchy of 'civilized' behaviors in the human realm.

Aesthetics

More than simply a pragmatic or ecologically moral argument, however, conservationists also leveraged this profound affinity between bird and human life into a specifically gendered argument about the connection between the reproduction and the aesthetics of display. Modern women inverted Darwinian reproductive dimorphism, according to this narrative, adopting the showy habits male birds employed to attract mates, and reconstituting them as necessary to the future survival of the human species. As fashion writer Mary Eliza Haweis cautioned, women were "not to be blamed for" their love of conspicuous display. "A woman's role is to attract," Haweis intoned, "and when she has attracted, to enchain" (Haweis 1883, 129). Birds and humans could thus both be seen as driven to ostentatious display – as natural in avian bipeds as it was in their mammalian counterparts. The very same affinities joining birds and humans in a sort of vital aesthetics of species-perpetuation could be seen as the source for humans' love of birds as adornment.

Writing for *Popular Science Monthly*, Canadian zoologist Grant Allen pursued this linkage, arguing for a strong evolutionary link between animal and human aesthetics. In birds, he reasoned, the color, pattern, proportion and visibility of plumage served the struggle for survival. Bird display was simply an earlier developmental manifestation of the common human capacity for all aesthetic appreciation. As he put it, "If ... we admit the reality and potency of sexual selection, in however modified a form, it must follow that birds, being on the whole the most ornamental of all classes in the animal world, are also the most aesthetic, with the exception of man." He asserted that birds demonstrate "extraordinary evidence of a taste for all that man considers lovely or artistic." And thus, sexual selection in both species was conveyed by superior display and aesthetic discernment: "as every individual is himself the product of countless thousands of prior individuals, all of whom have been in the main successful in the struggle for life and the search for mates," he said, "it must follow that he will have inherited from them, ... a healthy taste for that particular arrangement of limbs and features [or feathers] which best suits the essential conditions of the species" (Allen 1880, 662). This "particular arrangement," in fact, closely coordinated the purpose of human and animal display.

Further evidence of the complexities of animal-human hybridization arose from branches of modern science other than ornithology. As anthropologist Franz Boas would tell his students in an 1896 lecture concerning "Primitive Art," the feathered plume played an important role in the natural history of humankind. Human beings always decorate themselves, said Boas. They expend energy in self-adornment – in display – as part of mating and reproduction (Boas 1896). But lest we think that modern humans have a different standard of decoration than their "primitive" peers, Boas pointed out that

> feathers wave on the hats of primitive warriors as well as on those of generals of our armies. Bird-skins are worn in the hair of primitive women as well as on the hats of her civilized sister. There is one important difference, however [between primitive and civilized societies]: among primitive people, it is principally the man who adorns himself, while in civilized society, it is principally the woman (Boas 1896).

He went on to say that this was precisely because of the material conditions of modern society, which made it such that women were in the position of needing to attract a mate – rather than the other way around. Thus, although conspicuous plumes were widely thought to overturn the 'law' of Darwinian dimorphism, wearing feathers did not recapitulate 'primitive' behavior so much as update nature for the particular conditions of modern society.

Such musings on the nature of culture and the culture of nature were not limited to lecture halls and popular press. To combat bird death by raising public awareness, conservationists like Chapman – who became curator of Mammals and Birds at the American Museum of Natural History (AMNH) in 1901 – began drawing attention to the aesthetic affinities of birds and humans through the emerging medium of "habitat groups." These displays in natural history museums featured carefully taxidermied animals whose lifelike appearance was enhanced by painted and fabricated simulations of their surroundings.[2] Like windows into the natural world, these vivid presentations enabled visitors to imagine, even to experience, nature at its most relatable, as the posed animals fed, fought, courted and – perhaps most sympathetically – cared for their own preserved families.

Chapman's 1909 diorama for the AMNH depicting "the American Egret in a Southern Carolina Cypress Forest" (Figure 15.3) was an exemplar of didactic display, depicting a group of this species of heron in their annual mating plumage tending to their young. In the text which accompanied the exhibit, Chapman explained that egrets had been "brought to the verge of extinction" by the millinery trade. "So effectively, indeed, have the plume hunters done their work," wrote Chapman, "that it was feared the vanishing species could not be included" among the museum's dioramas. Fortunately, the museum found a group of egrets that had been protected from hunting – and was able, then, to "collect" the birds for themselves (Chapman 1909, 21).

Yet the egrets did not wear just any plumes; rather, as Chapman put it, they sported "their wedding costume." In the seasonal southern rookeries to which thousands of avid hunters flocked, both male and female birds flashed their beautiful feathers to attract one another, and the staged 'narrative' as constituted by Chapman showed the result: a domestic, nuclear family group. By their very position as sympathetic spectators, the middle-class viewers to whom Chapman addressed this drama were also

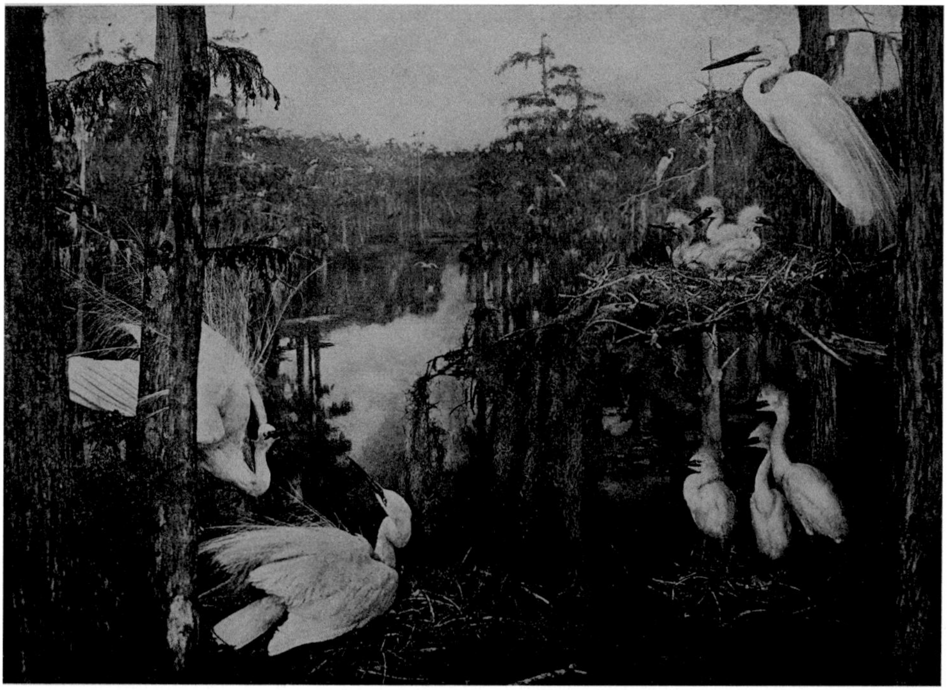

Figure 15.3 The American Egret in a South Carolina Cypress Forest. North American Bird Hall, American Museum of Natural History, ca. 1909. Image courtesy of the American Museum of Natural History Library Services.

implicated in the cycle it illustrated. Invited to admire the plumes for their aesthetic virtues, and to connect those qualities to hats, they were also asked to understand bird nature (in the form of avian matrimony) as similar to their own. Just as a gaudy display was the key to the underpinning of the egret "family" group, so, too, did modern humans use gaudy displays in pursuit of their own nuptials. Chapman's language urged modern viewers to understand their own socio-biotic substance as aligned with this vision of endangered nature.

The "wedding clothes" that aigrettes wore were more than mere ornamentation: they were a reminder to viewers – in particular, the middle-class women who comprised some of conservation's earliest leaders and advocates – of their duties at the intersection of nature and culture. May Riley Smith, representing the Audubon Society of the State of New York, argued emphatically that *aigrettes* especially should be "sacred in a woman's eyes, for it is the nuptial plume of the bird and its token of motherhood" exhorting women "In the name of humanity, of womanliness, of motherhood [...] to refuse to wear the aigrette, and to influence others to do so" (Smith 1906, 43).

Calling upon viewer's emotional affinities, the dynamic at play in the egret group posed humans as both implicated with and husbands of the natural world. From Chapman's vantage point, birds preserved for posterity in a natural history museum supported a view of conservancy that secured human paternalistic stewardship of nature: unmoored from natural law by the virtues of knowledge and management, human agency was both omnipotent and dependent. As Donna Haraway (and others) have shown, this 'view from nowhere' contained within in it the very notion of culture that was to supersede nature, by means of its advancement, indeed by the very 'civilization' that Merriam's sister championed (Haraway 1888, 1984–1985; Wonders 1993).

Monstrosity

Notwithstanding the aesthetic logic of Boas and Allen's observations, or their scientific credentials, conservationists served their cause by re-casting hat display as a violation of natural order. Yet accomplishing this rhetorical strategy meant reconfiguring the terms of being natural in culture – measured not in terms of display, but in terms of an essentialized nurturing role. What mattered to saving birds was not demonstrating the affinity between egret and human propensity for display, but rather stressing its end goal: the reproduction of the species. Motherhood could be equally naturalized as the fundamental mode of women, went the argument. In fact, to forego this natural maternity in favor of rapacious display was not to echo bird habits, but to perform a perverse amalgamation between species: to become a monster.

Thus in order to demonstrate their superior maternal, caring and nurturing behaviors, truly moral women had to cast off the deep connection to "nature" championed by Boas and Allen, and adopt a different configuration of "modern" nature. In these restructured priorities, superficial, visual affinities with the animal were sundered, in favor of deeper, underlying attachments to the promise of perpetuating desirable, even ennobled characteristics of the species that would further advance 'civilization.'

Of course, not only defenders of bird hats but even some of their staunchest critics admitted that the desire to display was perfectly natural. Edward Bok, longtime editor of the *Ladies Home Journal*, argued that the demands of human reproduction prevailed over other sensibilities: "When it comes to the question of her personal adornment, a woman employs no reason; she knows no logic. She knows that the adornment of her body is all that she has to match the other women and out do her, and to attract the male, and nothing that you can say will influence her... that is the feminine nature" ("Edward Bok: The American" 1920, 261). Even devoted ornithologists like Pearson conceded that the tendencies of fashion were inborn: "Nearly two decades ago a lady came into an evening gathering with a dead bird on her bonnet. This was something new, a novel idea, her admirers said. [...] with the hereditary impulse to adopt everything new, the fancy was seized upon by those we term the gentler sex, and only when Fashion itself shall be no more will the full extent of the mischief be known" (Pearson 1895, 102).

At the same time, it was possible to argue that even if woman's 'hereditary impulses' were natural, fashion itself was not only irrational, but also intransigent, unnatural or even inhuman. "Fashion has a more or less hypnotizing effect on us all, and we soon become accustomed to even its absurdities," said Mary Bannister Willard, writing for *Harper's Bazar* (Willard 1908, 580). Therefore, many conservation advocates, convinced of the seemingly obvious or indeed the 'natural' merits of preserving bird life, counterposed their moral arguments against such competing demands of fashionability. Bird enthusiast Mary Thatcher was outspoken on the matter, also writing for *Harper's Bazar*: "Fashion delights to set all the laws of nature at defiance, but she never showed more plainly her ignorance of the fitness of things than when she took the birds from their native haunts and perched their lifeless bodies upon the heads of our mothers and sisters and daughters" (Thatcher 1875, 338).

Celia Thaxter – celebrated nature writer, horticulturalist and bird lover – penned a scathing diatribe against "Woman's Heartlessness" in 1887, recalling how an acquaintance wearing "a charnel-house of beaks and claws and bones and feathers and glass eyes upon her fatuous head" had professed indifference to the wanton cruel destruction of birds. She told Thaxter that they "will soon go out of fashion, and there will be an end of it." Denouncing such callous disregard, Thaxter praised women who wore "birdless bonnets" in defiance of stylishness. As she claimed, women who instinctually followed trends "like a flock of sheep" were liable to "forget reason, forget the human heart within, forget everything but the empty pride of being 'in the fashion,'" and she implied that women only truly surpassed their underlying animality when resisting fashion's urgent pull (Thaxter 1887, 13–14).

Fashion's bird–human hybridity was not only hypnotic, steering entranced women toward mindless pursuit of perversity, but under its sway, women became hunters and predators, which, in the eyes of many, was a decided offense against the natural human order. In his full-page 1911 caricature 'The Woman Behind the Gun" (Plate 9), popular *Puck* magazine illustrator Gordon Ross shows just such a violation of roles, in the form of a slender, prominent and stylishly attired woman taking expert aim at a threatened egret colony in the southern American swamps. Sporting an immense *aigrette* spray on her hat, she targets the remaining wild birds, while the bodies of various other resplendent species lie at her feet. Her trained game dogs – identified as milliners – carry other bodies to join the pile, supporting the proposal that it was chiefly fashion-seeking women whose avidity fueled the ongoing market for plumes.

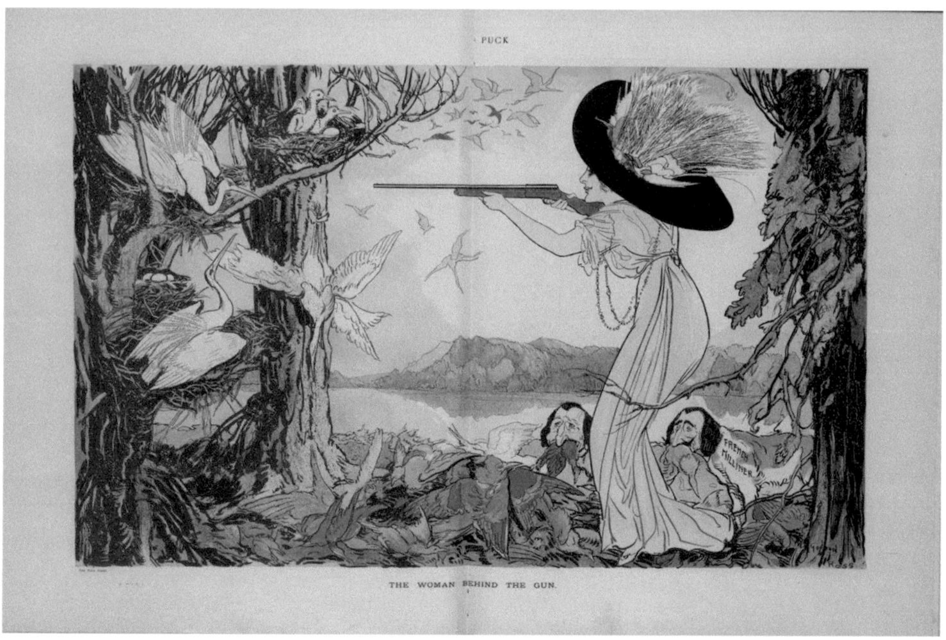

Plate 9 Gordon Ross, "The Woman Behind the Gun." *Puck Magazine*, Vol. 69, No. 1786, May 24, 1911. Image courtesy of the Library of Congress Prints and Photographs Division.

Moreover, if bird-bedecked hats mirrored nature in their display, the vision of nature they manifested was not merely inverted, but was also monstrous. The editor of *Our Animal Friends* argued fiercely against wearing birds in 1895, "A woman who would be guilty of such cruelty for the sake of ornament to her person would be no true woman, but a feminine monster"("Women and Egret Plumes" 1895, 4). With canny wit, British illustrator Linley Sambourne envisioned just such a creature: a plume-wearing 'Bird of Prey' invading the peace of a wetlands refuge, her arms flung wide to display a fearsome wingspan and her feet transformed into a raptor's claws (Figure 15.4). Her victim, a smaller bird, eludes her grasp, but she has evidently

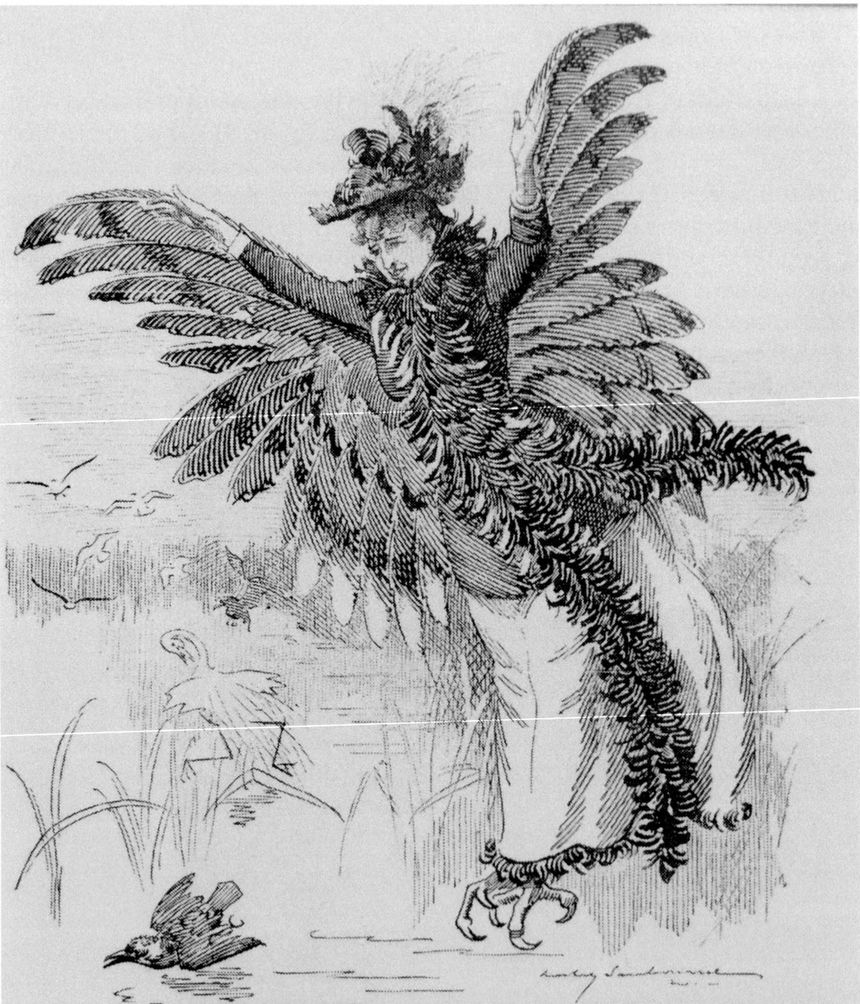

Figure 15.4 Linley Sambourne, "A Bird of Prey." *Punch, or the London Charivari*, 102, May 14, 1892. Image courtesy of Harvard University Libraries.

already preyed upon others: she sports an aigrette spray atop her bonnet, and a long feather boa swings from around her neck.

In another of his astute pictorial commentaries drawn for *Puck*, Ross showed how well such monsters could hide in plain sight (Plate 10). Here, he condemns fashionable women's ostentatious assumption of roles that violated their natural modesty, reserve and motherhood: a group of wealthy socialites wearing preposterous feathered picture hats surrounds a bareheaded nursing mother, gazing down at her with puzzlement and scorn. Carrying dogs and dolls instead of babies – and accompanied by a chimpanzee in a tuxedo – they embody the grotesque absurdity of refusing a purportedly 'natural,' biologically driven and socially necessary role in favor of fashionability.

The particular kind of monstrosity demonstrated by bird hats was made quite plain not merely because they implicated women in killing, but because that killing was proudly sported as a very prominent badge of fashionable womanhood. As reported in *The Sanitary Record*, "It is no more cruel for a woman to cause a bird to be killed for an ornament for her hat than for a man to shoot a bird for the pleasure of pursuing and killing it. Perhaps neither action is strictly defensible on the score of humanity, but ... [w]ith regard to the birds, it is not so much the actual cruelty as *the cruel look* of the fashion which is the ugly thing" ("Fashions of Dress and Their Wearers" 1888, 271). In this, it is not the moral opprobrium of bird death that was chiefly to be despised, but rather its visibility. The male hunter killed his birds out of the public view – and if he taxidermied his trophies, they were, by and large, displayed in his home, out of the public eye.[3] The "cruel look" of fashion, on the other hand, was

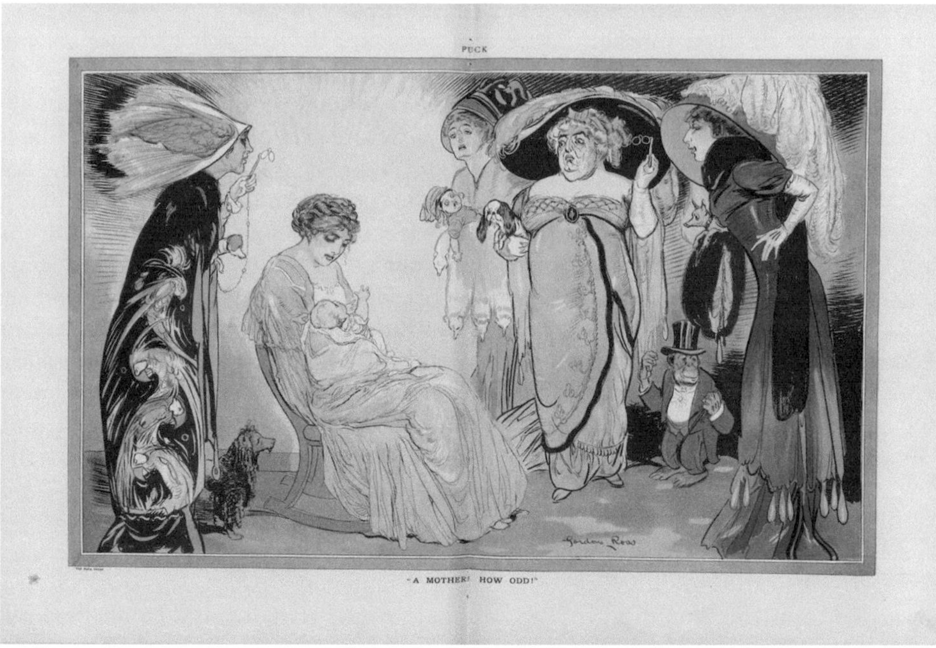

Plate 10 Gordon Ross, "A Mother! How Odd." *Puck Magazine*, v. 68, no. 1764, December 21 1910. Image courtesy of the Library of Congress Prints and Photographs Division.

very much in the public eye – as was, therefore, the fact of the visual stimulation (and accompanying prompt to reproductive desire) that accompanied the wearing and viewing of bird hats.

Mounting arguments against bird hats did not merely invoke practicality, or even progressive stewardship, then, but called upon the need to regulate the degree of affinity between the human and the animal, directing it away from monstrosity, and toward other forms of aesthetic rapport. To be a modern woman – the argument went – was to disaggregate the animal propensity for display, while compensating for its loss by embracing the alternate natural behaviors of maternal nurturing and compassionate care. "Women are supposed to be of gentler and more refined natures than men," wrote one conservation-minded observer, but the outrageous, vulgar presentation "of heads and wings, and often of entire birds on hats...would go far to make one doubt their superior and gentler natures" (O. W. R. 1886, 83). The implied solution, of course, was to remove all shadows of such doubt by refusing to sport feathery fashion.

Pearson rallied his bird-loving supporters by raising the venerable banner of 'true womanhood': "This terrible slaughter can only be checked by our true women who ... cease to wear aloft the mangled remains of the little innocents, and use their influence on behalf of those creatures who cannot speak for themselves" (Pearson 1895, 108). This essentialized model of behavior was naturally above moral reproach, but more importantly, it was nurturing, regardless of the material conditions of modern society. The cause of 'true womanhood' was, as it had ever been, also a class obligation: another way of articulating the needs of those who – supposedly – could not give voice to their own (Welter 1866). Indeed, as mothers were presumed to "speak for" babies, so, too, should all women speak on behalf of birds.

In fact, middle-class motherhood could also "speak" for other classes as well as (or even in the manner of) speaking for other species. As one Mrs. Charles Mallet explained, deftly mixing sentiments of emergent conservationism with *bourgeoise* entitlement, millions of songbirds suffered the same fate as more exotic species like Birds of Paradise, simply "because servant girls and factory lasses must, of course, follow the fashion set them by their leaders" – and thus set about decorating their hats with less expensive local birds, rather than more spectacular and pricier tropical plumes (Mallet 1900, 244). Celia Thaxter also affirmed, "let fine women lead the way, the rest will soon follow; the servant will not wear what the lady refuses to countenance; for curiously enough, fashion is respected as much by the ignorant as by the cultured" (Thaxter 1886, 186). Audubon Society members and other conservation advocates agreed that guidance, governance and moral tutelage had to come from women like Hemenway, educated to 'know better' than be led by fashion's fickle trends, and not only by means of strident editorials in fashion magazines, but by the legislation that they could lobby for via cooperative male associates.

Although fashion and millinery magazines noted that it was easily possible – and especially artful – to use the feathers of domesticated fowl or other nonlethal materials to fabricate facsimiles of exotic and threatened species, the staunchest conservationists in the Audubon movement maintained that display of all bird-related embellishment must be entirely eschewed, in order to meet the greater goal of protecting birds. "Even by wearing plumes that are genuinely artificial a bad example is set and an evil fashion kept in vogue," proclaimed Mrs. F. E. Lemon to the readers of *Women in Social Life* (Lemon 1900, 236). Such claims suggest that although no animal needs to suffer for fashionable display, "unnaturally natural" instincts were nonetheless stirred by the ongoing fusion and confusion of bird and human.

Conclusion

Beyond looking at the *Frank Leslie's* illustration merely as a clever pictorial argument against the self-evident 'Cruelties of Fashion,' analysis of its deeper – and fundamentally contradictory – visual logics reveals other dynamics at play in the service of saving birds. If these contradictions may seem overt, they were also subtly inconspicuous at the time. In fact, the decades it took for conservationists' arguments to sway public sentiment against bird hats attest to the complexity of the 'transcorporeal' imbrication we have investigated. What matters in considering how conservationists' arguments were framed visually is therefore as crucial now as it was for the egret in its time of dire threat: it shows how to think with and think through the tools that such aesthetic frameworks can provide.

In the long, drawn-out efforts to secure bird conservation – and constituting a counter-narrative lurking behind the caricatures and diatribes – a mode of aesthetic kinship requiring careful regulation was at stake. This model figured interrelationships between all animal species not just through the practical benefits of birds to humans, or the moral duty to care for nature, but rather through a shared affinity for display – a nature which underwrote culture. Yet constituting the 'nature' of display in favor of birds was crucially contingent on policing its boundaries. While conservation advocates sought to establish a framework in which humans are still, fundamentally, beholden to certain natural laws – such as (and especially) those of reproduction and kinship – ironically, the showy demonstration of those natural laws, and especially the threat of hybrid monstrosity, had to be sloughed off in favor of more deliberate distance from their shared potential for naturalness. After all, when regarded in light of Darwinian evolution, the difference between adaptation and monstrosity was merely one of degree, not kind.

Thus, the aesthetic terms in which 'human nature' was configured to argue for bird preservation involved a complicated negotiation of ecological relationships among animal attributes. If, in some ways, this conservationism recapitulated Boas and Allen's arguments, it shed the animal-aesthetic which too uncomfortably tied human and animal behavior to eroticized display. Instead, conservationists favored a peculiarly desexualized reproductive aesthetic, in which animals and their potentially uncontrollable urges were held at a safe remove, even if animal nature was still admittedly close.

Notes

1 Despite a promising start and the publication of *Audubon Magazine,* Grinnell disbanded his bird-protection group after only two years, due to declining enrollments and the lack of participation of women.

2 Although such dioramas became standard in many similar museums, the very first of them were devoted to birds (Wonders 1993; Gephart and Rossi 2017).

3 The exceptions to this, of course, were the vast trophy collections that served as the basis for natural history collections, like those that provided the foundation for the Field Museum's department of Ornithology – an instance of private gain turning public, but not one without an affinity for the public duty being leveraged upon women in their capacity as caretakers of the civilization (Directors Papers General Correspondence, Field Museum Archives).

Bibliography

Archives

Boas, Franz. 1896. *"Primitive Art." Lecture notes. Franz Boas Professional Correspondence (Mss.B.B61p).* American Philosophical Society, Philadelphia, PA.
Directors Papers General Correspondence, Box 1893–1907, COM-CUS; Folder: CB-Cory Ornithology, Field Museum Archives, Chicago, IL.

Published Sources

"A Bad Showing –for Missouri." 1913. *The Guide to Nature.* 6, 7: 173.
"A Few of the Novelties Shown in New York." 1902. *Illustrated Milliner.* 3, 9: 36.
Aikin, Charlotte Rankin. 1918. *Millinery.* New York: Ronald Press Co.
Alaimo, Stacy. 2010. *Bodily Natures: Science, Environment, and the Material Self.* Bloomington: University of Indiana Press.
Allen, Grant. 1880. "Aesthetic Feeling in Birds." *Popular Science Monthly.* 17, 37: 650–663.
"Birds and Bonnets." 1885. *Forest and Stream.* 25, 20: 385.
Cart, Theodore Whaley. 1973. "The Lacey Act: America's First Nationwide Wildlife Statute." *Forest History Newsletter.* 17, 3: 4–13.
Chapman, Frank Alvah. 1886. "Birds and Bonnets." *Forest and Stream.* 26, 5: 84.
———. 1899. "The Passing of the Tern." *Bird-Lore.* 1, 6: 205–206.
———. 1909. *Groups of North America: Guide Leaflet 28.* New York: American Museum of Natural History: 21.
Clemens, Jean. 1905. "Correspondence: The Passing of the Egret." *Harper's Weekly.* 49, 2524: 658.
Doughty, Robin W. 1975. *Feather Fashions and Bird Preservation: A Study in Nature Preservation.* Berkeley: University of California Press.
"Edward Bok: The American." 1920. *The Argonaut.* 87, 2274: 261.
"Fashions of Dress and Their Wearers." 1888. *The Sanitary Record.* 10: 270–271.
Gephart, Emily, and Michael Rossi. 2017. "Dovetailed Displays: Show Windows, Habitat Dioramas, and Bird Hats." In Anca Lasc, Patricia-Lara Betancourt, and Margaret Petty, eds., *Architectures of Display: Department Stores and Modern Retail.* London: Routledge: 202–216.
Graham, Frank Jr. 1990. *The Audubon Ark: A History of the National Audubon Society.* New York: Knopf.
Hansen, Thor. 2012. *Feathers: The Evolution of a Natural Miracle.* New York: Basic Books.
Haraway, Donna. 1984–1985. "Teddy Bear Patriarchy." *Social Text*, 11: 20–64.
———. 1988. "Situated Knowledges: The Science Question in Feminism and the Privilege of Partial Perspective." *Feminist Studies*, 14, 3; 575–599.
Haweis, Mary Eliza. 1883. *The Art of Beauty.* London: Chatto and Windus.
Lemon, Mrs. F. E. 1900. "Protection of Animal Life." In *Women in Social Life.* London: Unwin: 236–242.
Mallet, Mrs. Charles. 1900. "Dress in Relation to Animal Life." In *Women in Social Life.* London: Unwin: 242–245.
Mason, Kathy. 2002. "Out of Fashion: Harriet Hemenway and the Audubon Society, 1896–1905." *The Historian.* 65, 1: 1–14.
Merchant, Carolyn. 2010. "George Bird Grinnell's Audubon Society: Bridging the Gender Divide in Conservation." *Environmental History.* 15: 3–30.
"Midsummer Fashions." 1904. *Harper's Bazar.* 38, 7: 688.
Mitchell, John H. 1996. "The Mothers of Conservation." *Sanctuary: The Journal of the Massachusetts Audubon Society*, January–February: 1–20.

Moore-Colyer, R. J. 2000. "Feathered Women and Persecuted Birds: The Struggle Against the Plumage Trade, c. 1860–1922." *Rural History*. 11, 1: 57–73.

O. W. R. 1886. "Murder Most Foul." *Forest and Stream*. 27, 5: 83.

Patchett, Merle. 2012. "Fashioning Feathers: Dead Birds, Millinery Crafts, and the Plumage Trade." *2011 Exhibition*, University of Alberta, accessed February, 2015. http://fashioningfeathers.com/.

Pearson, T. Gilbert. 1895. "Echoes from Bird Land." *Natural Science News*. 1, 26: 102–103.

———. 1916. "The Traffic in Feathers." *The American Museum Journal*. 16, 4: 253–258.

Price, Jennifer. 1999. *Flight Maps: Adventures with Nature in Modern America*. New York: Basic Books.

"Reintroduction of Feather Millinery." 1888. *Audubon Magazine*. 2, 10: 207–208.

Smith, May Riley. 1906. "The Aigrette: An Appeal to Women." *By the Way- Side: Official Organ of the Wisconsin and Illinois Audubon Societies*. 9, 6: 43.

Thatcher, Mary. 1875. "Slaughter of the Innocents." *Harper's Bazar*. 8, 21: 338.

Thaxter, Celia. 1886. "The Badge of Cruelty." *The Cottage Hearth: A Magazine of Home Arts and Home Culture*. 12, 6: 186.

———. 1887. "Women's Heartlessness." *The Audubon Magazine*. 1, 1: 13–14.

"The Cruelties of Fashion." 1883. *Frank Leslie's Illustrated Magazine*. 57: 183.

Weidensaul, Scott. 2008. *Of a Feather: A Brief History of American Birding*. New York: Harcourt.

Welter, Barbara. 1966. "The Cult of True Womanhood: 1820–1860." *American Quarterly*. 18, 2: 151–174.

Willard, Mary Bannister. 1908. "Symmetry and Character in Hats." *Harper's Bazar*. 42, 6: 580–584.

"Women and Egret Plumes." 1895. *Our Animal Friends*. 23: 4.

Wonders, Karen. 1993. *Habitat Dioramas: Illusions of Wilderness in Museums of Natural History*. Uppsala: Almqvist and Wiksell.

16 Visualizations of "Nature"

Entomology and Ecological Envisioning in the Art of Willem Roelofs and Vincent van Gogh

Joan E. Greer

Introduction

This essay focuses on two Dutch artists who bring together the visual arts with natural history: Willem Roelofs (1822–1897) and Vincent van Gogh (1853–1890). In considering these two artists together – the former most commonly associated with the naturalist approach of Hague School Landscape painting, the latter with modernist, non-naturalist experiments of French Post Impressionism – I am consciously departing from an understanding of nineteenth-century art organized according to traditionally accepted styles of visual practice and associated with artistic "movements." Instead, I am attempting here to recognize a new category of art historical inquiry, one predicated on shifting artistic responses to the natural world that were beginning to be seen in the nineteenth century, responses that challenged rather than upheld dominant anthropocentric discourses. Concurring with a growing number of voices engaged in important revisionary work in this area, I posit that such shifts warrant study as an emerging nineteenth-century category of "ecological envisioning": a category with its own complexities and histories that opens avenues for a reconsideration of visual engagements with nature. As such, the project at hand is responding to current imperatives found in discourses coalescing around the term Anthropocene as well as some emerging within art history itself.

More specifically, this essay seeks to investigate these questions in relation to two examples of the intersection of art production and the study of insects. Using this entry point, I will argue for the recognition of the boundary-breaking figure of an "artist-natural scientist-environmentalist," putting forth both visual works and the thus expanded role of the artist as central considerations. Two works, in particular, will form the focus of this enquiry: Willem Roelofs's *In the Floodplains of the River Ijssel*, 1870–1897 and Vincent van Gogh's *Giant Peacock Moth*, 1889.

Roelofs's landscapes were appreciated by many, including aspiring young artists such as Vincent van Gogh, who shared Roelofs's love of the natural world, a fascination that included its entomological inhabitants.[1] Roelofs, as will be discussed, kept the two worlds of landscape painting and entomological study *visually* distinct, even while the two were mutually informative, resulting in a heightened, ecological understanding of both forms of representation. Van Gogh, on the other hand, took a different approach, at times including the insect as an important protagonist within his art as was the case with his representation of a large, rare moth in *Giant Peacock Moth*. In the following discussion, questions of ecological envisioning will be examined by first identifying a framework of theoretical underpinnings before turning to the artists and visual works.

Theory

This analysis is undertaken with the recognition that the term "nature," appearing in the title of this essay, is itself a multi-valent and culturally determined construct. "Ecology," a term coined by Ernst Haeckel in 1866, is used in its most basic sense (but also with reference to its more recent uses which will be indicated in the discussion), to indicate the interaction between living organisms and the systems in which they live. An "ecological envisioning," then, is visual representation which engages with and prioritizes these interactions. Finally, it should also be noted by way of preface that shining a light on insects seems particularly pressing at this point in time in particular, one in which an "ecological Armageddon" has been heralded as European scientists have sounded warning bells after drastic drops in insect numbers (Carrington 2017; Hallman 2017).

Fore-fronting visual studies and material culture as it does, this inter-disciplinary project is grounded in art historical methods: close visual and object-based readings of artworks rooted in formal, iconographical and iconological analyses as well as contextual analyses associated with the social history of art. Theoretically, it is informed by three key areas: Art and the Anthropocene; Applied and Philosophical Ethology; and North American First Nations ideas of a "Grammar of Animacy."

In the first of these, the use of the term "the Anthropocene," as Donna Haraway has said, may be seen in light of the need "for a word to highlight the urgency of human impact on this planet, such that the effects of our species are literally written into the rocks" (Haraway 2015, 259). 2016 marked the date the United Nations' 17 Sustainable Development Goals – which include "tackling Climate Change and Environmental Protection" – came into effect. Recognizing the disciplinary-crossing imperatives of addressing the current situation, I frame my discussion within discourses of the Anthropocene, thereby taking a political stance that speaks to the urgency of our situation vis-à-vis destructive anthropocentric modes currently affecting all aspects of the world in which we live. Seeking and privileging historical examples of alternatives to this hegemonic approach to the environment are a crucial part of this project. At the same time as situating this study within discourses of the Anthropocene (discussed in the introductory section of this volume), I am in agreement with many of the arguments outlining the shortcomings of the term itself, one of which will be returned to shortly. It remains useful, however, in its basic premises and accompanying sense of urgency.

At the same time as looking to discourses of the Anthropocene, it is relevant to identify a fundamental problem associated with the term that limits but does not eliminate its usefulness. As Eileen Crist has argued in "On the Poverty of our Nomenclature," identifying our period as the one defined and controlled by the human species – so, in an anthropocentric manner, that places rather than displaces the human at the very heart of the question – is the one that arguably undermines the project at hand of questioning and decentering this noncritical historical and current mono-species hegemony (Crist 2013).[2]

Second, and more specifically related to theories of "nature" and the environment, this examination of visualizations of nature in late nineteenth-century Dutch art is informed by questions of applied and philosophical ethology, the first being a branch of zoology examining nonhuman animal behavior; and the second a related field of inquiry bringing together anthropology, philosophy, environmental humanities and animal studies. Ethology focuses on the zoological, and is thereby important in that it provides a model for decentering and more appropriately situating the human species within the life-forms in the world we inhabit and cohabit. Philosophical ethology likewise shifts the human species away from the center. Vinciane Despret, for example, writing in the areas of philosophical ethology and animal studies demonstrates "the reciprocal agency of all participants involved, and the striving for an animated common world wherein both humans and animals coexist in meaningful and transformative relations" (Despret 2015).

Closely tied to such ethological positions is that of the botanist and member of the Citizen Potawatami Nation, Robin Wall Kimmerer, who combines academic scientific methods with native North American traditional knowledge. Kimmerer identifies the importance of developing "a grammar of animacy," such as that found in the Potawatami language:

> When I am in the woods with my students, teaching them the gift of plants and how to call them by name, I try to be mindful of my language, to be bilingual between the lexicon of science and the grammar of animacy. Although they still have to learn scientific roles and Latin names, I hope I am also teaching them to know the world as a neighborhood of nonhuman residents, to know that, as ecotheologian Thomas Berry has written, "we must say of the universe that it is a communion of subjects, not a collection of objects" (Kimmerer 2013, 56).

Kimmerer expresses and calls for an integrative critical awareness of the power of verbal languages. An analogy may be drawn with visual and material languages, in this case being discussed in respect to the artwork of Roelofs and Van Gogh. It raises questions of an alternate approach to other-than-human subjects – one attuned to deep observation; deep listening; deep feeling: to what, in short, is an immersive practice, one that is closely related to the "ecological envisioning" discussed in this essay.[3]

Willem Roelofs (1822–1897): Of Snout Beetles and Landscape Views

I begin with Dutch artist-scientist Willem Roelofs who was both an important early Hague School landscape painter and an active entomologist and authority on *circulionidae* or snout beetles.[4] Roelofs's position vis-à-vis landscape had a strong basis in Dutch seventeenth-century and Romantic landscape traditions and was firmly rooted in his careful study of the Dutch and Belgian countryside; it was also inflected with a solid understanding of contemporary directions in French landscape art including that of the Barbizon School, an area he himself visited and painted on a number of occasions, beginning in 1850 (Bodt 1995, 122). Indeed, one of the artists he greatly admired was the environmentalist-artist Théodore Rousseau.

The insect specimen collecting and open-air painting practices of Roelofs were not mutually exclusive, although they are generally treated as such in historical writings. The organization and analysis of his combined entomological–visual arts research findings – ones that resulted in both insect specimens and landscape sketches – were, without a doubt, mutually informative in the overarching project of collecting, identifying and disseminating knowledge of the natural world. After a day's work, Roelofs would return to his studio/study where the fruits of his labor as an entomological field scientist and as a landscape painter came together, continuing to inform each other in Roelofs's larger project of understanding and representing the different aspects of the natural environments he visited. Indeed, returning us to the insights of Kimmerer, Roelofs's enterprise seems conceptually "bilingual," bringing together a "lexicon of science" with the "grammar of animacy."

While Roelofs's entomological and artistic excursions were often combined, there are no known obvious attempts at the *visual* integration of the two worlds,[5] either in Roelofs's artworks or his scientific illustrations. Drawings made for scientific journals (Figure 16.1) reveal one aspect of this study and visual analysis, while the landscapes, such as the painting reproduced here of *In the Floodplains of the River Ijssel*, 1870–1897 (Plate 11), take a more macrocosmic position, rather than a close-up view that would facilitate an inclusion of the insect inhabitants of that natural world.

The sketching expeditions undertaken by Roelofs are documented in numerous books (many of which are housed in the Rijksmuseum print room).[6] Close examination of these sketch books reveals little in the way of *artistic*-entomological study. In fact, examinations seem to indicate that these insects do not feature in the sketch books at all. These books do, however, serve to underline Roelofs's rigorous attentiveness to and recordings of the landscape views with care given to rendering all but the entomological (and other) minutiae of the natural environment; the results, as seen in the Dutch landscape painting reproduced here, reveal deep observational methods that again resonate well with Kimmerer's ideas concerning experiential modes of understanding, resulting in strong and recognizable ecological visualizations of a very specific part of the countryside.

Roelofs's two practices of entomological and artistic envisioning, then, were not integrative in the sense of being folded into one another; however, they did lead to an overall ecologically heightened visual reading of the ecosystem at hand, resulting in landscapes that were informed by visual examination undertaken while engaged with collecting, studying and even visually rendering the smallest of the visible inhabitants of that environment, its insects, even though not incorporating them into the landscape scenes. Conversely, it seems likely that the resulting understanding of the insect life, given the close observation of their environment, was undoubtedly also sharpened.

As Greg Thomas has pointed out regarding the landscapes of Théodore Rousseau, a deep envisioning of the natural world in the case of Rousseau is paradigmatically different from earlier as well as much contemporary landscape painting (Thomas 2000). I would suggest that the same is true of Roelofs, even though an ecological envisioning is achieved in a different way. In the case of Roelofs, it is based most centrally on the painstaking, deep observation involved in his methods, not (or not *solely*) through distinct, immersive landscape practices in which he becomes more engaged with the natural world in an embodied way. Rather, an integrative understanding of his two practices allowed him to arrive at a form of landscape painting that is meticulously observed and ecologically envisioned. While Rousseau devoted himself most fully to visually investigating forest environments, Roelofs especially sets himself apart with his precisely envisioned wetlands, such as the floodplains of the Ijssel River depicted here.[7]

In the case of Roelofs, the role of the artist vis-à-vis the natural world is that of a temporarily "embedded" artist-scientist, separate from the object of his close observation and study even while regularly entering into this natural world to engage in fieldwork. This is still a position of being an "observer of" rather than "participant

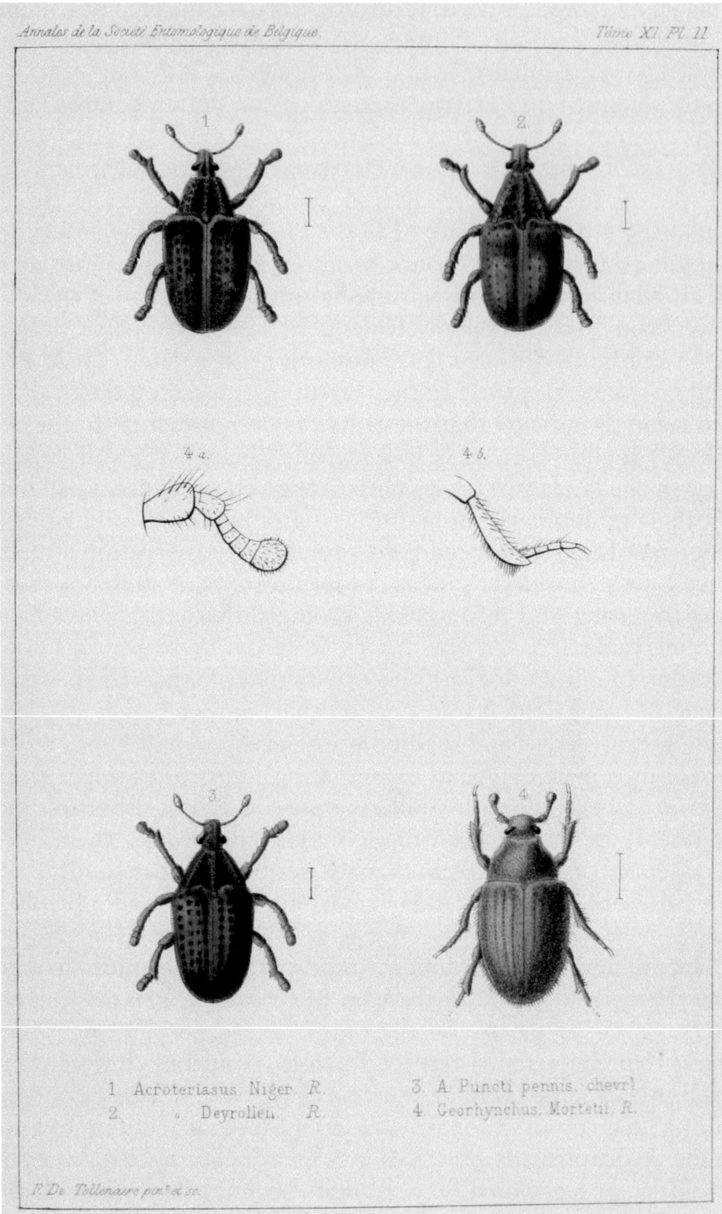

Figure 16.1 Willem Roelofs, illustration of Acroteriasus, "Notice sur le genre Acroteriasus, par W. Roelofs", in *Annales de la Société entomologique de Belgique*, 1867–1868, vol. 11, 75–77, pl. 2.

within" this world, in anthropological terms, an "etic" (outsider) rather than "emic" (insider) relationship to the object of study. Like the term Anthropocene itself, and as discussed at the outset of this paper, it needs to be made clear that this is a stance in which human activity still remains central and privileged. Nonetheless, this position

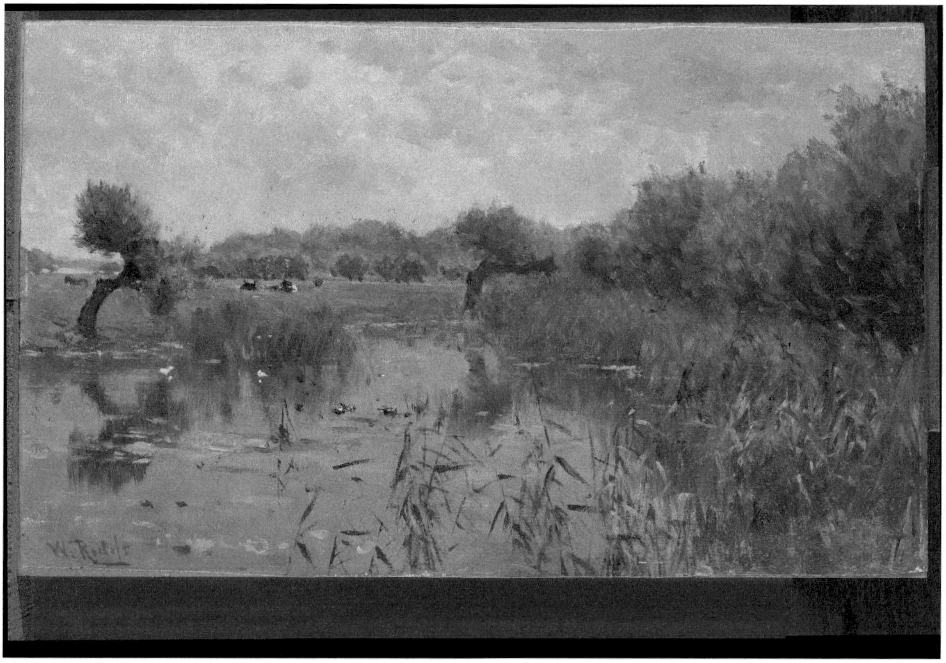

Plate 11 Willem Roelofs, *In the Floodplains of the River Ijssel*, 1870–1897, oil on canvas, Rijksmuseum, Amsterdam.

is beginning to be eroded and the emphasis has shifted: with its deep, engaged practice – one that combines to embrace the complexity and wonder found within this natural world – Roelofs's body of landscape art is the one that evokes strong, empathetic responses on the part of viewers.

Indeed, Roelofs's landscapes, devoted to the careful envisioning of the rich biodiversity of specific natural environments, including some at risk of destruction or severe degradation, gained widespread approval. Perhaps most notable is the recognition by early Dutch environmentalists occurring during the latter part of the century, such as that of the field biologist Jac P Thysse. Thysse, who would go on to found *Natuurmonumenten* (the Society for the Preservation of Nature in the Netherlands), made use of landscape art, including that of Roelofs, in his didactic environmentalist programs that included the publication of popular illustrated nature guides used to educate and garner support for environmentalist action.

Vincent van Gogh and Ecological Envisioning

In 1880, at the outset of Van Gogh's career as an artist, the paths of Van Gogh and Roelofs crossed briefly when Van Gogh sought out the older, well-established artist in Brussels for artistic advice. He referred to Roelofs and his artworks in five letters, and indeed seems to have owned a much admired pen sketch done by him.[8] Van Gogh was looking for mentorship and professional guidance during his visit to Roelofs, whose position of prominence in the art world and especially within the Hague School of landscape painting ensured him an esteemed position in the young Van Gogh's eyes. While there is no indication that insects, per se, played a role in Van Gogh's respect for the older landscape artist, the fascination for nature and natural history the two artists shared most certainly did. The close observation inherent within Roelofs' combined entomological–artistic pursuits that, as discussed, added a heightened element of deep observation of the natural world to his artistic practice undoubtedly held resonance to the young Vincent whose own early landscapes would also be based on a careful study of his natural surroundings.

The context of the Hague School landscape painting, more broadly, underpins these interests and artistic methods in the case of Van Gogh, a subject that has been explored in much scholarship.[9] The realist, *plein air* approach to the landscape, found within this group of painters working in the area around The Hague, was embraced wholeheartedly by the young Vincent in these early years and provided strong foundations for Van Gogh's nature imagery. This was reinforced by his knowledge of and sustained interest in the landscape art of the Barbizon painters.

It is of further relevance to mention two other components contributing to the late nineteenth-century nature discourses within which Van Gogh's own imagery was developing and functioning. That was the increasing production of didactic nature guides and illustrated publications, such as the slightly later ones by Jac P. Thysse just mentioned, and the importance of Japanese art including images of insects (Bealaram 2012, 259).

Van Gogh's engagement with nature predated his artistic ambitions, however, and was evident from an early age. As I have discussed elsewhere, it was a preoccupation that was importantly grounded in his early religious upbringing and studies (Greer 2013). A lifelong fascination with natural history and related experiential methods of exploring, collecting and sketching began with his boyhood scouring of Brabant gardens and countryside undertaken to augment his collection of beetles, birds' eggs and nests while simultaneously providing opportunities for examination of the ecosystems at hand. His sister Lies later commented on this period in Van Gogh's life and his fascination with natural history at this time, remembering that her brother knew the places where rare flowers grew as well as the names of the beetles he collected; the latter he preserved in cardboard boxes, like a junior entomologist at work, "placing small scraps of paper above each creature with their names, in Latin!" (Du Quesne-Van Gogh 1910, 18–19). These activities would translate into Van Gogh's later artistic expeditions in which his boyhood explorations would transform into a fully developed artistic envisioning of the countryside (Greer 2017).

As Van Gogh developed as an artist, his fascination with the natural world also continued to evolve, as is evident in numerous early landscapes of the Dutch countryside. In France,[10] especially during his period in Saint-Rémy (May 1889–May 1890), his approach increasingly shifted from an "observational" to an "experiential" practice – one that would include detailed engagements with the undergrowth in Provençal forests including root systems visible on those forest floors, or the acutely examined and tenderly rendered flora and fauna of the garden surrounding the asylum in which he was living. The famous painting made at this time of a large moth (Plate 12), situated in a carefully rendered natural habitat, provides opportunity for an investigation of Van Gogh's shifting practice of ecological visualizing.

In a letter to his brother Theo in May of 1889, Van Gogh said that he had drawn "a very large, rather rare night moth ... which is called the death's head," and included

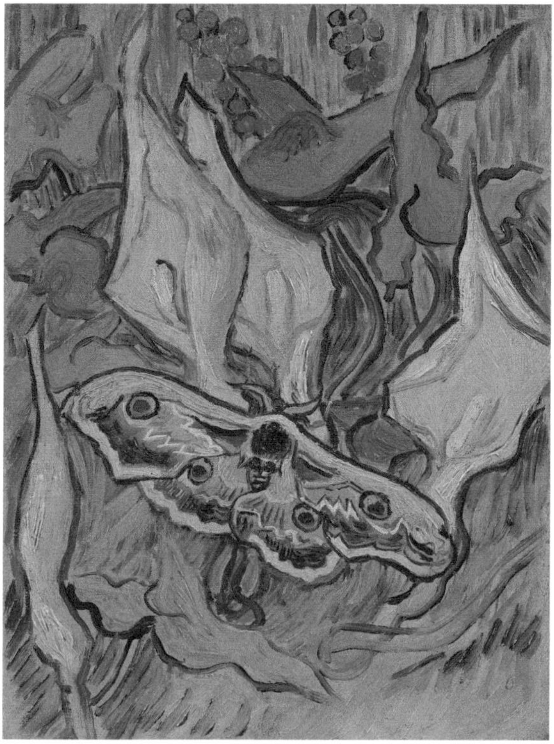

Plate 12 Vincent van Gogh, *Giant Peacock Moth*, 1889, oil on canvas, Van Gogh Museum, Amsterdam (Vincent van Gogh Foundation).

a sketch in his letter (Jansen et al. 2009, Letter 776. May 23, 1889). Van Gogh produced a drawing (Figure 16.2) and a painting of the same subject. The painting is a good example from Van Gogh's large body of work devoted to nature imagery that is clearly related to methods of visual enquiry found within the natural sciences and popular preoccupations with natural history. It is the one in which the artist zeroes in on an aspect of the natural world, taking a detailed, close-up view, like that of a field biologist.[11] In this body of work, Van Gogh's examinations range from views of entire ecosystems to those representing individual clumps of plant growth; and to those which move in on the subject yet more closely, for example, to the single sprig of a feather hyacinth, or a stalk of wheat; or to those taken from the animal world, for example, in the image of a kingfisher perched on a reed or, as here, moth on a leaf.

Both painting and drawing have since been entitled *Giant Peacock Moth* but, to Van Gogh, the large, striking moth with the central skull-like markings (especially visible in the painted version) was the mysterious Death's Head Hawkmoth or *Sphinx Atropos* – a creature that had captured the imagination of many. It was described in *L'Insecte* by Jules Michelet, an author whose works Van Gogh read and admired,[12] as "a frightful figure,—a great strong nocturnal butterfly, marked very plainly in tawny-gray, with a hideous death's head. This sinister being, which none had seen before, alarmed every countryside, and seemed an omen of the most terrible misfortunes" (Michelet 1875, 317).

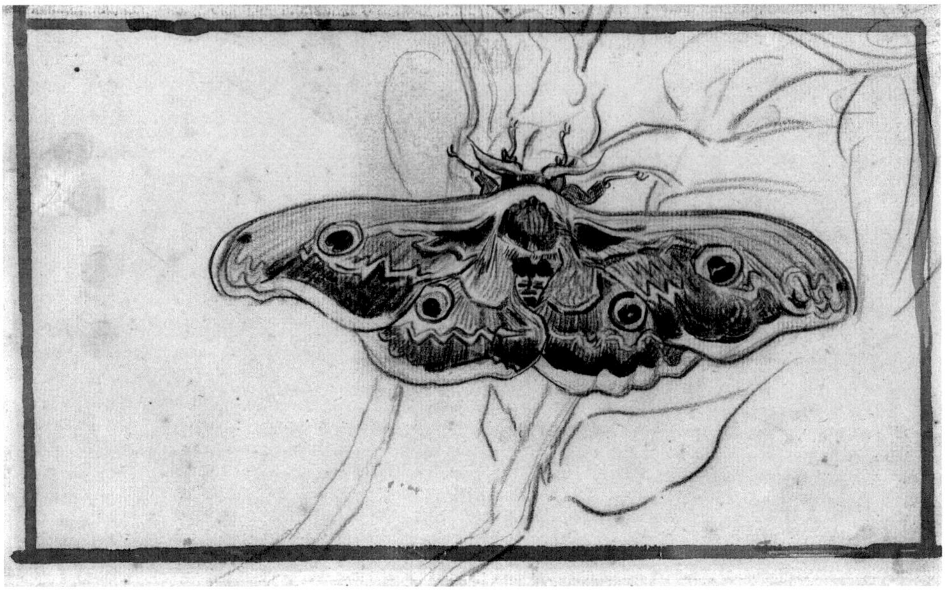

Figure 16.2 Vincent van Gogh, *Giant Peacock Moth*, 1889, chalk with pen, and brush and ink on paper, Van Gogh Museum, Amsterdam (Vincent van Gogh Foundation).

To Van Gogh, however, this night creature evoked none of this dread. To the contrary, he was full of wonder at the beauty of the moth and was excited that he had the opportunity to study this rare insect at close range. In a letter written to his brother Theo on May 23, 1889, Van Gogh wrote of its magnificence, especially noting the creature's astonishing coloration. The profound impact the encounter had on the artist speaks to a respect accorded to this nonhuman protagonist suggesting, in ethologist Despret's words, "meaningful and transformative relations" between artist and creature. Van Gogh wrote that he chose to sketch rather than to paint the moth as he would have had to kill it for the latter. No longer the junior entomologist-collector, Van Gogh, here, has become the naturalist-artist, one more fully engaged with a practice of heightened ecological representation in landscape art.

Earth Narratives and Ecological Envisioning[13]

In the drawing of the moth (Figure 16.2), the attentive investigation results in a detailed rendering of both the markings on the wings and back and the visible parts of the limbs and head protruding above the moth's wings. The branch and leaves on which the moth has alit, on the other hand, remain sketchy.[14] In the painting, however, the reverse is true. The limbs and head are more generally and simply suggested, while the vegetal growth, in which the moth all but disappears, is rendered with great interest and detail. It becomes a close-up, microcosmic view of this setting – akin to the undergrowth (*sousbois*) themes resulting from Van Gogh's intimate visual woodland studies (Homburg et al. 2016). The surrounding vegetation in the painting that Van Gogh has chosen for the habitat is the woodland Arum (*Arum maculatum*), a common plant throughout Europe and the one that Van Gogh sketched at this time in Saint-Remy.[15] The drawing reveals the direct experiential envisioning while the painting is a later encoded composite created by the artist.

In the painting, there is an element of careful examination combined with an understanding of popular lore that emerges in this ecologically informed but creatively conceived setting of forest undergrowth. Entomological accuracy is not the artist's sole concern. Van Gogh has chosen to feature more strongly the skull or "death's head," a well-known *vanitas* symbol of human transience and insignificance in the face of earthly existence and passing time, which was not readily discernible in the drawing (Greer 2017). The arum's association with male and female genitalia (the spathe and spadix) and especially its poisonous properties, and its strong scent and small hairs that enhance its function as an insect trap, make it a significant choice as a setting for this portentous moth. And it, too, is a choice. The insect and vegetation come together in this painting, subtly underlining the popular Death's Head Hawkmoth's sinister associations, even while functioning more significantly to create a bold, formally splendid and an ecologically envisioned close-up representation based on the acute observation of this magnificent moth and its creatively devised ecosystem.

We are no longer, as in the case of earlier landscapes by Van Gogh or Roelofs, confronted with a natural world predicated on a separation between human and nonhuman. This is moving toward a recognition of the "animated common world" of philosophical ethology outlined earlier in the paper, and one can speak as well of a move toward a holistically conceived visual "grammar of animacy." The approach in Van Gogh's painting considers the creature not merely as a discrete object of study but also as a subject in its own right. The moth itself is an integral part of this richly rendered habitat, but so too is the artist and, implicitly, the viewer. As such, the oppositional human/nature and related human/nonhuman animal binaries are disrupted. The resulting visual experiencing of the painting is "immersive," with the viewer having a sense of being placed within this richly rendered natural world; the intimate, close-up view, the compositional cropping of the plant forms and the great sensual beauty found in the sumptuous colors and thick buildups of impastos encompass and engage the viewer, evoking an embodied response – an integral part of Van Gogh's ecological envisioning.

Notes

1 Van Gogh sketched insects on a number of occasions. Other notable examples of paintings with insects include: *Butterflies and Poppies*, 1889, and *Wild Roses with Beetle*, 1890.
2 See also the critiques of the term "Anthropocene" made by T. J. Demos (2017) and Donna Haraway (2016).
3 Such acknowledgment of "other-than-human subjects" is part of new materialist positions such as that of Jane Bennett, whose writings on "thing power" argue for an understanding of the basic shared materiality of all things. See, for example, Bennett (2001, 347–372). Of further relevance is physicist and feminist theorist Karen Barad's theory of "agential realism" which sets forth a world of "entanglements" of social and natural forces and their intra-actions (Karen Barad 2007).
4 Saskia de Bodt has provided extensive and seminal work on Roelofs in her 1995 dissertation (De Bodt 1995).
5 This has been commented on by Rinnie Kooi of the Biology Institute of Leiden in her 2015 exhibition review "Een bezoek aan het Gemeentemuseum in Den Haag."

6 A good example is his 1851 sketchbook of landscapes, figures and animals (RP-T-1994-18) which includes close to 40 pencil sketches.
7 These approaches are variable in the case of different artists and need to be assessed with this complexity in mind. Rousseau's formal responses to the landscape environments, well outlined by Greg Thomas (2000), are not the same as those of Roelofs. What brings them together and allows them both to function in paradigmatically shifted ways in respect to an ecologically informed practice is the approach and deep engagement with the environment itself.
8 This sketch, referred to in the correspondence, has not been identified (Letter 252, Jansen et al. 2009).
9 Van Gogh was well aware of this art long before he met Roelofs. It would increasingly become more of a focus as he began his work as an artist; at the end of 1881, Van Gogh was taught briefly by Hague School artist Anton Mauve, who was a relative by marriage.
10 Van Gogh arrived in France in 1886.
11 Van Gogh's close-up views of nature have been brought to public and academic attention in the exhibition and accompanying catalog, *Van Gogh Up Close* (Homburg et al. 2012).
12 The Death's Head Hawk Moth captured the imagination of many, often due to its sinister associations. It appears in William Holman Hunt's *The Hireling Shepherd* of 1851. At the end of the century, Bram Stoker would include it in *Dracula*.
13 "Earth Narrative" is a term that has been used by Thomas (2000) as a counterpart to Barbizon paintings involving social narratives.
14 This plant has tentatively been identified as a periwinkle, based on the shape of leaves (Vellekoop and Zwikker 2007, 186).
15 Van Gogh drawing: F1613/JH 1703.

Works Cited

Barad, Karen. 2007. *Meeting the Universe Halfway: Quantum Physics and the Entanglement of Matter and Meaning*. Durham, NC: Duke University Press.

Bealaram, Rakhee. 2012. "Japonism and Nature." In Cornelia Homburg, ed., *Van Gogh up Close*, New Haven: Yale University Press: 259–261.

Bennett, Jane. 2001. "The Force of Things: Steps toward an Ecology of Matter." *Political Theory*. 32: 347–372.

de Bodt, Saskia. 1995. "Halverwege Parijs: Willem Roelofs en de Nederlandse Schilderskolonie in Brussel 1840–1890." PhD dissertation, University of Groningen.

Carrington, Damian. 2017. "Warning of 'Insect Armageddon' after Dramatic Plunge in Insect Numbers." *The Guardian* (17 October).

Crist, Eileen. 2013. "On the Poverty of our Nomenclature." *Environmental Humanities*. 3: 129–147.

Demos, T. J. 2017. *Against the Anthropocene*. Berlin: Sternberg Press.

Despret, Vinciane. 2015. "Editorial Introduction." *Angelaki: Journal of the Theoretical Humanities*. 20: 1–3.

Du Quesne-Van Gogh, Elisabeth Huberta. 1910. *Persoonlijke herinneringen aangaande een Kunstenaar*. Bussum: Van de Ven.

Greer, Joan. 2013. "Late Nineteenth-Century Visualizations of Nature and the Dutch Theologians' Culture." In G. Ulrich Grossman and Petra Krutisch, eds., *The Challenge of the Object: Proceedings of the 33rd Congress of the International Committee of the History of the Arts*. Nuremburg: Verlag des Germanisches Nationalmuseum: 240–245.

———. 2017. "'To Everything There Is a Season': The Rhythms of the Year in Vincent van Gogh's Socio-religious World View." In Sjraar van Heugten ed., *Van Gogh and the Seasons*. Melbourne: National Gallery of Victoria: 60–89.

Hallmann, Caspar A., Martin Sorg, Eelke Jongejans, Henk Siepel, Nick Hofland, Heinz Schwan, et al. 2017. "More Than 75 Percent Decline over 27 Years in Total Flying Insect Biomass in Protected Areas." *PLoS ONE*. 12: e0185809.

Haraway, Donna. 2015. "Donna Haraway in Conversation with Martha Kenney: Anthropocene, Capitalocene, Chthulhucene." In Heather Davis and Etienne Turpin, eds., *Art in the Anthropocene: Encounters Among Aesthetics, Politics, Environments and Epistemologies*. London: Open Humanities Press: 255–270.

———. 2016. *Staying with the Trouble: Making Kin in the Chthulucene*. Durham, NC: Duke University Press.

Homburg, Cornelia, ed., 2012. *Van Gogh Up Close*. New Haven, CT and London: Yale University Press.

———. 2016. *Van Gogh: Into the Undergrowth*. Cincinnati, OH: Cincinnati Art Museum.

Jansen, Leo, Hans Luijten, and Nienke Bakker. 2009. *Vincent van Gogh – The Letters*. Version: December 2010. Amsterdam & The Hague: Van Gogh Museum & Huygens ING.

Kimmerer, Robin Wall. 2013. *Braiding Sweetgrass. Indigenous Wisdom, Scientific Knowledge, and the Teachings of Plants*. Minneapolis: Milkweed Editions: 48–59.

Kooi, Rinnie. 2015. "Een bezoek aan het Gemeentemuseum in Den Haag." *Entomologische Berichten*. 75, 3: 85.

Michelet, Jules. 1875. *The Insect*. London: T. Nelson and Sons.

Thomas, Greg M. 2000. *Art and Ecology in Nineteenth-Century France: The Landscape of Theodore Rousseau*. Princeton, NJ: Princeton University Press.

Vellekoop, Marjie, and Roelie Zwikker. 2007. *Vincent Van Gogh and the Drawings*. Zwolle: Waanders.

Part 5

Agriculture and Resource Husbandry

In the *Botany of Desire,* nature and science writer Michael Pollan meditates on cultivated plants such as apples and potatoes as "partners in a coevolutionary relationship" wherein "we have spent the last few thousand years remaking these species ... [but] what is much less obvious, at least to us, is that these plants have, at the same time, been going about the business of remaking us" (Pollan 2002, xiv and xvii). Patterns of globalized food production and consumption had planetary consequences for nineteenth-century ecologies. If Voltaire in *Candide* (1759) had already shown to Europeans the misery suffered by slave labor on a Surinam plantation as the price by which they ate sugar, the Enlightenment-era "bioprospecting" (Schiebinger 2004) of exotic species (for agricultural, medicinal, domestic and commercial uses) and the domestication of wild species for food production continued to be complicit with taxonomic nineteenth-century sciences and the legacies of the transatlantic slave trade.

The three essays in this section consider visual culture's role in critically modeling modes of the anthropocentric production and consumption of "natural resources" in the form of food (via the technological remaking of plant and animal species) or as embodied, geo-political processes. Eating a Cincinnati-grown pineapple, drinking Brazilian coffee in New York or feasting on a haunch of English estate-grown venison implicates the diner in a complex network of conduits and mechanisms for the movement of organic entities: flora and fauna alike. As Caroline Gillaspie demonstrates, North American coffee consumption created both the burgeoning trade center of Lower Manhattan (just before the dawn of the nineteenth century) and the alternate and invisible sites of environmental disruption in the rapidly deforested landscapes of the Caribbean and Latin America; she tracks the agricultural labor of humans implicated in such coercive structures of oversight and control.

Kim Rhodes likewise examines a site shaped by consumption: the seemingly natural yet manicured estate of an aristocrat's deer park that enclosed, concentrated and farmed animals (that were threatened in the wild) for privileged delectation. Although the viewer may on first glance be convinced that they are glimpsing wild creatures looking back at them, the miniscule deer, rendered as "staffage" markers of the landscape, are revealed to be imprisoned livestock. In representing the "husbandry" or stocking of this captive herd in a rural landscape (that he himself had dined upon), the painter Constable entered into an intriguingly visual communion with this place and this reserve of animals.

Food consumption is bound up in the constructive performances of cultural identities; as the early nineteenth-century French gastronome Jean Anthelme Brillat-Savarin (1825) famously pronounced, "Tell me what you eat and I will tell you who you are."

It also can be construed as political action: as Gillaspie considers, the consumption of coffee in nineteenth-century New York had cosmopolitan or even subversive overtones in an American context and yet it financially supported slave labor in the Global South. Gillaspie and Shana Klein both connect global commodity demand in the U.S. for coffee and exotic fruit with larger ecological and social justice issues of enslaved or forced labor and environmental costs. Klein further considers how the still life representation of abundant, cultivated food in Robert Duncanson's paintings participated in the broader cultivation of American empire. As a working-class African-American artist, Duncanson had unique access to agricultural products (typically inaccessible to marginalized communities in antebellum society) because of the horticultural engineering that fostered the growing of a tropic fruit in the hothouses of his Ohio hometown. Through visual analysis and contextual research in the history of agriculture, Klein finds singular capacity in Duncanson's paintings to contradict racial prejudice, to embody ideas about equality and democracy and to validate America's burgeoning horticultural industries. Klein's close reading of images exemplifies how all three essays draw insight from acute, art historical analysis of the artists' choices of form, vantage point and socio-cultural perspective; their images consequently reveal important stories about the circulation of organic entities and the ways that human communities live within, interpret, affect and maintain their material environments.

Works Cited

Brillat-Savarin, Jean Anthelme. 1825. *Physiologie du goût, ou méditations de gastronomie transcendante* (*The Physiology of Taste*). Paris: Sautelet.

Pollan, Michael. 2002. *The Botany of Desire: A Plant's-eye View of the World*. London: Bloomsbury.

Schiebinger, Londa L. 2004. *Plants and Empire: Eolonial Bioprospecting in the Atlantic World*. Cambridge, MA: Harvard University Press.

Voltaire (pseud. François-Marie Arouet). 1759. *Candide, ou l'Optimisme*. Paris.

17 Coffee House Slip

Ecocriticism and Global Trade in Francis Guy's *Tontine Coffee House, N.Y.C.*

Caroline L. Gillaspie

The eighteenth century witnessed the meteoric rise of North American coffee house culture, and a demand for imports served in these establishments. Images from the Early Republic representing coffee houses and their city surroundings celebrated the role of these public spaces in the formation of early financial systems and increasingly global commerce. Francis Guy's (1760–1820) *Tontine Coffee House, N.Y.C.* (c. 1797) is named for the establishment depicted at the composition's left, partially obscured by the canvas edge (Figure 17.1). Although only half-visible, the Tontine dominates the skyline of Wall Street, its imposing architectural presence indicating its central importance to business activity at this Lower Manhattan commercial hub and the flag that adorns its roof suggesting its national significance to politics and economic growth during the Early Republic.

Guy's painting foregrounds the Lower Manhattan intersection of Wall Street and Water Street, on the northwest corner of which the Tontine Coffee House, established

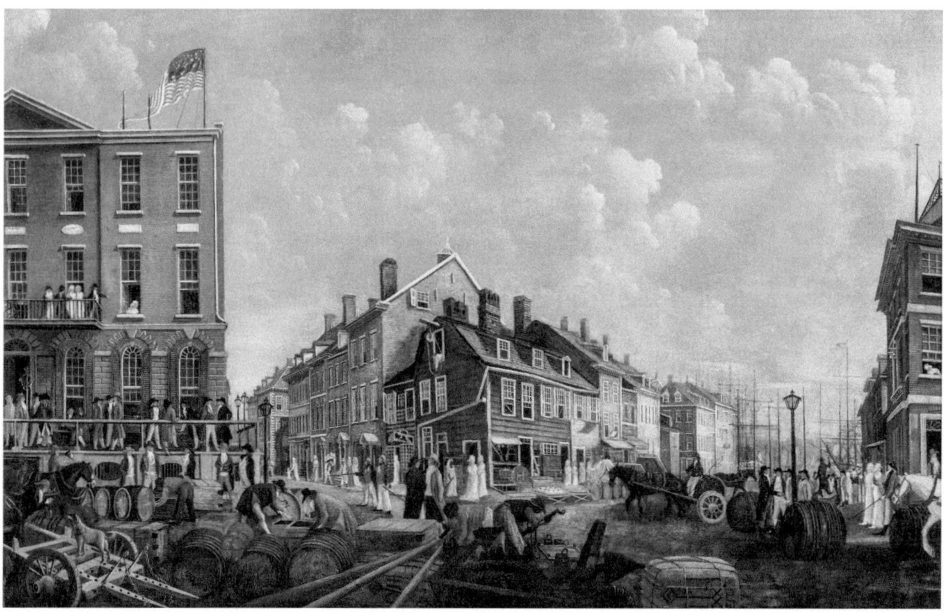

Figure 17.1 Francis Guy, *Tontine Coffee House, N.Y.C.*, ca. 1797, oil on linen, lined to fiberglass, New-York Historical Society.

by merchants, was constructed in 1793. This intersection was a major site of global trade and referred to as Coffee House Slip for the additional presence of the Merchant's Coffee House, located at the far right of this composition (Stokes 1928).[1] The presence of cargo, laborers and well-dressed businessmen surrounding the Tontine is indicative of this location's significance for merchants who learned of shipping news and made commodity trades at the establishment. By the eighteenth century, Caribbean and Latin American sugarcane and coffee beans were steady imports to North American cities, and rum and coffee were served in coffee houses amid business dealings and political discourse.

My analysis of Guy's painting calls attention to the artist's references to shipping in the distant piers, the inclusion of unloaded cargo and slave laborers, and their juxtaposition with the Tontine's merchant clientele. I argue that the depiction of this intersection, framed by coffee houses and culminating in ships docked in the East River, is layered with indications of the global transport of products whose cultivation resulted in brutal exploitation of the landscape and slave labor force. In comparing this painting with a Caribbean plantation view, I unite Guy's allusions to commodity demand and exchange that fueled economic and political development in the U.S. with the larger ecological history of the coffee and sugar trades. My work draws upon postcolonial ecocritical scholarship such as literary scholar Rob Nixon's discussion of "slow violence," or incremental ecological destruction with delayed yet devastating consequences on human and nonhuman agents, particularly in the Global South (Nixon 2011). I explore the imbrication of imperialist attitudes and globalizing exchanges with severe environmental damage in regions where coffee and sugar were sourced (DeLoughrey, Didur and Carrigan 2015; Gómez-Barris 2017).[2] Utilizing anthropologist Arjun Appadurai's model that explores the "trajectory" of objects in circulation to illuminate their social contexts, I view the Tontine Coffee House of Guy's painting as an end point in the journey of sugar and coffee from plantation to cup (Appadurai 1986). This study, therefore, entangles the operations of Coffee House Slip painted by Guy with the exploitative labor practices and environmental damage in plantation regions of the Global South that arose from the European and American demands for luxury commodities.

As Nicholas Mirzoeff has argued in his discussion of Anthropocene visuality, art produced in the Anthropocene era tends to aestheticize human dominance over nature (Mirzoeff 2014). I posit that images of Latin American and Caribbean plantations that harvested coffee and sugar for export to the U.S. normalize scenes of land cleared for cultivation, portraying it as an amelioration of the landscape as opposed to devastation. Amid colonial control in the Caribbean, European nations sought to subjugate nature so that extractive commodities could be circulated. In such artworks, the wilderness of the tropics appears tamed, and slaves depicted harvesting the land become additional, even naturalized signs of ownership and dominance over both human and natural resources. Charmaine Nelson's recent scholarship on the landscapes and geographies of slavery informs my analysis of cityscape and plantation views, both of which primarily focus on the spaces inhabited and worked by slaves, as opposed to slave bodies. Building on this, I also examine the marginalized presence of slaves that simultaneously acts as representation and omission of the injustices of slavery (Nelson 2016).

Although Guy's painting title alludes to the popular establishment, the central focus of the composition is the intersection outside the Tontine's doors, populated with merchants, dockworkers and slaves who examine recently unloaded cargo. The striking arrangement of perpendicular roads provides a dual vantage point, as the viewer's gaze is led both to the left along Water Street and the right along Wall Street. Extending along the east side of the Tontine Coffee House, the representation of Water Street follows the northward trajectory of development in Manhattan as the population began to expand uptown. However, the dramatic diagonal leading to the right represents the lively economic center of Wall Street terminating at major shipping piers. This dynamic juxtaposition of orthogonal lines enhances the sense of bustling movement and excitement that pervades the scene, and suggests momentum as the U.S. was propelled into political and economic independence.

The Tontine's impressive redbrick Georgian-style structure featuring gray rusticated stones and tall arched windows on the first story stabilizes the composition's left side, suggesting an ending point for cargo brought from the piers. Guy portrays the building's porches filled with male and female patrons, all of whom are white, well-dressed, merchant-class New Yorkers who converse while overlooking the busy intersection. Merchants actively used the Coffee Room, first floor Exchange Room where ship sales were made, and upstairs Long Room that hosted auctions (Tontine Coffee House Records, January 31, 1824). Coffee house images like Guy's celebrate the role of these businesses in the formative years' post-independence, and some (even created after the coffee house was demolished or burned down) seem to reflect nostalgically on their engagement with Revolutionary politics.

Coffee consumption in colonial North America became a crucial act of rebellion against British control, and some coffee houses were gathering places for patriots. This new public sphere emerged as a place of political activism where meetings often centered on opposition to the presence of British soldiers stationed in colonial cities (Conroy 1995, 158). Seeking to demonstrate elevated status as cosmopolitan citizens, coffee house patrons accessed newspapers, pamphlets, broadsides and shipping news at the establishment, which was often nearby or connected to a print shop (Cowan 2005; Kamensky 2014, 190). Coffee houses served the stimulating brew often in lavishly decorated Coffee Rooms, suggesting a desire in the U.S. for cosmopolitan habits and luxury imports (Conroy 1995, 96; Minutes, 1805–1813; Dividend receipts, 1805–1814, Vol. 4; and Account Books, 1791–1814, Vol. 7, Tontine Coffee-House Records.).[3]

Coffee houses, however, served more liquor – especially rum – than coffee (Conroy 1995, 95; Kamensky 2014, 189). As the British crown levied taxes on tea and sugar, such commodities (including the rum distilled from sugar) became controversial amid the political discourse of colonists, and coffee became the drink of choice among patriots. Criticism of rum was tempered as it was consumed with gusto, and the import of sugarcane continued despite political debates. But the consumption of coffee as an increasingly cosmopolitan act – and eventually as one of political subversion – laid a foundation for the emerging American coffee culture. However, desire for these libations also strengthened imperial control and expansion of plantation culture in the Caribbean and Latin America, exacerbating environmental degradation and the slave trade.

Created for a European travel book, Frédéric Mialhe's (1810–1881) image of La Ermita, a coffee plantation near Havana, Cuba, represents a common type employed by artists in representing Caribbean and Latin American plantations (Figure 17.2). My use of this print is also intended to highlight the expansion of coffee plantations from other islands such as Hispaniola and Jamaica in the eighteenth century, to Cuba in the nineteenth century. Like the strong perspective lines that draw the viewer into the midst of Guy's image of bustling commerce, a sense of forward momentum and expansion is also visualized in Mialhe's landscape. In place of Guy's street-level view,

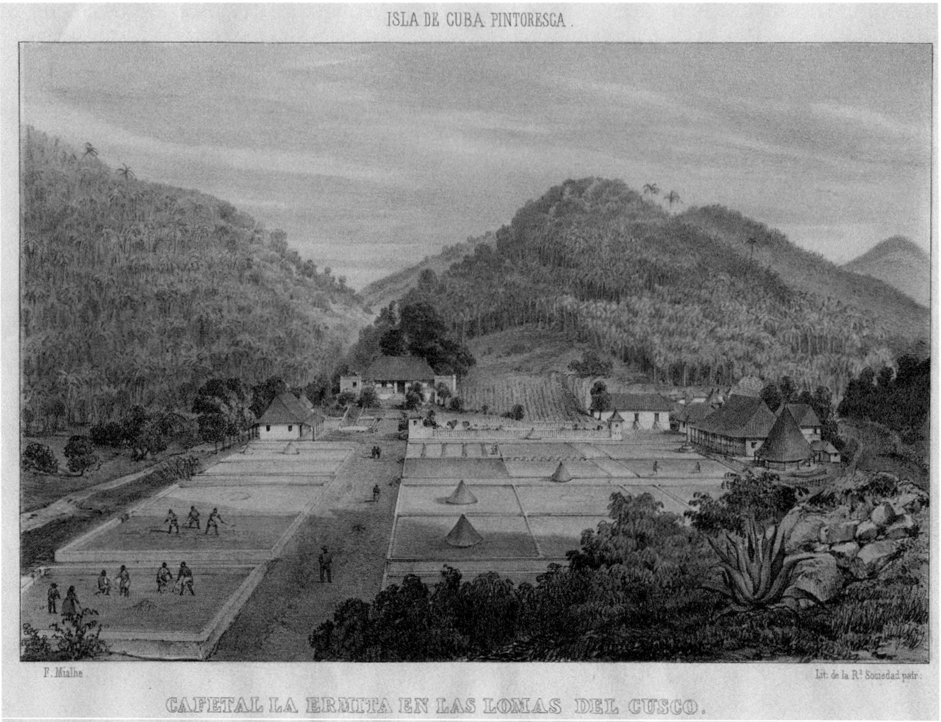

Figure 17.2 Frédéric Mialhe, *Cafetal La Ermita en las Lomas del Cusco*, from *Isla de Cuba Pintoresca*, 1839, lithograph, Cuban Heritage Collection Books, University of Miami Libraries.

here a commanding, elevated vantage point allows the viewer to survey the cultivated land, although Mialhe similarly emphasizes a prominent central avenue that extends from foreground to the distant manor house set against rolling hills. Flanking the avenue on either side is an orderly arrangement of square yards where coffee was raked and dried in the sun. Small coffee trees appear on the slope at the far left interspersed with taller palm trees, a technique that provided shade for the coffee crop (Van Norman, Jr. 2005). In a compelling detail, the cleared and partly plowed field on the hill behind the estate suggests the plantation's growth and the necessity for additional territory. Art historian Kay Dian Kriz's study of Jamaican landscape images argues that artists "detoxified" the environments they depicted to counteract the stereotypical assumptions about the extreme tropical climate of the Caribbean and the immoral and irrational inhabitants it produced (Kriz 2008, 157–160). Evident in Mialhe's image, the highly regimented French system of plantation organization, imposed earlier in Saint-Domingue before its adoption in Cuba and Brazil, implies a system of human rationality dominating the chaos of the tropical landscape (Marquese 2009). Only a small hint of frontier wilderness prevails in the background of this composition.

Although few signs of aestheticized nature distinguish Guy's coffee house painting, contemporaneous Caribbean plantation views record a characteristic picturesque outlook while also normalizing the Euro-American domination over landscape and human resources. The "magisterial gaze" Mialhe employs – common in traditional American landscape paintings to suggest expansion and ownership – provides an authoritative command over the estate (Boime 1991). However, historian and anthropologist John Michael Vlach has argued that U.S. plantation pictures often incorporated a lower vantage point to enhance the dominance and prestige of the manor house. Even Guy's paintings of Maryland estates, created after his 1798 move to Baltimore, feature sprawling landscapes with impressive manor houses elevated above the viewer's position (Vlach 2002, 49–65). Guy employs this compositional technique in *Tontine Coffee House* as the establishment's impressive architecture overwhelms the surrounding figures and buildings, commanding its corner of Wall Street and amplifying its importance as an economic center.

Entrance into this bustling crossroads at the receiving end of a cargo ramp secures the viewer's participation – or perhaps complicity – in an extensive exchange network. Yet the sanguine flurry of economic activity captured in this composition obscures the concomitant devastation inflicted upon Caribbean and Latin American nations subjected to European and U.S. capitalist ambitions and commodity extraction. As the first home of the New York Stock and Exchange Board, the Tontine Coffee House was constructed within the city's economic and political center on Wall Street, midway between the shipping piers and Federal Hall. Guy's careful delineation of the vicinity heightens the establishment's centrality to capitalist concerns that had lasting destructive effects due to the exploitative practices of sugar and coffee exchange.

Guy's representation of Wall Street visually links the port of New York with both the origins of the arriving commodities, and the operations of the Tontine Coffee House where they were consumed. More conspicuous visual depictions of the environment from which these resources derived may be missing from Guy's composition, yet these locations would have been familiar, readily recognized in the minds of merchants and subscribers to the Tontine Coffee House. At the time of the establishment's founding – and Guy's painting – the slave uprising and anti-colonial efforts of the Haitian Revolution (1791–1804) commanded the attention of merchants and consumers of coffee and sugar, as the export of these commodities from Saint-Domingue (or Haiti after 1804) was disrupted.

The bales and barrels occupying the foreground of Guy's image would have fostered such associations. The cultivation of sugarcane dominated the Caribbean landscape from the mid-seventeenth century when sugar plantations were introduced to Hispaniola and the islands of the Lesser Antilles, particularly Barbados. By the following century, Saint-Domingue, Jamaica and Cuba competed to be the largest exporters of sugarcane. Coffee also became a major export from Caribbean islands, particularly Hispaniola, Jamaica and Puerto Rico (Watts 1987; Monzote 2008; Leal, Pádua, and Soluri 2013). Mainland Latin American countries likewise grew these crops, and by the mid-nineteenth century, Brazil became the largest exporter of coffee in the world (Fausto 1999; Dean 2008).

Economic growth generated by plantation culture fed the bustling activity at Coffee House Slip, but also fueled environmental degradation in the Antilles and the import of slaves to Caribbean islands. Forested land replete with rich soils was rapidly cleared for cultivation. Coffee crops planted on hillsides in cooler climates required the clearing of forests in mountainous areas, leading to soil erosion. The lowland of Saint-Domingue, for example, was cleared for sugar plantations, while coffee estates cleared the forested hills, making Haiti's landscape arguably the most environmentally devastated in the Antilles (Tucker 2000, 181; Leal et al. 2013). Yet the imperialist attitude of subduing wild nature, and the ostensible necessity of an efficient plantation structure reliant upon slave labor to satisfy demands for coffee and sugar, led to continued expansion of such estates into frontier lands.

Both Guy and Mialhe staff their images with busy slaves, alluding to the ongoing exploitation of slave labor for the coffee and sugar trades, while also denying reference to the brutality of slavery. Typical of his cityscapes, Guy includes a cross section of classes and races working in Lower Manhattan (*Picturing Place: Francis Guy's Brooklyn, 1820*, 2006). Two white figures directly to the right of the Tontine Coffee House inspect and mark barrels of cargo, possibly puncheons of rum ('*The Slave Market in Wall Street; Puncheons of Rum*' 1907). To their right, wooden poles form a ramp from the viewer's space leading toward two black figures. Possibly enslaved, they are tasked with the manual labor of hauling cargo. An additional African laborer appears directly in front of the coffee house, while a black woman trails behind in attendance of two white women at the street intersection. The artist's characterization of these figures further highlights racial inequalities. Guy does not directly reference slave sales in this image, but these figures allude to the import of slaves to the city and their sale at this precise intersection until New York abolished slavery in 1827. In his coffee plantation landscape, Mialhe also foregrounds the slave population of La Ermita who rake coffee beans to dry. If this image obscures violent realities of slave punishments, it nonetheless reveals the backbreaking labor and the continued subjugation of a marginalized population who are portrayed here as picturesque staffage and a sign of the common ownership of land and people. Visual references to the "slow violence" of resource extraction are interwoven with the injustices of slavery, serving as picturesque tropes that further marginalize the lasting effects of suffering on vulnerable populations in the Global South (Nixon 2011).

In the eighteenth and nineteenth centuries, some environmental degradation caused by plantation culture was combatted through the creation of forest reserves in Caribbean colonies and later reforestation efforts in Brazil. Yet these environmental protections were primarily authorized to preserve the livelihood of elites (Drummond 1996; Leal et al. 2013). Meanwhile, the demand for sugar and coffee intensified the forced migration of Africans to the Americas. My ecocritical reading of Mialhe's print alongside Guy's painting, which identifies the beginning and end point in the trajectories of slave-harvested commodities, illuminates a convergence of social and environmental issues. Deforested land and the presence of slave laborers in Mialhe's plantation were normalized amid capitalist gain and imperial control, suggested by the bustling activity of Guy's Coffee House Slip. As the picturesque tropes and anecdotal charm in their work aestheticized natural resource extraction, transport and consumption, beholders should attend equally to the visual cues of shipping routes, globalizing commerce, land clearing and human oppression that bring to light other narratives: ones that reveal the sustained and catastrophic violence committed against colonized landscapes and people.

Notes

1 The Merchant's Coffee House, established in 1737 on the corner that the Tontine eventually occupied, was later moved to the southeast corner in 1772 before burning in 1804.

2 These works in form my examination of extractive industries that disproportionately impacted marginalized populations and the environment of Latin America.
3 With fine furnishings, coffee house patrons modeled themselves on cosmopolitan Londoners. In 1808, the Tontine was furnished with mahogany tables, Windsor chairs, chandeliers, a telescope and more.

Bibliography

Archives

Minutes, 1813–1844. Volume 10. Dividend receipts, 1805–1814, Vol. 4. Account Books, 1791–1814, Vol. 7, Tontine-Coffee House Records. New-York Historical Society, New York.
'*The Slave Market in Wall Street; Puncheons Of Rum*.' Art and Picture Collection, The New York Public Library. New York Public Library Digital Collections. 1907. Accessed October 12, 2017. http://digitalcollections.nypl.org/items/510d47e1-06cc-a3d9-e040-e00a18064a99.

Published Sources

Appadurai, Arjun, ed. 1986. *The Social Life of Things: Commodities in Cultural Perspective*. Cambridge: Cambridge University Press.
Boime, Albert. 1991. *The Magisterial Gaze: Manifest Destiny and American Landscape Painting, c. 1830–1865*. Washington, DC: Smithsonian Institution Press.
Conroy, David W. 1995. *In Public Houses: Drink & The Revolution of Authority in Colonial Massachusetts*. Chapel Hill: University of North Carolina Press.
Cowan, Brian. 2005. *The Social Life of Coffee: The Emergence of the British Coffee House*. New Haven, CT: Yale University Press.
Dean, Warren. 2008. *With Broadax and Firebrand: The Destruction of the Brazilian Atlantic Forest*. Berkeley: University of California Press.
DeLoughrey, Elizabeth, Jill Didur, and Anthony Carrigan, eds. 2015. *Global Ecologies and the Environmental Humanities: Postcolonial Approaches*. New York: Routledge.
Drummond, José. 1996. "The Garden in the Machine: An Environmental History of Brazil's Tijuca Forest." *Environmental History*. 1, 1: 83–104.
Fausto, Boris. 1999. *A Concise History of Brazil*. Translated by Arthur Brakel. Cambridge: Cambridge University Press.
Gómez-Barris, Macarena. 2017. *The Extractive Zone: Social Ecologies and Decolonial Perspectives*. Durham, NC: Duke University Press.
Kamensky, Jane. 2014. *The Exchange Artist: A Tale of High-Flying Speculation and Americas First Banking Collapse*. New York: Penguin Books.
Kriz, Kay Dian. 2008. *Slavery, Sugar, and the Culture of Refinement: Picturing the British West Indies, 1700–1840*. New Haven, CT: Yale University Press.
Leal, Claudia, José Augusto Pádua, and John Soluri, eds. 2013. "New Environmental Histories of Latin America and the Caribbean." *RCC Perspectives*, 7. doi: 10.5282/rcc/5921
Marquese, Rafael de Bivar. 2009. "Luso-Brazilian Enlightenment and the Circulation of Caribbean Slavery-related Knowledge: The Establishment of the Brazilian Coffee Culture from a Comparative Perspective." Translated by Glenn Ellefson. *História, Ciências saude-Manguinhos*. 16, 4: 855–880.
Mirzoeff, Nicholas. 2014. "Visualizing the Anthropocene." *Public Culture*. 26, 2: 213–233.
Monzote, Reinaldo Funes. 2008. *From Rainforest to Cane Field in Cuba: An Environmental History since 1492*. Translated by Alex Martin. Chapel Hill: The University of North Carolina Press.
Nelson, Charmaine A. 2016. *Slavery, Geography and Empire in Nineteenth-Century Marine Landscapes of Montreal and Jamaica*. London: Routledge.

Nixon, Rob. 2011. *Slow Violence and the Environmentalism of the Poor.* Cambridge, MA: Harvard University Press.
"Picturing Place: Francis Guy's Brooklyn, 1820." 2006. Exhibition at The Brooklyn Museum.
Stokes, Isaac Newton Phelps. 1928. *The Iconography of Manhattan Island 1498–1909.* Vol. 6. New York: Robert H. Dodd.
Tucker, Richard P. 2000. *Insatiable Appetite: The United States and the Ecological Degradation of the Tropical World.* Berkeley: University of California Press.
Van Norman, Jr., William C. 2005. "Shade Grown Slavery: Life and Labor on Coffee Plantations in Western Cuba, 1790–1845." PhD Dissertation, University of North Carolina, Chapel Hill.
Vlach, John Michael. 2002. *The Planter's Prospect: Privilege and Slavery in Plantation Paintings.* Chapel Hill: University of North Carolina Press.
Watts, David. 1987. *The West Indies: Patterns: Development, Culture and Environmental Change since 1492.* Cambridge: Cambridge University Press.

18 "A Haunch of a Countess"

John Constable and the Deer Park at Helmingham Hall

Kimberly Rhodes

Helmingham Hall and its adjacent deer park in Suffolk, England, about 20 miles north of John Constable's home territory of East Bergholt, provided the artist with creative, bodily and economic sustenance from 1800 until his death in 1837. During this period, Constable drew the park's majestic oak trees and painted its distinctive topography; ate the meat of the fallow and red deer cultivated there; and benefited from the patronage of the Tollemache family, particularly the Countess of Dysart, who provided him with the venison he consumed. Indeed, Constable used alimentary language to refer to his drawing practice at least once, writing to Maria Bicknell (later his wife, Maria Constable) from Suffolk in 1814, "I took several beautifull [sic] walks in search of food for my pencil this summer when I hope to do a great deal in landscape" (Beckett 1964, vol. 2, 121). *The Dell at Helmingham Park* from 1830 (Plate 13) is one of several images he painted of the exact site late in his career; it was exhibited at the Royal Academy in 1830 but remained in the artist's possession until his death.[1] Constable's rendering of deer as staffage in the 1830 version sets it apart from his other depictions of Helmingham Park by clearly indicating it as a site of animal husbandry. The inclusion of the deer also loosely associates it with his concurrent representations of Jaques and the wounded stag from William Shakespeare's *As You Like It* as well as with his painting *The Cenotaph* from 1836.[2] Most significantly for the purposes of this essay, Constable's addition of deer as staffage to his 1830 canvas offers a novel, ecocritically sensitive account of his practice that conjoins the immediate and remembered sensations of the artist's body – his hand's touch of brush to canvas, his eyes' gaze upon the picturesque scenery, his mouth's taste of venison – with the flora, fauna and owners of Helmingham's deer park in a trans-corporeal network of communion and exchange over more than 30 years.[3]

In his influential Marxist study of Constable's representation of rural workers as diminishing figures in the landscape, John Barrell has observed that "Constable … is not simply a landscape painter … but one clearly concerned to express in his landscapes a social vision – the image of a productive and well-organised society. It's remarkable, then, how little *social* his landscapes are. The nobility and gentry, of course, almost never appear in them" (Barrell 1980, 133). While this is true on the level of direct representation, nobility is indirectly present in *The Dell at Helmingham Park*. Constable emphasized the private, aristocratic associations of his painting during its exhibition at the Royal Academy in 1830, showing it under the title *Dell scene, in the park of the Rt. Hon. The Countess of Dysart, at Helmingham, Suffolk*. Connecting the depicted deer with an aristocratic landholder underscores the animals as captive in a deer park despite their appearance as inhabitants of the

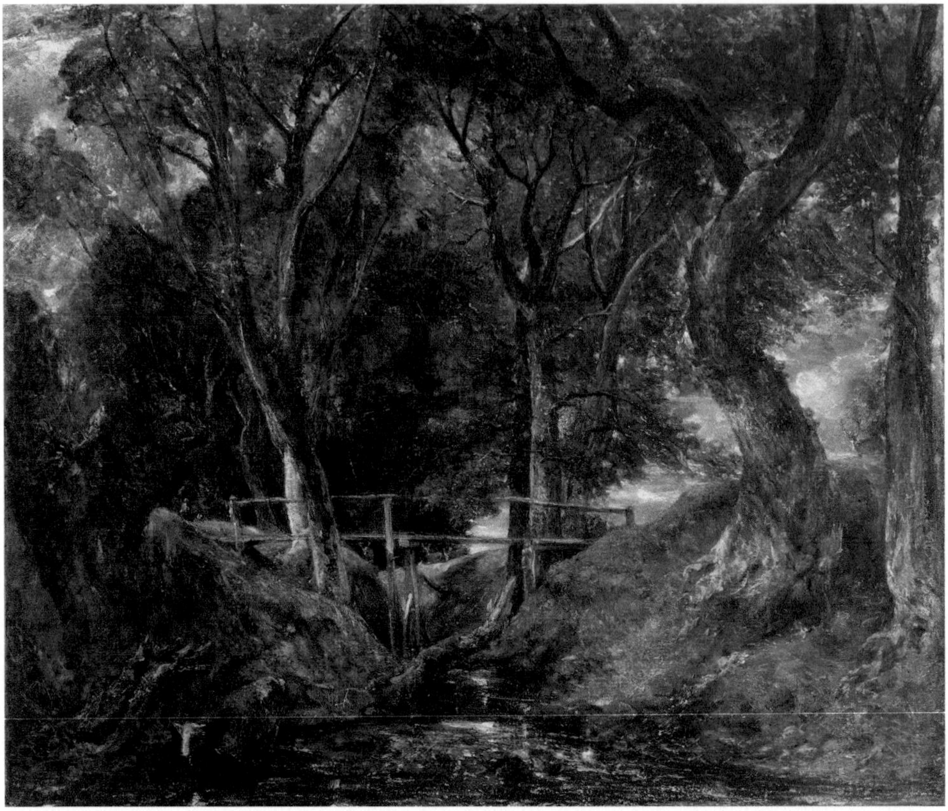

Plate 13 John Constable, *The Dell at Helmingham Park*, 1830, oil on canvas, The Nelson-Atkins Museum of Art, Kansas City, Missouri. Purchase: William Rockhill Nelson Trust. Photo courtesy of Nelson-Atkins Media Services/E.G. Schempf.

wild and emphasizes the function of the aristocrat as a patron providing food and artistic support in Constable's "well-organised society." Constable's titular reference also links him directly to the Countess of Dysart and the deer park: denoting the person who gave him venison and the place from which it originated. The small deer, then, are both symbolic extensions of the aristocrats who have contained and farmed them and "workers" on their land, providing food for the Tollemache family and their intimate circle of friends, Constable among them, while supplies lasted. Indeed, the number of deer in Helmingham Park diminished greatly over the nineteenth century. In 1867, Evelyn Philip Shirley reported "herds of 30 red, and 450 fallow-deer" (Shirley 1867, 120). About 30 years later, Joseph Whitaker reported 72 red deer and 260 fallow deer (Whitaker 1892, 143). Thus, by centralizing the significance of the deer park and utilizing animal studies' theories in my analysis of *The Dell at Helmingham Park*, I demonstrate how the implications of aristocratic patronage extend beyond financial transactions and unpack the potential of nonhuman agency (particularly that of animals as living beings and meat) in the making of Constable's landscape paintings.

The original title of Constable's painting triangulates the relationship between artist, aristocrat and animal and moves it beyond the subject/object, artist/animal, patron/artist binaries of power. The composition of the painting also suggests triads rather than binaries. For example, the points of four triangular wedges converge toward the bottom of the painting where the center support of the wooden bridge intersects with a tree limb severed from the twisted oak on the left side of the composition. Formally, this configuration creates a space generously opening up on all sides to the viewer like a blooming flower. Figuratively, the triangular spaces repeatedly signify the interdependent triad of artist, aristocrat and animal at the heart of this analysis. To make this move from the binary to the triangulated, it is necessary to join theoretical positions about human/animal relations expressed three decades apart by two different writers: John Berger's "Why Look at Animals?" originally written in 1977 and Stacy Alaimo's "Trans-Corporeal Feminisms and the Ethical Space of Nature" from 2008. Berger's essay considers the marginalization of animals and the progressive alienation of humans from nature in the context of modern, Western capitalist society, a state he traces back to the nineteenth century and detects in landscape painting: "the treatment of animals in nineteenth-century romantic painting was already an acknowledgment of their impending disappearance. The images are of animals *receding* into a wildness that existed only in the imagination" (Berger 1980, 17). Reading Constable's painted deer through Berger's lens as symbolic objects subservient to the human gaze yet, as the quotation at the beginning of this essay suggests, capable of moving the imagination is useful in situating *The Dell at Helmingham Park* in a broader social context acknowledging power relations precipitated by capitalism and industrialization as well as the aristocratic associations of the landscape depicted in the painting. Alaimo, on the other hand, proposes an urgent ethics of animal agency that also relies on the analysis of power dynamics, yet runs counter to Berger's pessimistic assessment of the asymmetry of power between animals and humans: "acknowledging the agency of the more-than-human world is crucial for environmental ethics because it challenges the prevalent practice of 'thingification' (in [Karen] Barad's terms), which, in this case, means the reduction of lively, emergent, intra-acting phenomena into passive, distinct resources for human use and control" (Alaimo 2008, 249).

Admittedly, Berger's and Alaimo's generational and theoretical differences create friction. However, by combining their theories, I am able to position animals as objects (Berger) and agents (Alaimo) simultaneously to provide an emblematic framework for Constable's complex relationship with Helmingham's park and patrons. The artist, himself, was both an object of patronage and an agent of creation at Helmingham; the Countess, as a woman, had limits placed on her powers of patronage and land ownership and was at least once an object of ridicule for Constable. The Countess did not have full rights to Helmingham in 1830. At the time, she held not its title, but a life-interest in the property (Beckett 1964, vol. 4, 65). Within the realm of art practice and representation, Berger's approach offers a compelling social historical account of why paintings of animals look the way they do and what animals symbolize in the nineteenth century. Alaimo's notion of trans-corporeality, on the other hand, suggests ways to think about Constable's practice in a complementary way, wherein the animals he depicts assert their own agency and shape their own form as they pass before his eyes, through his body and onto the canvas. She asks, significantly, "What are some of the routes from human corporeality to the flesh of the other-than-human and back again? How are both terms transformed by the recognition of their interconnection?" and answers that "perhaps the most palpable example of trans-corporeality is that of food, whereby plants or animals become the substance of the human" (Alaimo 2008, 253).

While only one of Constable's versions of *The Dell at Helmingham Park* includes deer in the composition, each is sited in a space that served as a deer park, a space of illusionistic wildness wherein animals seem to roam freely despite being enclosed in approximately 400 acres of land and cared for by humans in exchange for providing hunting fodder, meat and picturesque beauty for aristocratic pleasure. Around the time Constable composed his painting, Helmingham Hall and its park were described as such:

> This stately residence is situated in the hundred of Bosmere and Claydon, four miles south-east from Debenham, eight miles from Ipswich, in a beautiful park comprehending four hundred acres, which contains some of the finest oak-trees in the county, many of them of great age, and which is abundantly stocked with deer, there never being less than seven hundred head in the park, amongst which are a few remarkably large stags. (*Jones' Views of the Seats, Mansions, Castles, Etc. of Noblemen and Gentlemen in England, Wales, Scotland, and Ireland* 1829, 3-Q)

Constable visited Helmingham Park for the first time in 1800 and described the experience to his friend John Dunthorne, echoing remarks about the magnificent trees: "Here I am quite alone amongst the oaks and solitude of Helmingham Park.... There are [sic] abundance of fine trees of all sorts; though the place upon the whole affords good objects rather than fine scenery" (Beckett 1964, vol. 2, 25). A subsequent visit to Helmingham invoked in an 1814 letter to Maria Bicknell culminated with a "long walk in the park" with Lady Dysart and Mary Constable (Beckett 1964, vol. 2, 121). The artist fails to mention observing deer in either account, which is curious considering how others attest to the powerful presence of the animals in the park. However, a charcoal drawing from 1803 identified as a depiction of Helmingham Park (Figure 18.1) includes deer in a composition similar to the 1830 painting. The drawing, however, does not feature a man-made structure to indicate its location in a deer park; the deer seem to be roaming freely in a sylvan environment. Two drawings from 1800, *Helmingham Dell* and *A Tree at Helmingham Dell*, provide inspiration for the overall landscape composition with the bridge and the sinuous, expressively twisted tree on the right side of the painting. Neither includes deer, but they have clearly been "digested" by Constable for use in the later painting.

As a composite product of both direct observation recorded in drawings made 30 years earlier and sensory memory summoned during its execution in the studio, *The Dell at Helmingham Park* is not unique in Constable's oeuvre. The deer notably signal specific historical and personal events from 1800 to 1830 and simultaneously collapse time through trans-corporeal unity. Features of the landscape that appear in Constable's drawings but had disappeared since he visited in 1800 are present in his painting. The rustic wooden bridge, for example, had been replaced with a stone bridge by 1815 (Reynolds 1984, 217). By extension, the deer Constable painted were plentiful in Helmingham Park between 1800 and 1830, but were disappearing in the wild during this period, conjuring Berger's notion that "the image of a wild animal

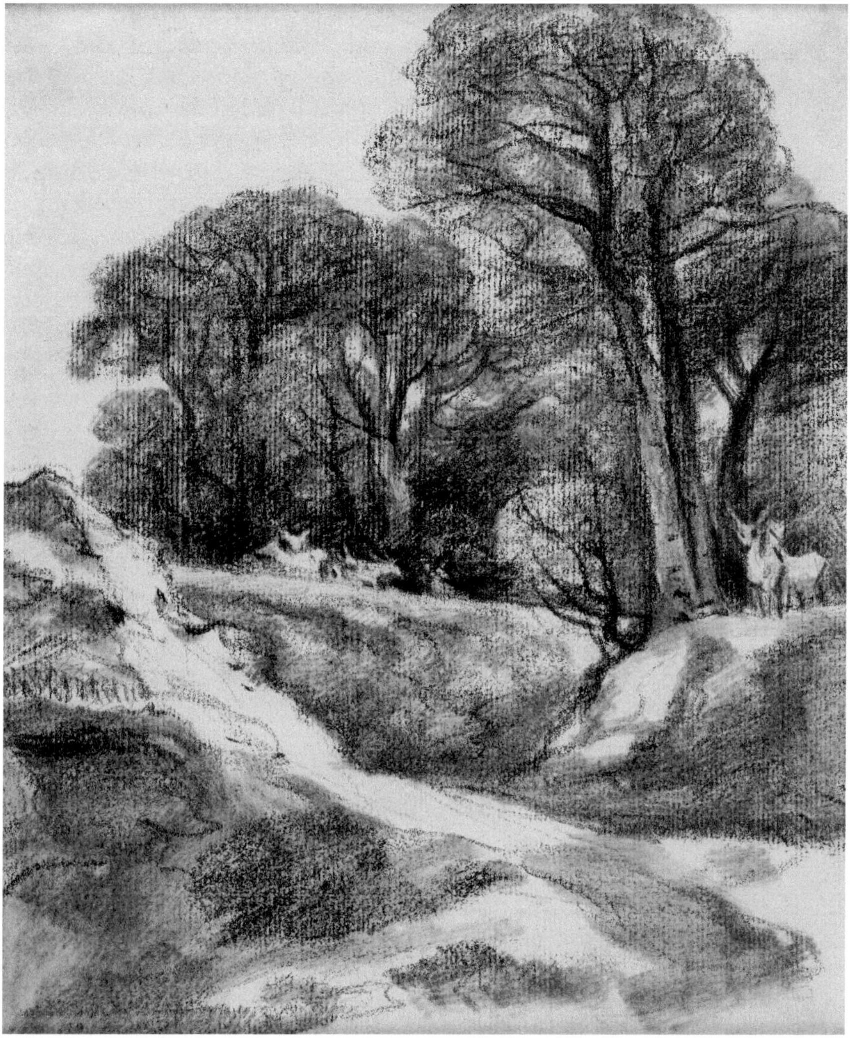

Figure 18.1 John Constable, *Helmingham Park*, 1803, charcoal on paper, Gallery Oldham.

becomes the starting-point of a daydream: a point from which the day-dreamer departs with his back turned" (Berger 1980, 17). According to *The Hunting Directory*, opportunities to encounter deer outside the confines of private parks were limited by 1830: "In regard to Stag Hunting … it has gradually given way to the increasing cultivation of the country; and as the object of pursuit has nearly ceased to exist in a state of unlimited freedom, this noble and princely diversion has, of course, in a great degree subsided" (Johnson 1830, 231). P.B. Munsche has traced this process through analysis of the enactment of enclosure and game laws between 1671 and 1831, stating that after 1692, "deer and rabbits were protected by statutes which forbade anyone, qualified or not, to hunt or take these animals without the permission of the person on whose land they were found.... As a result of enclosure, they had become a type of private property and were entitled to legal protection as such" (Munsche 1981, 5). In accordance with these facts about wild deer and Berger's description of the minimized depiction of wild animals in romantic landscape painting cited earlier, the five deer (all of which appear to be stags) arrayed along the ridge in the middle ground of *The Dell at Helmingham Park* are quite small, and the centrally located members of the herd are camouflaged against a backdrop of dense forest. The cow wading in the stream in the left foreground emphasizes the diminishing figures of the deer above it and appears, at first, to counter the "wildness" of the deer with its domesticated status. The wooden footbridge at the center of the composition and the woman about to cross it, however, belies the built environment of the deer park, enabling the deer and the cow to harmoniously occupy the space together. The four deer in the center appear to be behind a fence, framed as they are by the bottom edge of the bridge; by contrast, a fifth deer stands majestically at the top of the hill on the right framed by two naturally growing trees, rather than the man-made structure of the bridge, to signal its relative freedom. The low vantage point from which the viewer looks at the deer emphasizes both the deep natural hollow of the dell and the human distance from the animals, per Berger's description of animals receding into an imaginary wilderness. Symbolically, looking up at the deer suggests their higher status than farm animals like the cow, even though the deer are being "farmed" as well. Placed above the cow, the deer are more akin to the aristocrats who own them and the land they live upon.

The bridge, seen from below, also creates an artificial horizon marking the space between the viewer and the deer. Constable's use of the bridge to structure the space of his painting and speak to the duality of the deer park as both wild and cultivated (accentuated also by the juxtaposition of the broken tree limb and the wooden split rails of the bridge at the bottom center) is reminiscent of Berger's observation that "animals came from over the horizon. They belonged *there* and *here*" (Berger 1980, 6). The deer, in other words, are in a space of the past that is retreating. Finally, all of the deer are looking out at the viewer as if they have been startled. This exchange of gazes between the partially enclosed deer and the viewer is reminiscent of animal/human encounters in zoos, described by Berger as similar to art museum visits: "Visitors visit the zoo to look at animals. They proceed from cage to cage, not unlike visitors in an art gallery who stop in front of one painting, and then move on to the next" (Berger 1980, 23). Deer parks resemble zoos in their confinement, objectification and ownership of wild animals, yet are private rather than public spaces. On the other hand, the animals appear to be able to retreat at any time to the wooded space behind them, rendering their gazes elusive and transitory, and suggesting their agency in capturing and then escaping human attention.

One way to consider the simultaneous role of deer as objects of representation and agents of artistic transformation is through Constable's consumption of venison. The remainder of this essay follows Alaimo's logic about the trans-corporeal effects of food, considering deer and their meat as creative agents, muses if you will, for the idiosyncratic corporeal, sensory, temporal and social dimensions of Constable's painting of Helmingham Park. Constable and his extended family received gifts of venison from Helmingham's deer park throughout the 30 years under consideration here and the artist often shared the meat with fellow artists at the Royal Academy in the 1820s and 1830s.[4] Lady Dysart seemed to take these transactions very seriously, at one point in 1828 nervously asking Constable's brother Abram to inquire as to whether the artist had received her gift. Of the many references to venison from Helmingham Park in Constable's letters, a bawdy joke made to C. R. Leslie at the expense of Lady Dysart stands out as emblematic of the complex matrix of artist, aristocrat and animal in Constable's career:

> It would not be quite fair perhaps to ask you to leave home at a time so interesting—but I am anxious you should enjoy a haunch of a Countess, during your "gander month." Lady Dysart has sent half a buck which is unusually fine. I hope to see the President, & Mr. Bannister, Howard, & a few more to partake of it, about the middle of next week (Beckett 1961, vol. 3, 42).

The sexualized play on words here conflates Lady Dysart with her gift of venison, slated for consumption by Royal Academicians, confirming Alaimo's suggestion that animal and human bodies intermingle, sexually in this case, through ingestion. Lady Dysart is objectified here, despite her active and necessary role as patron, providing Constable with food and money. Thus, Constable, Lady Dysart and the deer all simultaneously inhabit roles as both objects and agents in this transaction.

Comparing two different versions of Helmingham dell helps deepen consideration of the trans-corporeal effects of the 1830 painting. Constable's 1825 or 1826 painting of Helmingham dell (Plate 14) was completed before the death of his wife Maria in 1828,

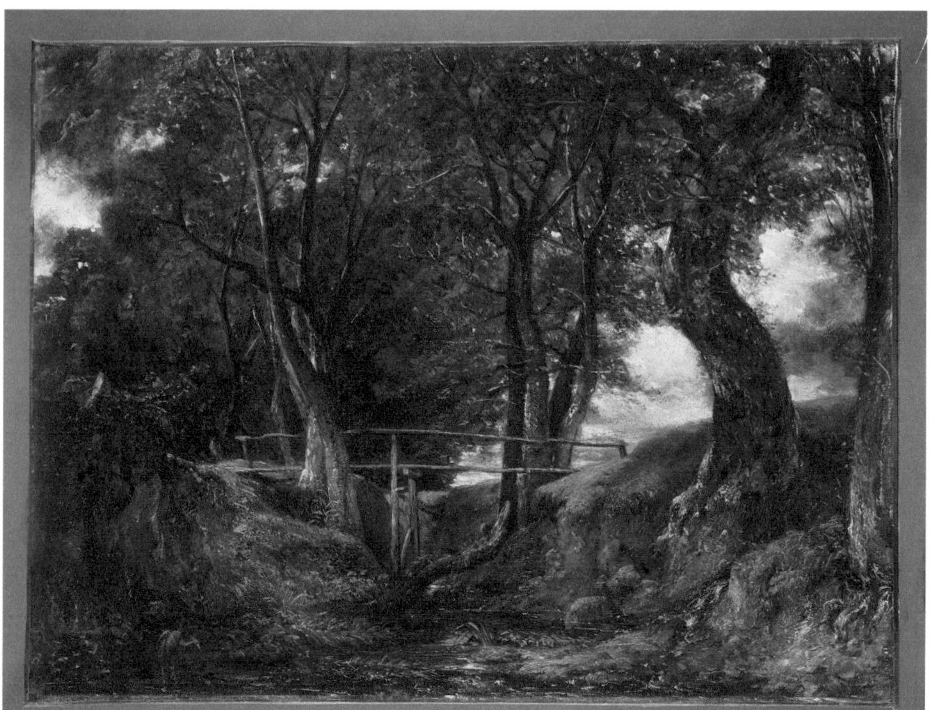

Plate 14 John Constable, *Dell at Helmingham Park*, 1825 or 1826 (retouched 1833), oil on canvas, Philadelphia Museum of Art, John G. Johnson Collection.

which marked an affective turning point in his life and played a role, perhaps, in his turn after that event toward romantic subjects of melancholy (*Jaques and the Wounded Stag*, for example) and memorial (*The Cenotaph*), and nostalgic return, through all of the avenues of memory available to him, to earlier, more sanguine subjects.[5] Interestingly, as a type of land-form, a dell suits both modes of feeling. It is frequently portrayed as a secluded, quiet and pleasant space: a safe haven. However, it is simultaneously a physically depressed spot in the landscape, bespeaking that psychological mood as well. The 1825–1826 painting is compositionally almost identical to its 1830 companion, save for its smaller size, absence of deer and the heron in the lower left-hand corner.[6] Like the deer in the 1830 painting, the heron may function as a locational signifier for the larger Helmingham estate: in this case as a typical occupant of the moat populated by water birds surrounding Helmingham Hall.

Surface facture and detail vary greatly between the two paintings, however, privileging different sensory engagements. The 1825–1826 painting is rendered with tight brushwork, appealing primarily to the viewer's sense of sight through inclusion of detailed water plants together with the heron in the foreground that call for close inspection. The emphasis on vision is appropriate to underscore the descriptive qualities of a primarily topographical work such as this one and to sustain the illusion of direct encounter and an eternal present, despite the temporal distance between the drawings from 1800 that form the painting's foundation and the finished work itself. The deer in the 1830 painting mediate the extended temporal gap in a slightly different way by transporting those who encounter them back in time and allowing them to occupy two temporal zones at once, as evoked by Lionel Tollemache's memory of encounters with deer at Helmingham Hall:

> In Helmingham Park there are red as well as fallow deer. The stags, in particular, are appropriate in the old place. During the rutting season their belling seems to take one out of the tame safeness of latter-day Europe, and to transport one bodily into Tudor or medieaval [sic] insecurity. I remember once feeling this strongly in my boyhood, when the menacing tones of a stag issued from the avenue between me and the house (Tollemache 1908, 5).

Constable may have sentimentally wished to return to his visit to Helmingham Park in 1814, which he described in his aforementioned letter to Maria during their courtship.[7]

The 1830 work is also much more painterly and tactile than its 1825–1826 counterpart, consistent with Constable's late style and suggestive, like the topographical qualities of the earlier painting, of immediate sensorial contact with the site as in one of the artist's sketches. However, the loose paint handling also quite literally "marks" the distance between the early drawings and the mature work with unsettled streaks and wayward daubs of paint that suggest gaps of memory filled in with visceral traces of the artist's hand in abstracted passages; one critic noted the "scratchiness in the execution" of the painting in their review ("Royal Academy" 1830, 444). Another significant mark on the surface of the painting is the visible crease running approximately two inches below the top of the work created when the artist inexplicably extended his painting area. This sign of a disrupted, reconsidered process, like the vigorous brushwork, alerts the viewer to temporal vagaries and the artist's inability to transparently collapse the time elapsed between his drawings and the present. By showing traces of his mark-making so vividly in his depiction of the deer park,

Constable completes a sensory, trans-corporeal cycle moving from sight to taste to touch and conjures the deer he has first gazed upon; then eaten; and, finally, painted for public consumption at the Royal Academy exhibition.

Analysis of place in relation to the personal and the social is a key feature of Constable scholarship. Animal studies, as practiced by John Berger and Stacy Alaimo, offers a novel lens through which to consider the significance of place in the artist's oeuvre by attending to his representation of nonhuman interaction with the landscape setting. When Constable first visited Helmingham in 1800, he remarked upon the "good objects" available for him to draw. In 30 years' time, he transformed these "objects" into creative agents in *The Dell at Helmingham Park*, using representational and trans-corporeal means. The painting is thus the product of a cyclical process of art making by which place is seen, taken into the body (as food, in this case) and conveyed onto canvas. This essay thereby proposes an alternative ecology of Constable's work that points toward a richer reading of place by insisting on the primacy and agency of its nonhuman inhabitants.

Acknowledgment

Funding for travel, research and image permissions for this essay were provided by the Art History Department Faculty Research and Development Restricted Fund and a Faculty Research Grant from Drew University. I would like to extend special thanks to Maura Coughlin and Emily Gephart for their guidance and the opportunity to publish this essay, to the anonymous readers who provided constructive feedback on it, and to Jeff Haddorff for being its first reader.

Notes

1 The most complete account of Constable's Helmingham paintings can be found in Beckett (1961). Together with providing a history of the production of these works, the author discusses the geographical significance of the dell to the Helmingham property.
2 This essay is part of a larger study in progress of the depiction of deer in Constable's work that will include these pieces.
3 For a notable ecocritical study of Constable's work, see Wood (2007).
4 An additional family connection to the Tollemache family is Golding Constable's employment as the warden of Bentley Woods, owned by the aristocrats (Beckett 1964, 67, vol. 4).
5 After Maria's death, Constable was also increasingly agitated about the Reform Bill of 1832, which shifted Parliamentary representation and power away from rural districts controlled by aristocrats toward growing urban centers. While this may be significant to discussions of other representations of deer by Constable such as *Jaques and the Wounded Stag*, it is outside the scope of consideration here.
6 The 1825–1826 painting was in a private collection until 1830, when Constable purchased it back from the owner. From 1830 to 1833, when the earlier painting went into a different private collection, the Helmingham paintings were therefore both in Constable's studio (Reynolds 1984, 174–175). Additionally, the mezzotint *A Dell, Helmingham Park, Suffolk*, which first appeared in part one of Constable's *English Landscape Scenery* in 1830, has been linked to the 1825–1826 version of the scene by Andrew Wilton. The inclusion of the cow in the lower-left-hand corner of the print, however, suggests that the mezzotint is an amalgam of the two Helmingham paintings (Wilton 1979, 34).
7 Ann Bermingham has convincingly suggested that during Maria and Constable's long and difficult courtship, local landscape subjects became more important to the artist and that he began to conflate the two in his mind. I am extending Bermingham's idea here (Bermingham 1986, 101–105).

Bibliography

Alaimo, Stacy. 2008. "Trans-Corporeal Feminisms and the Ethical Space of Nature." In Stacy Alaimo and Susan Hekman, eds., *Material Feminisms*. Bloomington: Indiana University Press: 237–264.

Barrell, John. 1980. *The Dark Side of the Landscape: The Rural Poor in English Painting 1730–1840*. Cambridge: Cambridge University Press.

Beckett, R. B. 1961. "Constable's 'Helmingham Dell'." *The Art Quarterly*. XXIV, 1: 3–14.

Beckett, R. B., ed. 1964. *John Constable's Correspondence*, vols. 2–4. Ipswich: Suffolk Records Society.

Berger, John. 1980. *About Looking*. New York: Pantheon Books: 3–28.

Bermingham, Ann.1986. *Landscape and Ideology: The English Rustic Tradition, 1740–1860*. Berkeley: University of California Press.

Fletcher, John. 2011. *Gardens of Earthly Delight: The History of Deer Parks*. Oxford: Windgather Press.

Johnson, Thomas Burgeland. 1830. *The Hunting Directory: Containing a Compendious View of the Ancient and Modern Systems of the Chase*. London: Sherwood, Gilbert, and Piper.

Jones' Views of the Seats, Mansions, Castles, Etc. of the Noblemen and Gentlemen in England, Wales, Scotland, and Ireland: And Other Picturesque Scenery Accompanied with Historical Descriptions of the Mansions, Lists of Pictures, Statues, Etc. and Genealogical Sketches of the Families and Their Possessions. 1829. London: Jones & Co.

Munsche, Peter Bernard. 1981. *Gentlemen and Poachers: The English Game Laws 1671–1831*. Cambridge: Cambridge University Press.

Reynolds, Graham. 1984. *The Later Paintings and Drawings of John Constable: Text*. New Haven, CT and London: Yale University Press.

———. 1996. *The Earlier Paintings and Drawings of John Constable: Text*. New Haven and London: Yale University Press.

"Royal Academy." 1830. *The Gentleman's Magazine and Historical Chronicle*. May: 444.

Shirley, Evelyn Philip. 1867. *Some Account of English Deer Parks*. London: J. Murray.

Tollemache, Lionel A. 1908. *Old and Odd Memories*. London: E. Arnold.

Whitaker, Joseph.1892. *A Descriptive List of the Deer-Parks and Paddocks of England*. London: Ballantyne, Hanson & Co.

Wilton, Andrew. 1979. *Constable's 'English Landscape Scenery'*. London: Colonnade Books.

Wood, Gillen D'Arcy. 2007. "Constable, Clouds, Climate Change." *Wordsworth Circle*. 38, 1 (Winter–Spring): 25–33.

19 Cultivating Fruit and Equality

The Still-Life Paintings of Robert Duncanson

Shana Klein

In 1849, Cincinnati artist Robert Duncanson depicted two still-life paintings with a pineapple at their center (Figures 19.1 and 19.2). In one painting, Duncanson showed the fruit is in its most vibrant stage of life, with prickly skin, tight cross-hatching and waxy green leaves. In the other interpretation, Duncanson's pineapple is rotting, with frayed leaves and balding skin, where the effects of time have carved valleys between each crosshatch. His careful and faithful study of the pineapple wilting over time suggests that he drew this fruit from real life rather than merely copying it from paintings or horticultural engravings. It might seem peculiar that Duncanson, a lower-class, African American artist located in the chilly climes of Cincinnati, Ohio, had access to the tropical pineapple. This fruit, in fact, was an exotic and expensive import from the Caribbean and Central America that cost a whole day's wage for most middle-class Americans at that time (Okihiro 2010, 136). Duncanson's location in Ohio might explain his access to tropical fruits that were cultivated by local, elite horticulturists in their hothouses. A number of these horticulturists were also Duncanson's art patrons, who helped cultivate land in what was at the time the western frontier and transform the Ohio region into one of the most vibrant fruit markets in the trans-Appalachian West. In this lively gastronomical environment, a still-life and landscape painter such as Duncanson took advantage of the opportunity to paint pineapples and other opulent fruits. The artist's romance with painting food was also strategic to appeal to Cincinnati's leading fruit growers, who doubled as the city's leading art patrons. A number of those patrons, moreover, were abolitionists and supporters of greater equality for African Americans like Duncanson. This essay focuses on the edible matter in Duncanson's art and examines the many ways that fruit connected him to local networks invested in art, horticulture and racial equality.[1]

The following essay also advocates for an environmental turn in scholarship on nineteenth-century art since many still-life artists and patrons in the United States were redefining their relationship to nature and using experiments in fruit cultivation to document, study and modify the natural world. Duncanson displayed this changing relationship in paintings that combined fruits from different seasons and regions, showing man's perceived ability to defy the limits of the natural world and locally grow seasonal and foreign fruit varieties at all times of the year. These horticultural advancements had real consequences on Duncanson, whose race and economic status historically excluded him from accessing luxury fruits, much less studying and illustrating them. Duncanson's botanically astute still-life paintings, consequently, raised questions around race and privilege and who had access to horticultural resources in the mid-nineteenth century. His paintings also confronted questions about slavery and racial equality by displaying politically charged foods strongly connected to abolitionist circles protesting slave labor.

Upon more closely examining the political conditions of agriculture, readers will discover how non-corporeal subjects like food triggered the nation's heated debates over a very corporeal subject: human slavery. The following text uses an amalgamation of methodologies across Eco-Art History, Environmental History and Food Studies to reveal how Duncanson's still-life paintings of food tapped into broader conversations about racial equalities and inequalities in antebellum society.

Fruit and Manifest Destiny

While Duncanson is best known for his pastoral landscape paintings of majestic mountains and verdant fields in the style of the Hudson River School, he won much praise for his early still-life paintings that also celebrated America's natural abundance (Figure 19.1). Pictures such as *Fruit Piece* show raspberries, cherries, strawberries, gooseberries, citrus fruits and two varieties of currants resting on a ceramic plate. Confined by a shallow foreground, the berries are squeezed between the soft skins of raisins and the less hospitable surface of a prickly pineapple.

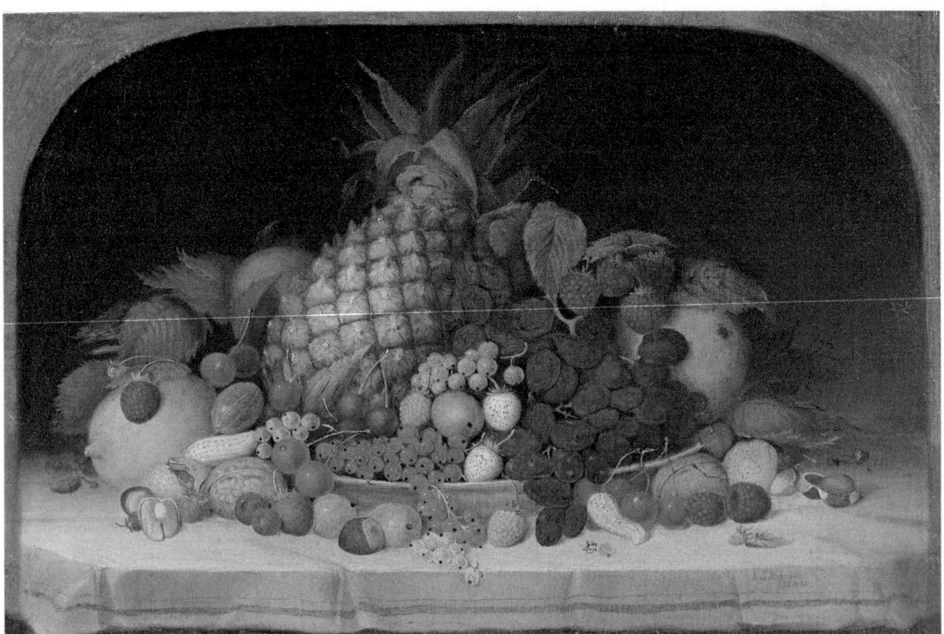

Figure 19.1 Robert Duncanson, *Fruit Piece*, 1849. Oil on canvas, Detroit Institute of Arts, Detroit, Michigan, gift of the estate of Miss Elizabeth Gray/gift of the estate of Mr. Henry Lyster Walker. Photo © Bridgeman Images.

The fruit is represented in such abundance that it spills off the plate's edge onto the table beneath it. No facet or feature of the fruit escaped Duncanson's keen eye; he even indented every strawberry with tiny dots of paint to look like seeds. The artist's meticulous regard for detail brings to mind the early nineteenth-century still lifes of the noted Philadelphia artists Raphaelle and James Peale whom displayed fruit with relentless naturalism, illuminating every dimple or blemish of foods by a sharp spotlight.[2] Duncanson similarly lavished attention on the unique characteristics of fruit, endowing each one with its own individual portrait. For his masterful still-life paintings, Duncanson won a premium at the Michigan State Fair in 1849 and gained admission to the esteemed American Art-Union with a still-life depiction one year later (*List of Premiums Awarded* 1849, 295; *Bulletin of the American Art-Union* 1850, 88). The artist excelled in still-life painting, an artistic genre that accelerated his career.

Duncanson's still lifes may have garnered favorable reviews because they translated period beliefs that America's fruits were blessings from God into paint. Such notions had gained currency earlier in the nineteenth century, in 1802, when the English clergyman and philosopher William Paley proclaimed in his influential book *Natural Theology* that fruits were evidence of God's handiwork and a reflection of the Creator who produced them (Paley 1802, 241). By the mid-nineteenth century, in 1850, the writer and editor of the *Cincinnati Chronicle* Edward Deering Mansfield delivered a lecture to the Cincinnati Horticultural Society that urged every grower to plant trees and build orchards to "take from the earth its blessings ... which Heaven ordained" (Mansfield 1850). Lectures celebrating the fruits of God's design carried particular significance in the frontier state of Ohio, where the nurseryman and apple missionary Johnny Appleseed (né John Chapman) famously distributed seeds to farmers so that they would cultivate the fruits of God's blessings (Pollan 2001, 3). Underpinning this call to cultivate America's God-given fruits was the concept of Manifest Destiny, a notion that divinely sanctioned American efforts to expand the nation's territory westward and reap the attendant financial rewards. Viewers may have embraced Duncanson's still-life paintings of natural bounty as an endorsement of this national agenda to cultivate the continent's providentially granted fruit and land.

The artist's *Still Life* with a honeycomb (Figure 19.2) is particularly resonant in this regard. Agricultural journals frequently described America as a land "flowing with milk and honey," drawing on the description in the Bible of the rich landscape God promised to the Israelites ("Ploughing Exhibition at Harlæm" 1839, 612; Fennell 1848, 468; "Address of Walter W. W. Bowiee" 1849, 282; Paulding 1849, 16). Mansfield took this comparison a step further by declaring in his lecture to Cincinnati horticulturists that America's "fields produce more than Canaan with its milk and honey!" (Mansfield 1850, 8). Such notions inform Duncanson's *Still Life* that shows honey oozing from every cell of the honeycombs onto the gold-rimmed plate beneath. The comb releases nectar all on its own, needing little human effort to retrieve it. Like many still-life artists, Duncanson portrayed nature as generous and accessible, awaiting human intervention to take advantage of its richness. By painting foods oozing forth and bursting open, Duncanson subscribed to artistic traditions that invited viewers to exploit the Godly abundance of the American landscape.

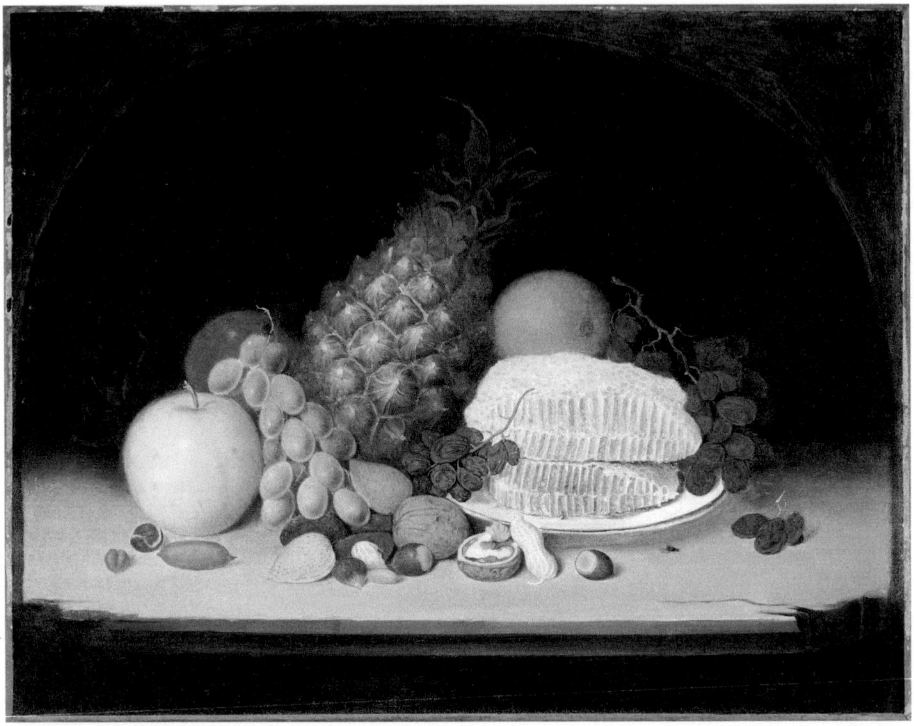

Figure 19.2 Robert S. Duncanson, *Still Life*, 1849. Oil on canvas, Los Angeles County Museum of Art, Gift of Mr. and Mrs. Robert B. Honeyman, Jr. Photo © Museum Associates/ LACMA.

The Wedding of Art and Horticulture in Cincinnati

More specifically, Duncanson displayed the rewards of extracting fruit from Cincinnati, the artist's resident city. Located on the Miami and Erie Canal and Ohio River, at the transportation nexus linking North, South, East and West, Cincinnati was one of the leading economic centers of the American West and a major depot for the transportation and manufacture of food. Galvanized by farmers, agricultural educators and members of the Cincinnati Horticultural Society, this city was a powerhouse of fruit cultivation in the mid-nineteenth century. A print from 1843 (Figure 19.3) shows the nine founding members of the Horticultural Society posing around a table laden with apples and grapes. The fruits are piled high but remain contained by the plates' borders, a metaphor for the horticulturists' ability to control wild nature. Although fruit traditionally accessorized portraits of women in the home, symbolizing ideas about social class and child-rearing, in this lithograph, the fruits function as masculine attributes that affirm the sitters' mastery over the landscape (Staiti 1998, 21–22). Such representations notably honored the creative and intellectual rather than the physical labor of fruit cultivation, the latter of which was typically relegated to immigrant or enslaved farmers. The members' direct gaze forced viewers to bear witness to their successful manipulation of nature and role in pioneering plant cultivation in the American West.

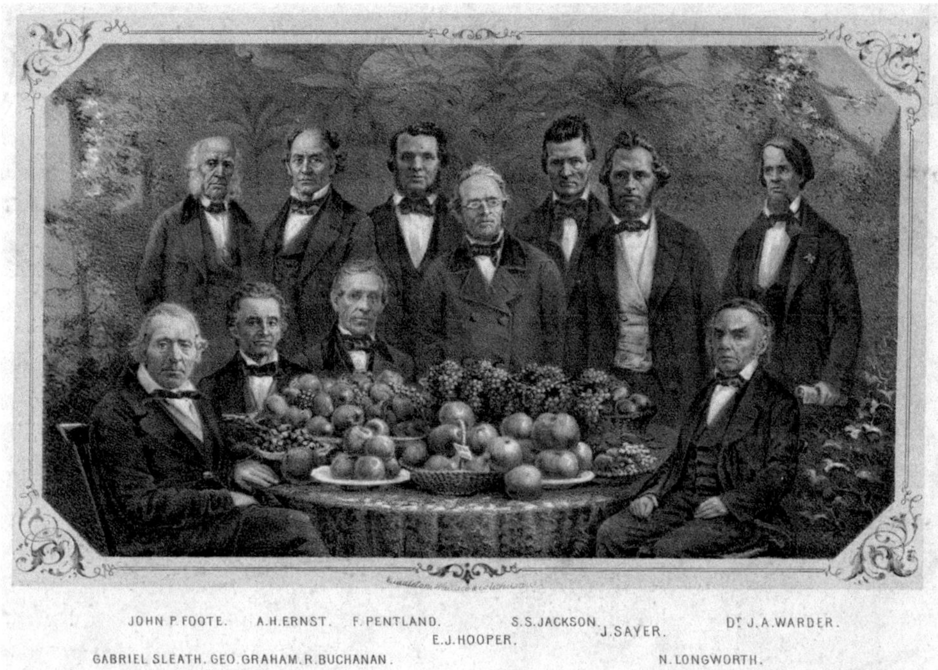

Figure 19.3 Middleton, Wallace & Co., Founding members of the Cincinnati Horticultural Society, 1843. Lithograph. Cincinnati Museum Center.

This engraving recording the members' likeness signals a more self-conscious generation of horticulturists who understood how visual images could help preserve their legacy and map America's participation in the enlightened arts of horticulture.

Members wanted to visually confirm their contribution to horticulture because it was a new discipline in antebellum America that promised to bring national distinction and advance national expansion. After the eighteenth century when European plantsmen began consciously growing plant varieties, botanists in the nineteenth century began to organize these species as part of a new enterprise called horticulture (Pauly 2008, 53). It spread to the United States where horticultural societies organized nomenclatural projects for new varieties and seed exchange programs and produced botanical engravings that detailed the biological and aesthetic features of each plant and fruit. Although American horticulture was a clear extension of European study, Americans sought out "horticultural independence" and the development of specifically American fruits, explains historian Philip Pauly (3). The process of classifying and illustrating fruit sought to naturalize foreign varieties in America and botanically colonize the indigenous land upon which they grew. Cincinnati writer Edward Mansfield specifically connected the cultivation of fruit to land in his 1850 lecture to the Cincinnati Horticultural Society:

> New nations, new trees, new fruits, new flowers have come to magnify and glorify the face of nature, with far more than its original magnificence; and you, ladies and gentleman, have come to testify to this revival of civilization… We are

> now entering the era of high and finished civilization... The race of Pontiac and Little Turtle has disappeared! ...a new race has come... A stronger race, a better destiny, a higher glory...will soon build the house of a new and better Garden (Mansfield 1850, 3).

In Mansfield's perspective, the transformation of Native American land into "a new and better Garden" was inevitable. Horticulture was a servant to building a "stronger race" and ushering the United States into a new chapter of civilization that depended on the theft of Indigenous land. The 1848 print of Cincinnati horticulturists posing with fruits reaped from the West asserted their destiny to cultivate America's frontiers and carry forward the legacy of civilization first started by European cultivators.

Cincinnati's leading fruit growers understood the power of visual images to communicate social progress since they were also the city's leading art patrons invested in boosting artistic production. Andrew Ernst (pictured in the engraving's top right), Nicholas Longworth (pictured at bottom left) and Freeman G. Cary were all members of the Horticultural Society and active art collectors and members of the Western Art Union – a subscription-based organization founded in 1847 in Cincinnati to promote and sell local art (*Transactions of the Western Art Union* 1847, 1848, & 1849). Subscribers could view art in the Western Art Union building, which contained artist studios and a gallery. Members of the Western Art Union were also entitled to engravings of select oil paintings by paying an annual subscription fee. Horticulturists such as Nicholas Longworth were not only subscribers, but founding members of the Western Art Union and belonged to the "Committee of Ten," helping to recruit new subscribers.

Longworth, Ernst and Cary all commissioned portraits in the 1850s from Duncanson, a contributing artist to the Western Art Union, who might have first painted still lifes of fruit in the 1840s to attract the patronage of horticulturists who doubled as art collectors. Duncanson's portraits of the three patrons embodied the marriage between art and horticulture in Cincinnati, showing all three cultivators posed with the fruits of their labors.[3]

Painting and horticulture were uniquely wedded in the antebellum period when both disciplines were considered equal art forms in the pursuit of civilizing society. Edward Mansfield explicitly linked the two fields, declaring, "The fine Arts have come. Painters, poets, sculptors, and not least, Horticulturists have come, each in his turn, to erect the Corinthian columns of the solid and stately structure of Society" (Mansfield 1850, 9). The strong connection between the cultivation of fruit and art was also evident in an art exhibition spanning five pavilions at the Cincinnati Horticultural Society Fair in 1856, which the *Cincinnatus* journal described in detail:

> The whole was converted into a splendid picture gallery, and nearly three hundred fine paintings, engravings, photographs, daguerreotypes and other works of art, were added, to beautify the display, and blend together the imitations of art with the productions of nature. Over the green-house plants, were arranged some of the best paintings of our native artists, such as Whittridge, Beard, Sontagg, Frankenstein, Lee, Cridlin, Spinning, and others... .Over the fruit tables there was also suspended a splendid collection of paintings and engravings, many of them imitations of the natural fruit below them (Graham, "Cincinnati Horticultural Society," (Graham 1857, 22).

By suspending paintings of fruit above tables of specimens, these pictures literally elevated fruit to an artistic form. Fruits were considered aesthetic objects in their own right as materials with properties in color and shape that horticulturists manipulated like artists in the soil. Fruits were similarly valued for their aesthetic properties that some horticulturists even collected fruit specimens and displayed them like paintings in personal galleries (Hannickel 2013, 41). A study of nineteenth-century fruit painting thus crucially parallels a study of nineteenth-century horticulture since art patrons in cities like Cincinnati fused the two disciplines together to legitimate cultural progress in the West.

Duncanson's still-life paintings of summer cherries, peaches and pears perched atop autumn and winter apples and quinces highlighted the advancements of local horticulturists.[4] While Duncanson's liberal fruit combinations uniting fruits from different seasons, climates and regions might have been an invention of his imagination, they more likely reflected concurrent fashions in hothouse technologies that enabled local horticulturists to surpass the limits of a region's climate and transform Ohio's Temperate Zone into the Tropics. His longtime patron Longworth, for instance, famously experimented with fruit cultivation in his hothouse, producing pineapples that normally had to be imported from the distant Caribbean ("Exhibitions of Horticultural Societies" 1843, 465; Lockwood 1931, 415). He also invented a hermaphrodite strawberry variety that ended a decade-long "Strawberry War" concerning which American horticulturist could produce the most prolific strawberry seedling (Buchanan 1854). Still-life paintings that paired fruits from different seasons and regions, and showed an abundance of fruits that were difficult to propagate, appealed to patrons who were agriculturally experimental and had an intimate knowledge of hothouse production. Rather than display horticulturists' failed attempts at modifying nature and the limits of human dominion, artists portrayed a diversity of fruits and plants in still-life paintings that validated America's burgeoning expertise in cultivating specimens.

Pomological Patronage and Fruit Philanthropy

Duncanson was no doubt knowledgeable about local agricultural innovations since his art patrons were profoundly improving the public's access to fruit. Longworth built vineyards, hothouses and greenhouses, allowing the public to visit his gardens and view exotic fruits for the first time (Young 1982, 30). The journalist Charles Cist applauded Longworth for democratizing fruit, writing, "he has made these fruits accessible to the means of purchase of every man, even the humblest among us" (Cist 1998, 297). Cist continued, "How much more manly and spirited is this, than tempting the poor man with the sight of luxuries he may look at, but can never expect to taste" (Cist 1998, 336–337). Duncanson belonged to the group of "poor" and "humble" men in society, who benefitted from the generosity of Longworth that Cist described.[5] Longworth specifically gave away his products to artists, sending his wines abroad to sculptor Hiram Powers ("Letters of Hiram Powers to Nicholas Longworth, 1906, cited by Hannickel 2013, 115). Andrew Ernst shared similar goals with Longworth, claiming that he founded his nursery "not merely to make money... but to produce something to add to the happiness of his fellow citizens" (Ernst 1843, preface). In this sense, Longworth and Ernst were fruit philanthropists who cultivated specimens so that a broad swath of the local community could enjoy the world's fruit. As fruit was being democratized in Cincinnati, Duncanson showed the results.

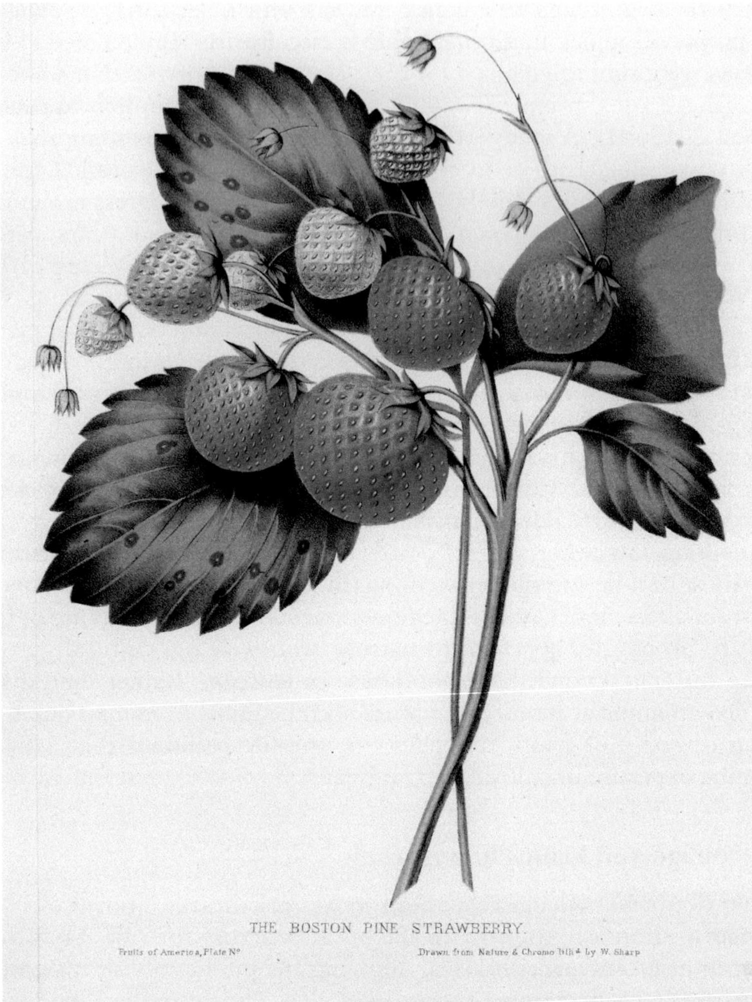

Figure 19.4 Charles Mason Hovey, *The Boston Pine Strawberry*. From *The Fruits of America, Containing Richly Colored Figures, and Full Descriptions of All the Choicest Varieties Cultivated in the United States* (C. C. Little & Jas. Brown, and Hovey & Co., 1847). Image courtesy of Crystal Bridges Museum of American Art Library, Bentonville, Arkansas.

Duncanson's knowledge of fruit also was likely enhanced by the growing circulation of illustrated horticultural books and journals in the 1840s. Between the 1820s and 1850s, horticultural literature dramatically increased, with newspapers regularly publishing articles and illustrations from horticultural journals. Duncanson may have drawn inspiration for the strawberries in his *Fruit Piece* from the engravings in Charles Hovey's multivolume tome *The Fruits of America,* the authority on American pomology in the antebellum period (Figure 19.4). In both cases, the berries dangle from a leafy vine, with their crowned leaves and calyx still intact. Duncanson even depicted the strawberries' hairy bristles and the horseshoe neck of the vine, features that Hovey's engravers highlighted to educate viewers on the anatomy of this fruit. Duncanson's portrayal of apples in *Fruit Still Life* (see Figure 19.2) also resembles Hovey's written description of the fruit:

> Skin, fair, smooth, glossy, bright yellow in the shade, but nearly covered with deep orange red, approaching to scarlet, indistinctly striped with crimson, and covered with prominent grayish specks, thickest near the crown; blotched with russet in the cavity around the stem, which, in some specimens, extends in irregular tracings over the base of the fruit (Hovey 1847, 11).

Hovey would have been pleased to find that his text inspired artistic replications, since he recognized how depictions of fruit in partnership with horticultural journals "have done much to spread a better knowledge of the many varieties which have been brought to notice" (Hovey 1847, prospectus). Hovey specifically sought to "give correct Drawings, and full descriptions" to fruits of American origin, reinforcing how visual images participated in a national program to produce a distinctly American discipline of horticulture. Only upon studying nineteenth-century attitudes toward nature do scholars discover how Duncanson's still-life paintings relayed the principles of a larger national educational project to showcase the colors, textures and anatomical features distinguishing the fruits of America.

Although Duncanson's paintings conformed to a greater visual and national formula, pictures displaying the horticultural intelligence of an African American man were no meek statement at this time. His studied depictions of fruit demonstrate a more inclusive field in which marginalized voices – and not only white, affluent, male landowners – contributed to the national agenda of recording America's fruits. Still-life paintings that display the pomological knowledge of an African American man also contradicted long-held beliefs that black Americans were slow, lazy and incapable of mastering the science of plant cultivation. In 1835, a writer in the *Farmers' Register* expressed this prejudice against a colony of African American farmers, writing, "they are so excessively lazy and stupid, that ... neighboring farmers will not employ them as work hands to any extent. They do not raise produce enough on their lands to feed their families, much less do they have a surplus for sale abroad" ("An Experiment of Emancipating Negroes" 1835, 430). Paintings such as William Sidney Mount's *Farmers Nooning*, which depicts an African American farmer asleep on a mound of hay during a pause from work, perpetuated these negative stereotypes.[6] Duncanson's skillfully rendered fruit paintings, on the contrary, demonstrate that African Americans possessed the intellectual capacity to master subjects like pomology and horticulture.[7] As the art historian Wendy Katz has argued, Duncanson's landscape paintings challenged derogatory imagery more directly by depicting African American laborers diligently working the land (Katz 2002, 103–104). What better way to convey the intellectual cultivation of African Americans than through the representation of cultivated fruit and landscapes? The use of fruit and land to undermine racial prejudice encourages a closer look at the ways artists – specifically artists of color – depicted food to challenge systems of power in the nineteenth century. Duncanson's proficient depiction of the anatomy of fruit not only displayed the altruism and aptitude of Cincinnati's horticulturists, but also the aptitude of their black visual interpreter.

Equality among Fruit and Men: Debates over Slavery in Still-Life Painting of Food

The capacity for Duncanson's paintings to contradict racial prejudice and embody ideas about equality and democracy added to their value for local fruit-growing art patrons who were also abolitionists, sympathetic to the plight of black Americans. Andrew Ernst, for example, partnered with his wife, Sarah Otis Ernst, in the organization of the Cincinnati Anti-Slavery Convention from 1851 through 1855. He was well known for housing fugitive slaves at his Spring Garden nursery and reportedly never turned away a person in need (Douglass 1850). Cary, too, was an abolitionist and advocate for African Americans, using the Farmers' College that he founded as a station for the Underground Railroad (Dwight 1955, 207). Although there are ambiguous reports about Longworth's attitude toward abolition, he funded the construction of a school and an orphan asylum for African Americans in Cincinnati and was even rumored to have saved a runaway slave by hiding the man in his wine cellar (Woodson 1916, 19–20).[8] In such cellars and nurseries where fruit was cultivated, so too were freedom and equality for African Americans.

By the 1840s and 1850s, abolitionists of the Free Soil Party had already established a deep connection between slavery and the agricultural environment, believing that the practice's evils and inefficacies caused withered, barren fields when labored by slaves. One politician asked, "Shall unpaid, unwilling toil, inspired by no hope and impelled by no affection, drag its weary, indolent limbs over that State, hurrying the soil to barrenness and leaving the wilderness a wilderness still?" (Salter 1876, 48). Nature's decay, however, was not irreversible. With the end of the Civil War and the abolishment of slavery, horticulturists praised the effects of free labor on the landscape, saying, "now that we have perfect peace and quiet again….it will not be long before land will rise to treble its value now; flourishing farms and vineyards will be where everything is wilderness yet; and oh! Most glorious thought of all, they will be worked by *free* and happy people" (Husmann 1865, 267). The cultivated land, in this case, was no longer a victim of man's unethical actions but a partner to abolitionists in celebrating the dissolution of slavery.

Period discourses surrounding the cocultivation of fruit and equality shed new light on Duncanson's painting of honeycomb, a rare still-life subject before the Civil War, which may have been a means for the artist to engage contemporary, anti-slavery arguments (Figure 19.2). While the iconography of this picture may have referenced contemporary descriptions of the United States as a blessed nation "flowing with milk and honey," abolitionist newspapers – often the same publications that reviewed Duncanson's artwork – cast a shadow over this metaphor by publishing poems pointing to the incongruity that "In the fairest countries earth provides where fruits forever grow, whose rocks and hills, and plains besides, with milk and honey flow, are poorer than the poorest free, because by slavery cursed" ("The War of Slavery" 1839, 100). Perhaps Duncanson placed the plate of honeycomb next to a dead fly on a cracked table in *Still Life* to signal the broken system of slavery and its violent consequences. The rotting pineapple and brittle raisins further signaled decay, symbolizing human suffering in a still-life genre that long used foods as human proxies to comment on mortality (Gerdts 1976, 5). As much as Duncanson's still-life paintings seem to celebrate the democratization of food and sweetness of the American landscape, his rotting fruits and poisoned fly bite back at these uplifting notions by symbolizing the corruption of a "slave-cursed" country.

Honeycomb took on yet another political dimension around mid-century as an honorable replacement for the white sugar and molasses produced under the oppression of slavery. Advocates of "free produce" sugar hoped that the consumption of honeycomb and the boycott of slave-labored foods would diminish the profits of industries reliant on slavery. Renowned supporters of free-produce food included Harriet Beecher Stowe, Frederick Douglass and the Grimké sisters, one of whom famously served "free-sugar" desserts at her wedding (Glickman 2012, 71). Duncanson was undoubtedly familiar with the political connotations of sugar and honey in Cincinnati – a hot spot for the free-produce movement, where slave-free sugar was manufactured in the city's sorghum mills and sold in its free-produce grocery stores ("New Helps of Agriculture," 1862, 274).[9] But unlike the ethical honeycomb, the tropical pineapple Duncanson painted was typically harvested by slave labor on plantations in the Caribbean and Central America (Okihiro 2010, 112–113). Many abolitionists viewed pineapples as polluted by the systems of slavery, detailing the evils of tropical fruit in poems, including the following verse about a mother who warns her daughter against eating bananas, oranges and citrons:

> I told you those fruits, which you valued so much,
> Though tempting their look—'t was pollution to touch;
> . . . .
> They were raised in a land where the labor of slaves
> Has borne many thousands away to their graves!
> Alas! when I place the sweet cup to my lip,
> The tears of the Afric are all that I sip!
> Oh! who for a moment's enjoyment would crave,
> Did they think it was mixed with the blood of a slave? (A., "Slave Labor," 91).

Duncanson's juxtaposition of the golden honeycomb oozing natural sweetness next to the decaying pineapple in his painting expressed the values of the free-produce movement and resentment toward the rotten labor system producing tropical fruit.[10] Yet, if Duncanson's pineapple was produced domestically in a hothouse by a local patron such as Longworth, then his fruit is not an adversary to the honeycomb but another example of a slave-free product, albeit not the most vibrant one. More important than the labor source of Duncanson's pineapple is the fact that his painting raises such questions in the first place. The crisis over slavery demanded that viewers in the Civil War Era re-evaluate their relationship with agriculture and track how their personal eating habits intersected with those who grew or depicted foods that had been cultivated under controversial circumstances.

New Directions in Scholarship

Duncanson's fruit paintings ultimately demonstrate how art, racial equality and nature collided in nineteenth-century discourses. The artist's immersion in the world of horticulture invites scholars to consider the fruit and plants in his art. Examining his representations of food alongside antebellum perspectives on socially responsible cultivation practices shows how Duncanson's images, like many of the time, supported national efforts to cultivate western frontiers and promote a uniquely American identity. But an eco-critical approach that places fruit at the center of this investigation also reveals how Duncanson's still lifes displayed a unique connection between food and racial equality. He exploited the pictorial conventions of still-life painting in ways that expanded the inclusivity of horticulture, combated racial prejudice and rejected slave-labored products. Inserting messages about racial justice into still-life painting was a clever strategy to command the attention of horticulturists, abolitionists and art patrons in Cincinnati who spearheaded initiatives that advanced this very goal. Exploring the intersection of race and food through the history of a marginalized artist is especially productive since it is in these historically overlooked narratives that scholars often find threads of resistance to normative traditions oppressing both people and the natural environment. An eco-critical approach to food brings these dialogues to the surface, throwing into sharp relief the political importance of food in the antebellum period and the stakes of rendering it in paint.

Notes

1 A version of this essay previously appeared in *American Art*. 29.2 (Summer 2015). Many thanks to the leaders of the journal for allowing me to re-use this material for this anthology.
2 Duncanson may have seen the Peales' still-life paintings in the collection of art patron, Nicholas Longworth, who likely collected still lifes by the Peales for the short-lived National Portrait and Historical Gallery, located in the Fireman's Hall in Cincinnati.
3 In Duncanson's 1858 portrait of Longworth, the artist positioned the patron next to a grapevine in front of his vineyard; in an 1858 portrait of Ernst, the artist displayed the patron holding fruit on his lush property; an 1856 portrait of Cary showed the horticulturist at Farmer's College, one of the first agricultural colleges in the United States. Longworth was an especially important patron to Duncanson, funding his trip abroad in 1853 and giving him a large commission to paint a suite of eight murals in his Belmont Mansion.
4 See: *Fruit Still Life* (1849, Corcoran Collection, National Gallery of Art, Washington).
5 Author James McCabe Jr. (*Great Fortunes, and How They Were Made* 1871, p. 160) similarly admired the way in which Longworth "manifested no selfishness with respect to his fruits. He was anxious that their cultivation should become general, and his discoveries and improvements were always at the service of any and every one who desired to make use of them."
6 See: William Sidney Mount, *Farmers Nooning*, 1836 in the collection of The Long Island Museum of American Art, History & Carriages.
7 Slave gardens represent another way in which African Americans showed diligence in farming. It does not appear that Duncanson included any distinct references to slave-garden practices or African American foodways in his still lifes.
8 The strength of Longworth's commitment to abolition, however, is debated. While Joseph Ketner (*Emergence of the African-American Artist*, 50) argued that Longworth possessed "strict antislavery principles," Denny Carter Young ("Longworths," 33) classified Longworth as a political moderate on the issue of abolition.
9 There were at least two free-produce stores in Cincinnati: one operated by the famous abolitionist Levi Coffin and another in Mount Pleasant, Ohio, where Duncanson may have lived at one point.
10 A decade later, artist Eastman Johnson also acknowledged the free-produce movement in his Sugaring Off paintings, which celebrate the gathering of maple sap from which maple sugar was made, a product long hailed as an ethical alternative to the slave-produced cane variety. Read: Brian Allen's text *Sugaring Off* at the Clark Art Institute.

Bibliography

A. "Slave Labor." *The Liberator*. June 8, 1833.

Allen, Brian T. 2004. *Sugaring Off: The Maple Sugar Paintings of Eastman Johnson*. Williamstown, MA: Sterling and Francine Clark Art Institute.

"An Experiment of Emancipating Negroes, Under Very Favorable Circumstances." *The Farmer's Register, Monthly Publication, Devoted to the Improvement of the Practice, and Support of the Interests of Agriculture.* 3 (1836): 430.

Breidenbach, Paul. 2001. *Art Patronage and Class Identity in a Border City: Cincinnati, 1828–1872*. Dissertation, University of California, San Diego.

Buchanan, Robert. 1854. *The Culture of the Grape and Wine Making with an Appendix Containing Directions for the Cultivation of the Strawberry by Longworth, 5th Edition.* Cincinnati, OH: Moore, Anderson and Company.

Bulletin of the American Art Union 313 (December 1850): 88.

Cist, Charles. 1998. *Sketches and Statistics: Cincinnati in 1851*. Cincinnati, OH: W.H. Moore and Co.

Douglass, Frederick. 1850. "The Western-Anti-Slavery Tour." *Anti-Slavery Bugle*, August 24.

Dwight, Edward. 1955. "Robert Scott Duncanson." *Cincinnati Historical Society Bulletin.* 13 (July).

Ernst, A. H. 1843. *Catalogue of Fruit and Forest Trees: Ornamental Shrubs, Plants, &c.; Cultivated for Sale at the Spring Garden Nursery*. Cincinnati, OH: Ernst and Thorp Printers, preface.

"Exhibitions of Horticultural Societies." *Magazine of Horticulture, Botany, and All Useful Discoveries and Improvements in Rural Affairs.* 9 (December 1843): 465.

Fennell, James H. 1848. "The Farmers' Note-Book." *Journal of Agriculture.* 3 (July): 468.

Gerdts, William. 1976. "Introduction." In John V. Bridle and Sally Secrist, eds., *American Cornucopia: Nineteenth-Century Still Lifes and Studies*. Pittsburgh, PA: Carnegie Mellon.

Glickman, Lawrence. 2012. *Buying Power: A History of Consumer Activism in America.* Chicago, IL: University of Chicago Press.

Graham, George. 1857. "Cincinnati Horticultural Society." *Cincinnatus.* 1, 1–2: 22, Cincinnati Museum Center, Pamphlets Collection, Reference Number 143.

Hannickel, Erica. 2013. *Empire of Vines: Wine Culture in America*. Philadelphia: University of Pennsylvania Press.

Hovey, Charles. 1847. *The Fruits of America: Containing Richly Colored Figures and Full Descriptions of the Choicest Varieties Cultivated in the United States.* Boston, MA: C. C. Little, Jas. Brown, and Hovey.

Husmann, George. 1865, June 22. "The New Era in Grape Culture, No. II." In *The Horticulturist, and Journal of Rural Art and Rural Taste*. New York: George E. & F. W. Woodward: 267.

Katz, Wendy. 2002. *Regionalism and Reform: Art and Class Formation in Antebellum Cincinnati*. Columbus: Ohio State University Press.

Ketner, Joseph. 1994. *The Emergence of the African-American Artist, Robert Duncanson, 1821–1872*. Columbia: University of Missouri Press.

"Letters of Hiram Powers to Nicholas Longworth, Esq., 1856–1858." *The Quarterly Publication of the Historical and Philosophical Society of Ohio*. 1.2 (April–June 1906).

"List of Premiums Awarded by the Michigan State Agricultural Society." *Michigan Farmer.* 3, 19 (1849, October 1): 295, 299.

Lockwood, Alice B. 1931. *Gardens of Colony and State: Gardens and Gardeners of the American Colonies and of the Republic Before 1840* v.1. New York: Smallwood and Stewart.

Lyons, Maura. 2015. "Nature Defamiliarized: Picturing New Relationships between Humans and Nonhuman Nature in Northern Landscapes from the American Civil War." *Panorama: Journal of the Historians of American Art.* 1, 1 (Winter 2015). doi: 10.24926/24716839.1502

Mansfield, Edward. 1850. *Address Delivered before the Horticultural Society*. Cincinnati, OH: Wright, Ferris, and Co.

McCabe Jr., James. 1871. *Great Fortunes and How They Were Made, or the Struggles and Triumphs of Our Self-made Men*. Cincinnati, OH: E. Hannaford and Co.

"New Helps in Agriculture in Aid of War." *Scientific American.* 1862: 274.

Niblack, Rita Steininger. 1985. *Nicholas Longworth, Art Patron of Cincinnati.* Cincinnati, OH: University of Cincinnati.

Okihiro, Gary. 2010. *Pineapple Culture: A History of the Tropical and Temperate Zones.* Berkeley: University of California Press.

Paley, William. 1802. *Natural Theology or Evidences of the Existence and Attributes of the Deity.* New York: American Tract Society.

Paulding, James K. 1849. "The Illinois and the Prairies." *Graham's American Monthly Magazine of Literature and Art.* 34, Philadelphia: Samuel D. Patterson and Co.: 16.

Pauly, Philip. 2008. *Fruits and Plains: The Horticultural Transformation of America.* Cambridge, MA: Harvard University Press.

"Ploughing Exhibition at Harlæm." *Journal of the American Institute: A Monthly Publication, Devoted to the Interests of Agriculture, Commerce, Manufactures, and the Arts....* 4, 11 (August 1839): 612.

Pollan, Michael. 2001. *The Botany of Desire: A Plant's Eye View of the World.* New York: Random House Publishing Group.

Salter, William. 1876. *The Life of James W. Grimes.* New York: D. Appleton and Co.: 48.

Staiti, Paul. 1998. *Reading American Art.* Marianne Doezema and Elizabeth Milroy, eds., New Haven, NJ: Yale University Press: 12–38.

"The War of Slavery." *Liberator.* 9, 25 (1839, June 21): 100.

"Transactions for the Western Art Union 1847–1849." Cincinnati, OH: Cincinnati Museum Center.

Whittredge, Worthington. 1969. *The Autobiography of Worthington Whittredge 1820–1910.* New York: Arno Press.

Woodson, Carter G. 1916, January. "The Negros of Cincinnati Prior to the Civil War." *The Journal of Negro History.* 1, 1: 1–22.

Young, Denny Carter. 1982. "The Longworths: Three Generations of Art Patronage in Cincinnati." In Kenneth R. Trapp, ed., *Celebrate Cincinnati Art.* Cincinnati, OH: Cincinnati Art Museum: 29–39.

Index

Note: *Italic* page numbers refer to figures and page numbers followed by "n" denote endnotes.

For Product Safety Concerns and Information please contact our EU representative GPSR@taylorandfrancis.com Taylor & Francis Verlag GmbH, Kaufingerstraße 24, 80331 München, Germany

Batch number: 10777126

Printed by Printforce, the Netherlands